RADIOISOTOPE TECHNIQUES FOR PROBLEM-SOLVING IN INDUSTRIAL PROCESS PLANTS

RADIOISOTOPE TECHNIQUES FOR PROBLEM-SOLVING IN INDUSTRIAL PROCESS PLANTS

Edited by

J. S. CHARLTON

Physics and Radioisotope Services
Imperial Chemical Industries PLC
Billingham, Cleveland, UK

Leonard Hill

Glasgow and London

Published by Leonard Hill
A member of the Blackie Group
Bishopbriggs
Glasgow G64 2NZ

Furnival House
14–18 High Holborn
London WC1V 6BX

British Library Cataloguing in Publication Data

Radioisotope techniques for problem-solving
in industrial process plants.
1. Process control   2. Radioisotopes
I. Charlton, J.S.
670.42′7   TS156.8

ISBN-13: 978-94-010-8306-5   e-ISBN-13: 978-94-009-4073-4
DOI: 10.1007/978-94-009-4073-4

Photosetting by Thomson Press (India) Limited.

Contents

Contributors

J. S. Charlton, BSc, PhD
Manager, Plant Services and Instruments, Physics and Radioisotopes Research Group

E. A. Edmonds, BSc, MSc, PhD
Process Applications Manager, Physics and Radioisotope Services Group

K. James, BSc, PhD
Nucleonic Instruments and Development Manager, Physics and Radioisotope Services Group

P. Johnson, BSc, FInstP
Former Group Manager, Physics and Radioisotope Services, and Past President, Tracerco Corporation

T. L. Jones, BSc, PhD
North-west Area Manager, Physics and Radioisotope Services Group

G. Reed
Radiological Protection Manager, Physics and Radioisotopes Research Group

R. Roper
Radiation Scientist, Physics and Radioisotope Services Group

1 Radioisotopes in industry

J. S. CHARLTON

1.1 Introduction

Radioactive materials, as sealed sources of ionizing radiation and as radioactive tracers, are used extensively throughout industry. The field of application is extremely wide: this book is concerned with the application of radioisotope techniques to process investigation on full-scale industrial plant. Our objective is to explore the many ways in which radioisotopes can be used to help industrial plant to operate more efficiently. Because of the sheer volume and diversity of radioisotope applications, a selective approach has been adopted. We have concentrated upon those applications which have proved to be the most useful in terms of economic benefit, realized either as savings or as improved production efficiency.

As with any technology it is, of course, possible to achieve the benefits without a detailed understanding of the basic principles, just as one can make good use of an automobile with little or no knowledge of the workings of the internal combustion engine! Some understanding of the basics is nevertheless essential if one is to use the technology to its full effect and, equally importantly, appreciate its limitations. This background information is presented, in condensed form, in Chapters 2–4.

Safety is clearly a further important consideration. It is well known that unrestricted exposure to radioactive materials can lead to health detriment. However, it also needs to be appreciated that these hazards are well understood and that through appropriate precautions they can be reduced to a negligible level. Chapters 5 and 6 discuss health physics and radiological protection and describe briefly the practical measures which are taken to protect both workers and public.

The remainder of the book is devoted to the process applications of radioisotopes. Each chapter covers one technique or class of techniques. Both theory and experimental approach are described and, in addition, case histories are presented which, as well as illustrating the versatility of the technology, demonstrate the economic benefits which can be realized.

In describing case histories and in examining several other aspects of radioisotope applications technology we have drawn upon the experiences of two organizations—the Physics and Radioisotope Services Group (PRS) of Imperial Chemical Industries PLC, and its associate company, Tracerco Corporation of Houston. PRS was established some 30 years ago solely to

exploit radioisotope technology, and, together with Tracerco, provides
service to a broad spectrum of industry world-wide. It is the world's largest
organization specifically devoted to providing contract problem-solving
services using radioisotopes, and its activities will exemplify what radioisotope
technology can achieve.

1.2 Historical perspective

It is now almost ninety years since Becquerel[1] discovered the phenomenon of
radioactivity. For fifty years thereafter, the use of radioactivity was virtually
confined to medical, research, military and power-generation applications.
However, in the late 1940s and 1950s the increasing availability of man-made
radioisotopes produced in nuclear reactors resulted in a greatly expanded
sphere of application. The oil industry, in particular, was quick to appreciate
the potential of radioisotope techniques—the first recorded industrial use of
radioisotopes involved oil-well tracer studies [2].

Research institutes and industrial companies world-wide began to explore
potential uses of radioisotopes and ionizing radiations. To examine the factors

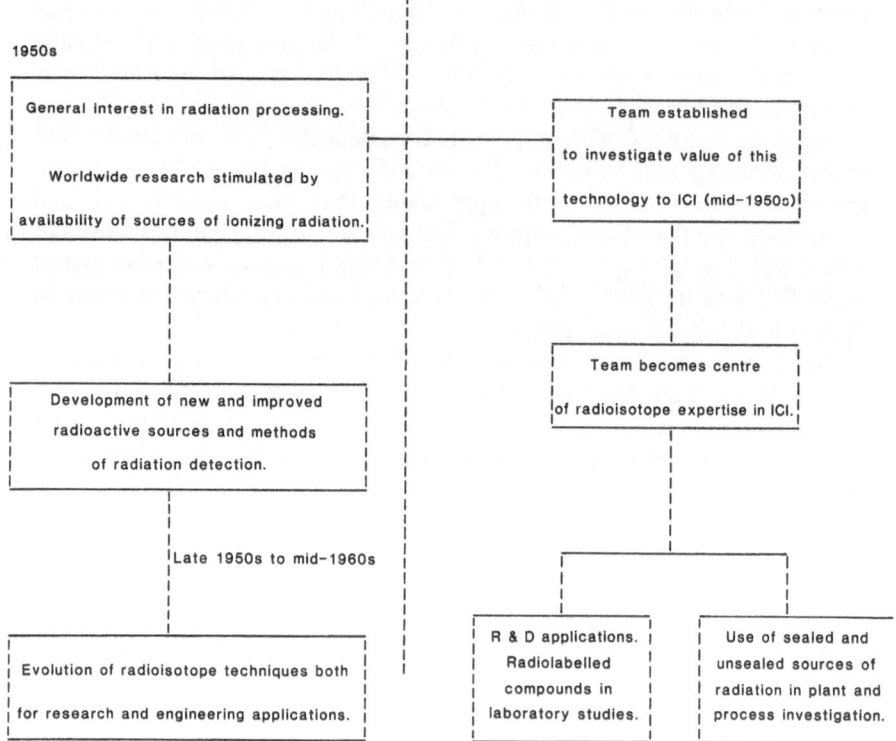

Figure 1.1. The origin of a radioisotope applications service.

which shaped the development of the technology, it is instructive to refer to our example of ICI's Physics and Radioisotope Services Group. Within ICI, interest was initially centred upon the possibility of using sources of ionizing radiation to induce chemical reactions (Figure 1.1). The findings here were not as encouraging as had been hoped. Though radiation processing possessed unique advantages in a limited number of situations, in the majority of cases large-scale radiation sources could not compete with conventional process technologies. However, while radioisotopes as a processing tool appeared to have limited use, it was already clear that there was considerable scope for them in process investigation, and that it was this sphere of activity which possessed the greatest potential for economic benefit.

The ICI Group was fortunate in being located at Billingham in the north-east of England, at the centre of one of the largest chemical complexes in Europe, and benefited greatly from close association with production personnel—a classic case of 'solutions in search of problems' and 'problems in search of solutions' being brought together in close proximity.

Considerable savings resulted through using radioisotope techniques to investigate operational characteristics and to diagnose faults on full-scale process plant. An important feature was the unique ability conferred by the properties of ionizing radiation to investigate problems without disrupting the process in any way. In particular, shutdown time could be avoided or reduced to a minimum. Process applications activity was therefore considerably stimulated by the construction, accelerating throughout the 1960s and 1970s, of very large single-stream production units (Figure 1.2). The financial consequences of shutdown of a unit of size comparable (say) with that of the 500 000 te p.a. No. 5 Olefines Plant at Wilton, Cleveland, were so large that a considerable investment of time and resources was made in on-line surveillance and fault-finding techniques, among which radioisotope methods were found to be by far the most versatile.

By 1970, the use of radioactive tracers had become routine to the extent that PRS Group could economically justify the purchase of its own radioisotope production facility, a TRIGA Mark I nuclear reactor. This was installed at the Group's headquarters in 1971 and continues to function as a safe and reliable source of radioisotopes. As is often the case, meeting the existing need (by the purchase of the nuclear reactor) stimulated an even greater demand: tracer applications which had hitherto been impossible because they depended upon the availability of a particular radiotracer (for example, material of very short half-life) became feasible, and further expansion of process applications resulted.

Continued development of this technology throughout the 1970s was directed more by economic than technical influences. Concerns about the continued availability and conservation of fossil fuel resources (especially oil and gas) led to the extensive use of radioisotope techniques to provide the basic information for energy-conservation studies. Additionally, and very

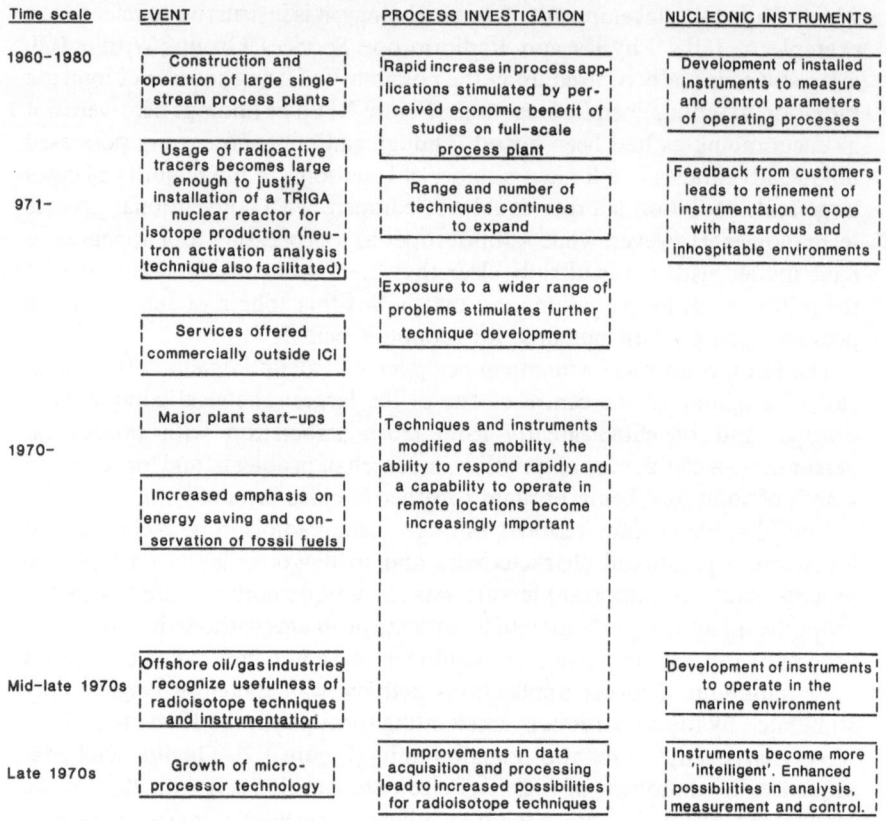

Time scale	EVENT	PROCESS INVESTIGATION	NUCLEONIC INSTRUMENTS
1960–1980	Construction and operation of large single-stream process plants	Rapid increase in process applications stimulated by perceived economic benefit of studies on full-scale process plant	Development of installed instruments to measure and control parameters of operating processes
971–	Usage of radioactive tracers becomes large enough to justify installation of a TRIGA nuclear reactor for isotope production (neutron activation analysis technique also facilitated)	Range and number of techniques continues to expand	Feedback from customers leads to refinement of instrumentation to cope with hazardous and inhospitable environments
	Services offered commercially outside ICI	Exposure to a wider range of problems stimulates further technique development	
1970–	Major plant start-ups	Techniques and instruments modified as mobility, the ability to respond rapidly and a capability to operate in remote locations become increasingly important	
	Increased emphasis on energy saving and conservation of fossil fuels		
Mid–late 1970s	Offshore oil/gas industries recognize usefulness of radioisotope techniques and instrumentation		Development of instruments to operate in the marine environment
Late 1970s	Growth of microprocessor technology	Improvements in data acquisition and processing lead to increased possibilities for radioisotope techniques	Instruments become more 'intelligent'. Enhanced possibilities in analysis, measurement and control.

Figure 1.2. Development of a radioisotope applications service.

importantly, in recent years the offshore oil and gas industry has increasingly recognized the value of radioisotope technology in assisting with process problems both on the production platforms and on onshore installations. This has certainly led to increased use of radioisotope techniques, but also has stimulated technical developments aimed at facilitating applications in remote and environmentally 'difficult' locations.

However, the greatest influence on radioisotope applications in recent years continues to be the advance of microprocessor technology. This has manifested itself primarily in advances in data acquisition and processing, but it is already clear that the speed with which raw data can be manipulated will enable radioisotope techniques to be used for applications which hitherto would have been considered impractical.

It is worth pointing out that, in parallel with the growth of radioisotopes in process applications there has been a continuous development of instruments which utilize the properties of radioactive materials in process measurement and control. The external stimuli which governed the course of the development of these two activities are broadly similar (Figure 1.2) and the usage of so-called 'nucleonic' instruments is similarly widespread throughout the chemical process industry[3].

1.3 Current uses of radioisotopes in problem-solving

The current use of radioisotopes in problem-solving on processs plant is not, in general, undergoing dramatic change. The technologies involved are reasonably mature, and major breakthroughs which might give rise to radically new techniques are not in evidence. The use of radioisotope techniques is, however, increasing steadily as more and more potential applications are identified. In large measure, this growth stems from the increasing awareness on the part of industry of the versatility of the technology and the fact that the benefits which it confers can be obtained at relatively modest cost. The problem-solving applications of radioisotopes, though very numerous, can be divided into two broad categories: techniques which utilize *sealed sources* of radiation and those which utilize *radioactive tracers*.

The essential feature of all sealed source techniques is that the radioactive isotope remains permanently sealed within the source capsule and makes no contact either with the plant or with the process material. Radiations from the source are directed at the item of interest (perhaps a process vessel), and by analysing variations in the intensity of the transmitted or scattered radiation beam, it is possible to draw conclusions about the vessel and its contents. Sealed-source techniques are discussed in Chapters 12–15. It will be seen that, because full-scale chemical plant is of substantial construction, the radioactive sources which are most useful in problem-solving are those which emit penetrating radiations: generally, gamma-ray and neutron sources.

In contrast to the sealed-source technique, the essential feature of a radiotracer application is that radioactive material in appropriate physical or chemical form is injected into the process material. A portion of the material thus becomes 'labelled' with radioactivity and, provided that the radiotracer faithfully follows the behaviour of the process material, its subsequent movement through the plant can be monitored using external radiation detectors. This gives rise to a range of techniques for studying the dynamics of process streams, and these are described in Chapters 7–11.

The utilization of radioisotopes in plant investigation is usefully demonstrated by the work spectrum of Physics and Radioisotopes Services Group (Table 1.1). Over 1500 applications are carried out each year, and though each is unique it is possible to divide them broadly into the several categories listed. Although this spectrum shifts from year to year, the problem associated with operating process plant remain broadly the same and so the applications spectra still possess many points of similarity. The case histories described in later chapters therefore possess a generality which should make them of interest to most production engineers, whatever type of plant they operate.

Table 1.1 Work spectrum of Physics and Radioisotopes Services Group, Imperial Chemical Industries PLC: a typical year's work

Technique	Number of applications
Flow measurement	604
Leak detection	92
Residence-time studies	30
Liquid carry-over studies	15
Activable tracer techniques	5
Level measurements: gamma-ray absorption	74
gamma-ray scattering	165
neutron backscatter	380
Blockage detection and deposition measurements	47
Corrosion and thickness measurements	19
Entrainment and voidage measurements	16
Distillation column scans	105
Miscellaneous: analysis using portable XRF, valve orientation studies, blending studies, scintillography of refractory-lined components	14

Which industries make most use of radioisotope technology for plant problem-solving? One might deduce the answer by considering the special benefits which the technology has to offer. These benefits all stem from the fact that measurements can, in general, be carried out on line with little or no disruption of the operating process and with minimum inconvenience to production personnel. The information obtained in this way facilitates better plant operation by providing the input for crucial decisions. Let us consider a fairly typical example.

A manager suspects that loss of process efficiency is due to internal damage inside a distillation column. To shut down the column and perform a check of the internal structure is a costly operation involving (possibly) several days of lost production: not a decision to take lightly. However, radioisotope technology permits an assessment of the internal condition of the column to be made *on-line*. Not only that, the column can be checked under various operating conditions and the effect of changing process flows associated with the system can be observed.

It is, of course impossible to predict what the results of the measurement will be. It may be that the original supposition of internal damage is confirmed and a shutdown *is* necessary. If this is the case, then the radioisotope measurement will have identified in advance the area of the column which is damaged, so that maintenance effort can be directed immediately at the problem area, and time-consuming off-line checks minimized. On the other hand, the radio-isotope measurement may indicate that, by making modest changes in process flowrates, the column can be made to operate perfectly well. In this case, an unnecessary shutdown will have been eliminated. In either case, the result is savings in terms of reduced down-time. On modern, large-scale plant, such savings may be very large indeed, outweighing by many orders of magnitude the cost of the radioisotope investigation.

The main users, then, are those who stand to benefit the most: the operators of large-scale continuously-operating plant. The major user continues to be the oil and gas industry (where radioisotopes first found industrial use). Production, refinery operations and petrochemical processing account for more than 50% of all applications. Large-scale chemical plant (agrochemicals, general inorganic chemicals, dyestuffs, organic chemicals and pharmaceuticals) represents the other main sphere of application, accounting altogether for some 30% of use. There are many smaller users: the steel, mining, utilities (electricity, water, gas and sanitation), paper-making, plastics, minerals processing and automobile industries are among those which have benefited from the application of radioisotope technology.

1.4 Growth trends and the future

So far, we have seen that radioisotopes are cost-effective and that they are of proven use in a wide variety of industries. Nevertheless, radioisotope applications are not as commonplace as their universal applicability might suggest. Some production managers are unaware of this technology, or, if they *have* encountered it, are unwilling for it to be applied on their production unit. This unwillingness may stem from several considerations.

(a) 'Radioisotopes are unsafe'. This is an understandable concern, but is ill-founded. Radioactivity is a natural phenomenon and, like other natural phenomena (electricity, for example), it can be harnessed to beneficial use. In expert hands radioisotopes are no more hazardous than many chemicals in routine use. On the contrary, because the hazards *are* well appreciated, precautions are correspondingly well-developed and the chances of health detriment are very small indeed.

(b) 'There are a lot of legalities to overcome before one can use radioisotopes'. The use of radioactive materials is (rightly) closely controlled by law. However, expert advice is available from government departments, research organizations, radioisotope applications contractors, or independent health physics advisers. If a radioisotope service is purchased, then the legal and safety aspects should be included as part of the contract.

(c) 'Radioisotopes are expensive'. Cost is purely relative: compared with the benefit, radioisotope applications are usually very good buys.

(d) 'I don't want to make the capital outlay to purchase equipment' and

(e) 'My organization lacks people expert in the use of these materials and in the interpretation of the data'. The answer to these latter two problems is, of course, to employ a contractor with a special expertise in this technology.

World-wide, many organizations offer radioisotope applications services. A few of the major companies possess their own in-house organizations. Alternatively there are government organizations and private contractors who can be called upon as the need arises. There are certainly opportunities for companies who wish to exploit radioisotope technology to do so.

The use of radioisotope techniques for problem-solving is undoubtedly increasing. It is difficult to be precise about the growth-rate world-wide. Our own organization is experiencing an upward trend of about 30% per annum, and given that this is taking place in a part of the world which has had ready access to the technology for many years, there is every reason to suppose that the overall growth-rate is significantly greater. This should come as welcome news to all who are concerned with greater industrial efficiency, greater safety and with making best use of the world's energy and feedstock resources. We hope that this book will help to promote this beneficial trend, a task well worth undertaking.

References

1. Becquerel, H. (1896) *Comptes rend.* **122**, 422.
2. Mott, W. E. *et al.* (1967) Review of radiotracer applications in geophysics in the United States of America. In *Radioisotope Tracers in Industry and Geophysics*, IAEA, Vienna, 111–113.
3. *Radioisotope Instruments in Industry and Geophysics*, Proc. Symposium, Warsaw 1965, IAEA, Vienna (1966).

2 The basic physics of radioactivity

E. A. EDMONDS

2.1 Introduction

This chapter presents the main features of the physics of radioactivity. The mathematical aspects of the subject are avoided as far as possible and the stress placed on the general concepts with the aim of clarifying the ideas involved.

2.2 The structure of the atom

Radioactivity and the emission of radiations are properties of matter and matter is composed of atoms. An atom is composed of three basic building blocks, electrons, protons and neutrons arranged systematically as shown in Figure 2.1. There are a number of electrons orbiting a central core or *nucleus*. The electrons are arranged in layers or *shells* of different orbital diameters. The size of the atom defined by the outermost electrons is about 10^{-10} m. Electrons carry a negative electrical charge. This is the smallest unit of electrical charge which has ever been identified in nature and is, in the SI system of units, -1.6×10^{-19} C (for *coulombs*). The mass of the electron is 9.1×10^{-31} kg. The electron is thus seen to be a small particle with a tiny and negative electrical charge.

The nucleus of the atom is composed of protons and neutrons, and is about 10^{-15} m across. Most of the mass of the atom, however, is concentrated in it.

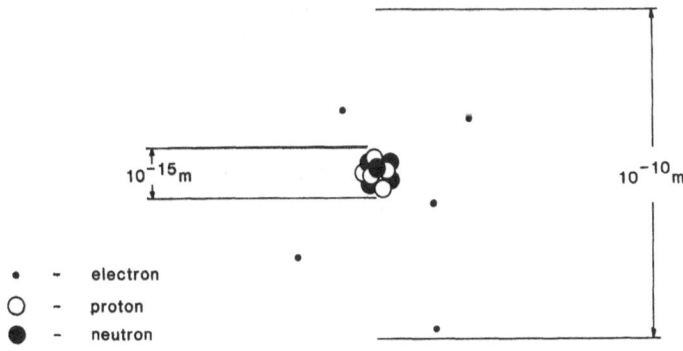

10^{-15} m 10^{-10} m

• - electron
○ - proton
● - neutron

Figure 2.1. The atom.

Protons are much heavier particles than electrons, with a mass of 1.7×10^{-27}kg (roughly two thousand times heavier than the electron). Protons also have an electrical charge and it is $+ 1.6 \times 10^{-19}$C; that is, a positive charge which exactly balances the negative charge of the electron. Neutrons appear to be very similar to protons. They have about the same mass as a proton but unlike protons they are electrically neutral. Protons and neutrons are collectively referred to as *nucleons*. Since they account for most of the mass of the atom, the total number of nucleons in the nucleus is called the *mass number*, usually written as A. The *atomic number* (Z) is the number of protons in the nucleus of an atom, which is the same as the normal complement of electrons for the atom.

Once the atomic number is defined, the chemical behaviour of the atom is also defined and thus its identity is fixed. In the example shown in Figure 2.1, where there are five protons and five electrons, $z = 5$, this atom will behave chemically as boron. An atom with $z = 6$ would be carbon; with $z = 26$, iron.

2.3 Isotopes

Nuclei of a given atomic number which differ in the number of neutrons they contain are called *isotopes*. Isotopes which are radioactive, that is, they are unstable combinations of neutrons and protons, are *radio*isotopes. Consider the example shown in Figure 2.1. This is a stable isotope of boron. It has five protons, by definition for boron, plus five neutrons in the nucleus. There is another stable isotope of boron, with six neutrons. In addition, there are several unstable, radioactive configurations of the boron nucleus. (An unstable configuration is here defined as one which will decay radioactively into another configuration but which persists long enough to be identified. The decay period of unstable nuclei varies from millions of years to fractions of a second. Between these extremes, there are many radioisotopes which are manufactured routinely for industrial and medical use, for research and as byproducts of the operation of nuclear reactors which last for periods from minutes to thousands of years, which are easily identified and which can be manipulated, used and stored before they decay away.)

At this point it is worth considering some of the nomenclature for atoms and isotopes. The general notation for an isotope is $^{A}X_{Z}$ where X is the chemical symbol for the element, Z is the atomic number and A is the mass number. The superscript A may be transposed right or left of the chemical symbol, and the atomic number subscript is redundant, given the presence of the chemical symbol of the element and so it is almost invariably dropped. The symbol for the isotope shown in Figure 2.1 is thus B^{10} or ^{10}B; in speech this would be referred to as 'boron ten', and is also often written boron-10.

Some isotopes have their own names, in particular, the isotopes of hydrogen. The most abundant isotope of this element is ^{1}H, Hydrogen -1, and there is one other stable isotope of hydrogen, however, much rarer (0.015%), ^{2}H,

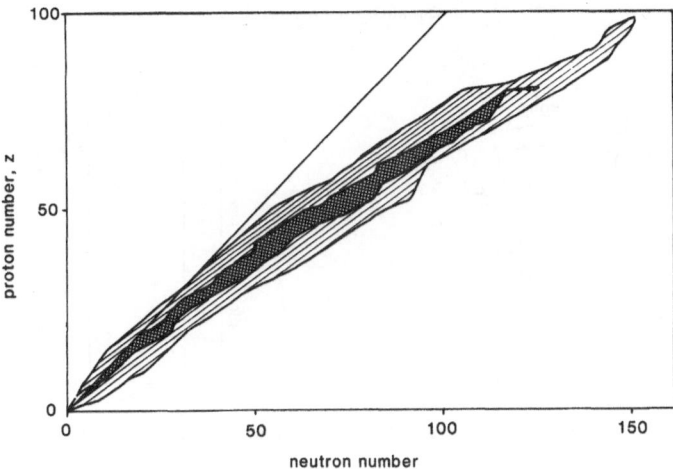

Figure 2.2. Stable and unstable isotopes.

hydrogen − 2, known as deuterium, and another, radioactive, isotope, ^{3}H, or tritium.

There are far more unstable isotopes, radioisotopes, than stable ones; a map of all the currently identified isotopes is shown in Figure 2.2, where proton number, Z, is plotted against neutron number, $(A − Z)$. The region of stability is shown in black. It can be seen that the stable isotopes tend to larger values of neutron number as Z increases.

2.4 Ionizing radiations

Ionizing radiations are of four main types, X-rays and gamma-rays, alpha-rays and beta-rays. X-rays and gamma-rays are electromagnetic waves, like light and radio waves, microwaves and ultraviolet, but of much shorter wavelengths. Alpha-rays and beta-rays are particulate radiations: that is, the 'rays' are actually composed of alpha-particles and beta-particles. All of these radiations are emitted from matter as the result of natural processes. Gamma-rays, alpha-rays and beta-rays are produced as a result of *nuclear* transformations associated with radioactive *decay*. X-rays are produced from transformations in the *electronic* structure of atoms.

2.4.1 *X-rays*

To understand what X-rays are and how they come from the electronic structure of an atom it is necessary to consider in more detail how the electrons are organized in atoms. There may be many electrons in an atom and their

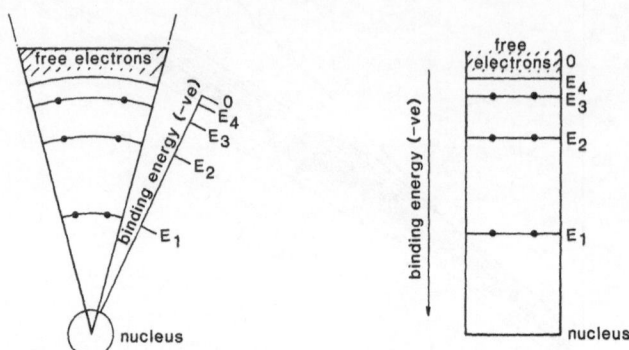

Figure 2.3.   Energy levels for electrons in an imaginary atom.

number is characteristic of that element. They are all bound to the nucleus, some quite loosely, others more tightly. Electrons in atoms exist in orbital shells, or levels, of differing effective radii. Physically, the more tightly bound electrons are closer to the nucleus while the more loosely bound electrons are those further from the centre of the atom. (These shells have historical labels, by which they are designated K, L, M... and so on). The electron shells of a particular nucleus can only have certain allowed amounts of binding energy which apply to that kind of nucleus only. The binding energies are said to be *quantized*. Figure 2.3 shows the electronic energy level scheme for a simple and imaginary atom. The electrons can occupy the horizontal positions with energies E_1, E_2 and so on but nowhere in between. Consider the analogy of a football and a hole in the ground. The football becomes 'bound' to the hole when it falls into it. The binding energy of the football can only be equal to the depth of the hole, as shown in Figure 2.4a. It cannot be at an intermediate

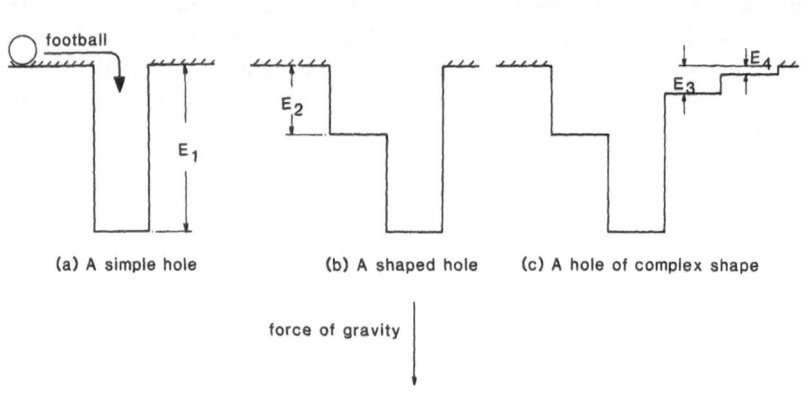

Figure 2.4.   A football falling into a hole in the ground.

value unless there is a 'shelf' such as in Figure 2.4b. Given the shelf, the football could have a binding energy of E_2 instead of E_1, but it could not have any intermediate value. To an electron, a nucleus looks like a hole with a complex shape offering certain energy levels but no others, as in Figure 2.4c. Different nuclei have differently-shaped holes. The analogy cannot be taken much further since electrons do not behave like footballs: numerous footballs would all tend to pile up in the bottom of the hole on top of each other, but electrons drop neatly into the available 'shelves' or levels of energy, two by two. It takes two electrons to fill an available energy level; then the next level begins to fill. This is because electrons have a property called *spin* and there are two spin states available, up and down. For each energy level there is room for one electron spinning 'up' and one spinning 'down'.

The lowest energy state that the electronic structure of an atom can attain is when the electrons lie in the lowest available energy levels. This is the natural state for the atom and it is called the *ground state*. Any other state, where electrons occupy higher levels leaving vacancies at lower levels, is called an *excited* state, as shown in Figure 2.5. When atoms absorb energy, the electrons move so that the atom is raised to an excited state. When electrons fall back naturally into the vacancies created in the electronic energy level scheme, energy is liberated in the form of *photons* of electromagnetic radiation. The electrons which fall into the vacancies may be from within the atom itself or they may be from the 'sea' of free electrons drifting in the spaces between atoms. These free electrons simply have the normal thermal energy of their environment, which is, in effect, the zero energy in Figures 2.3 and 2.4. Free electrons may be released when atoms combine and the outermost electrons receive sufficient energy to drift off. If an electronic transition fills one of the outermost energy levels, so that the energy difference between the excited state and the ground state is small, then a low-energy photon is emitted. This would correspond to visible light or infrared radiation. If an electron falls into a deep vacancy, so that the energy difference between the excited state and the ground

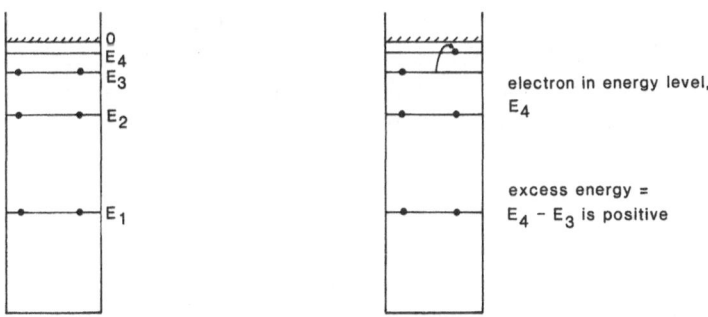

(a) Electrons in ground state   (b) Electrons in excited state

Figure 2.5. Ground and excited states of an imaginary atom; (*a*) Electrons in ground state; (*b*) electrons in excited state.

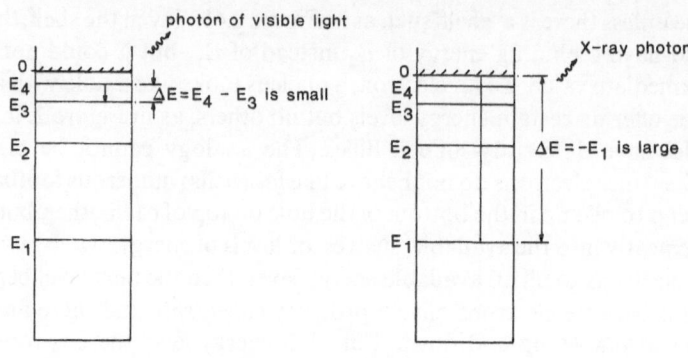

(a) Light emission (cf. Fig. 2.5b)   (b) X-ray emission

Figure 2.6. Decay of excited states—emission of photons, (*a*) Light emission (cf. Figure 2.5*b*); (*b*) X-ray emission.

state is large, then the photon which is emitted is very energetic—in fact, it is an X-ray photon (Figure 2.6).

To create an excited state, energy must be put into the system in some way. A block of copper, heated over a bunsen flame, can be made to glow red and then white-hot, as sufficient energy is being put into the copper atoms to create excited states in the outermost energy levels. When these excited states decay away, infrared (heat radiation) and visible light are emitted. Heating the block over a bunsen flame, however, will not remove an electron from a tightly bound energy level which would allow the transition which generates an X-ray—the copper block will probably melt or evaporate first. But X-rays can be obtained from a copper block by firing into it a beam of energetic electrons. The electrons from the beam are capable of knocking out tightly-bound electrons from the copper atoms, leaving the copper atoms in highly excited, ionized states. The X-rays which emerge are *characteristic* of copper, and so they are called characteristic X-rays. If a different material had been placed in the beam of electrons, then X-rays characteristic of that material would have been obtained. These X-rays correspond to the energy changes which are available to the copper atoms. All the atoms in the block are exactly the same and behave in exactly the same way when electrons are fired at them, generating many exactly similar X-ray photons.

There is another kind of radiation which comes out of the block, however. This is also X-ray radiation, but it is not characteristic of the material in the target block. By firing energetic electrons into the copper block we have created in it a population of fast-moving charged particles: the *primary* electrons from the beam itself, of energy E_0, the energy of the beam, and any *secondary* electrons of diminished energy from the beam or which have been knocked off the copper atoms during the process of ionization and excitation. The free electrons in the block can have any energy whatsoever up to the

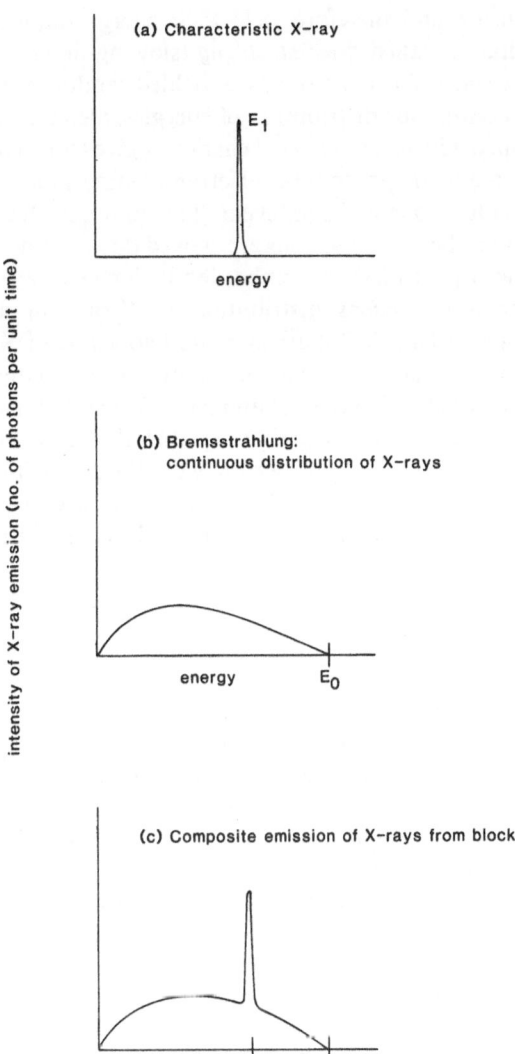

Figure 2.7. X-rays from a block of material composed of imaginary atoms, (a) Characteristic X-ray; (b) bremsstrahlung—continuous distribution of X-rays; (c) composite emission of X-rays from block.

maximum energy of an electron from the beam. (This is reasonably self-evident. If the beam contained electrons of energy E_0 then the most energetic secondary electron one of those can generate from those already in the block is one with an energy E_0. This occurs when an electron from the beam gives up all its energy in one go to an electron which was already in the block.) These free but energetic electrons come to thermal equilibrium in the block by a process of random collision and therefore yield their energy randomly, generating photons of radiation called *bremsstrahlung* (slowing-down radiation), often called 'brem' for short. Since the energy is yielded randomly, the photons are emitted with a continuous distribution of energies, also up to the maximum value, E_0, obtained when an electron of energy E_0 gives up all its energy in one go as an electromagnetic photon, the electron coming to a 'dead halt' in the block as the result of one collision event. This bremsstrahlung radiation is quite unavoidable whenever electrons are slowed down inside solid materials, and so when the copper block is bombarded with energetic electrons, it will always produce a *continuous* distribution of X-rays in addition to its characteristic ones. Figure 2.7 illustrates the two kinds of X-rays obtained from a block of some simple, imaginary material, the material having the energy levels presented in Figures 2.3 and 2.6. (Notice that had the energy of electrons in the beam, E_0, been less than E_1, then the beam would have been unable to stimulate the excited state which led to the production of the X-ray. The beam would have been unable to supply sufficient energy to overcome the binding energy E_1, and it could not have removed the inner electron from the atom.)

2.4.2 *Beta-rays*

Beta-rays are emitted from a radioactive nucleus when it undergoes a spontaneous radioactive decay of a kind known as beta-decay. A nucleus which has an unstable 'mix' of neutrons and protons (too many neutrons or too many protons) changes its internal composition to attain a stable configuration. At first it might seem likely that a proton-rich nucleus would simply emit a proton or two until a stable configuration were reached, or that a neutron-rich nucleus would throw out neutrons with the same end in view. The unstable nucleus would then progress vertically or horizontally across the map shown in Figure 2.2 until a stable configuration for the nucleus was reached, at which point, by definition, the process would cease. In fact this kind of process is very rare, and light nuclei which are neutron- or proton-rich never emit the fundamental particles of which the nucleus is composed. For a nucleus to emit a neutron, for instance, it would have to find the energy to overcome the nuclear binding forces and physically eject a fairly large object. It is energetically much easier and much less disruptive to the nucleus simply to convert an excess neutron into a proton or vice versa: this process is called beta-decay. Thus

$$\text{neutron} \rightarrow \text{proton} + \text{electron} + \text{antineutrino}$$

$$\text{proton} \rightarrow \text{neutron} + \text{positron} + \text{neutrino}.$$

These simple equations show that the single particle on the left becomes three particles on the right when it undergoes beta-decay. The produced proton or neutron merely remains inside the nucleus although that nucleus now contains one fewer neutron and one more proton, or vice versa. The original nucleus is usually referred to as the *parent* and the nucleus which is the product of the decay is usually called the *daughter*. Thus, when caesium-137 changes by beta-decay into barium-137 a neutron in the original caesium-137 nucleus turns into a proton to yield the barium-137 daughter nucleus.

Why are the other particles produced? In fact, when a neutron turns into a proton, an uncharged particle is changing into one which carries a positive charge. The electron, which is negatively charged, is produced to conserve the charge, in effect to balance up the equation. A positron is an electron which is positively charged. It is also there to balance the equation. When a proton changes into a neutron, the charge it carries must go somewhere. The electrons, whether the common negatively-charged variety (sometimes called *negatrons*) or the rather more exotic positively-charged positron, are emitted from the nucleus, carrying away some of the energy of the transformation. (Bear in mind that if there had been no excess energy to carry away, there would have been no instability in the parent nucleus.) It is these electrons which are called beta-particles. Streams of beta-particles emerging from radioactive material are beta-rays.

The neutrino and antineutrino are also emitted from the nucleus but they need concern us no further. They interact very weakly with matter of all kinds and are very difficult to detect. They do not ionize matter nor do they constitute any kind of hazard. They are in place in the equation to conserve another property of all these particles, which is spin. It is quite acceptable to think of a neutrino as nothing, spinning; and of an antineutrino as nothing, spinning the other way. In simplified texts these particles are often ignored altogether, but they have been mentioned here because it is important to appreciate that the neutrino and the antineutrino also carry away some of the energy of the radioactive decay. The energy of the decay is therefore divided three ways; between the beta-particle itself, the neutrino and the nucleus. This is why all beta-particles emitted as a result of a particular decay do not emerge from the nucleus with the same energy. The energy of the decay can be unevenly divided between the three bodies involved. Thus, beta-particles emitted from a particular kind of nucleus have a range of energies up to some maximum value.

2.4.3 Alpha-rays

In atoms with large nuclei another factor which renders them unstable becomes important. The nuclear force binding the nucleons together has a very short range, such that only particles which are physically touching are bound by it to any great extent, whereas the electrostatic repulsion amongst the protons is a long-range force. As a nucleus gets bigger, the repulsive force increases strongly, whereas the nuclear force between particles does not effectively increase. A large nucleus can be pictured as a clump of billiard ball-like objects fairly loosely tacked together at the points where they touch each other. They remain able to move and rotate and tumble around each other. Larger nuclei are more unwieldy. The binding energy per nucleon passes through a maximum when the atomic number, Z, is 26, (iron), and then progressively diminishes through elements of increasing atomic mass as the electrostatic repulsion builds up. Thus, instability increases due to the sheer size of the nucleus and it becomes energetically possible for the nucleus to emit large, heavy particles.

The particles emitted are not simply the neutrons or the protons of which the nucleus is composed. In fact, when alpha-decay occurs and a large particle (alpha-particle) is emitted from a nucleus, that particle is a bound state of two neutrons combined with two protons. This structure is equivalent to the nucleus of an atom of helium-4, and is about seven and a half thousand times more massive than a beta-particle. This configuration of two protons bound to two neutrons is particularly stable, and actually represents the basic building block for nuclei, rather than single protons and neutrons. Inside a large nucleus, many of the protons and neutrons will already be bound together in pairs in the configuration of the alpha-particle.

After an alpha-particle is emitted by a large unstable radioactive nucleus, the daughter nucleus is smaller and lighter. The process of alpha-decay might not, however, have resulted in a stable daughter. There could be a long chain of alpha-decay processes, punctuated by beta-decay transformations, before a stable end product is reached. For instance, the so-called uranium/radium decay series by which uranium-238 decays to stable lead-206 involves a branching series of eight alpha-decay steps with several different beta-decay intermediate stages.

2.4.4 Gamma-rays

Gamma-rays are electromagnetic photons emitted from the nucleus as a result of radioactive decay. The emission of gamma-rays almost invariably accompanies beta-decay and alpha-decay. This is because the daughter nucleus, the product of radioactive decay, is almost always produced in an excited state. Decay of this excited state involves a change of energy of the nucleus and the excess energy is carried out of the nucleus as a photon, a gamma-ray. The mechanism for the production of gamma-rays is quite similar to the mechanism which generates X-ray photons, but instead of the energy level scheme of the electrons being involved, as in X-ray production, it is the nuclear energy levels. Just as electrons in the atom can only exist in quantized energy levels, so there are quantized energy levels for the protons and neutrons in the nucleus. These decay from excited states to the ground states appropriate for the arrangement of protons and neutrons in the daughter nucleus.

An example is the decay of caesium-137 to barium-137 by beta-emission, followed by the emission of a gamma-ray. It is important to note that the gamma-ray is in fact characteristic of the energy-level scheme in the nucleus of barium-137. It is nothing to do with the energy-level scheme in the nucleus of caesium-137, despite the fact that a radioactive source which contains caesium-137 always emits gamma-rays and hence caesium-137, which actually decays by beta-emission, is commonly employed as a source of gamma-rays.

2.5 Important concepts in radioactivity

2.5.1 *Energy units*

Until now, no units have been given for the energy associated with nuclear processes. The unit generally employed to describe the energy of a particle emitted by a nucleus or of an X-ray photon, is the *electron volt* (eV), the amount of energy imparted to an electron when it passes through a potential difference of one volt. The electron volt is a very small unit of energy, only 1.6×10^{-19} joules.

Energy transitions which give rise to the production of X-rays, gamma-rays, beta-particles and alpha-particles are typically in the range from a few thousand electron volts (keV) to a few million electron volts (MeV). For example, the gamma-ray emitted by barium-137 (which as we have seen is usually associated with the beta-decay of the parent nucleus, caesium-137) has an energy of 0.66 MeV. Caesium-137 itself emits a beta-particle of 0.51 MeV. (Interestingly, a radioactive source which contains caesium-137 also emits beta-particles with an energy of 1.17 MeV. This is because some of the

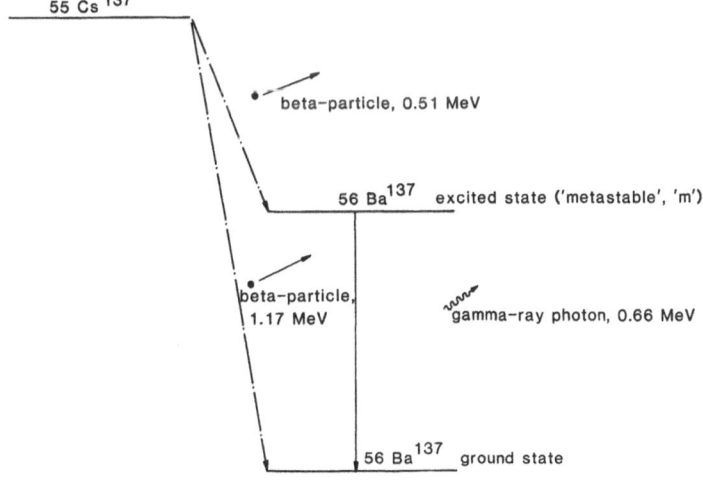

Figure 2.8. Beta-decay of caesium-137 and gamma-decay of barium-137m excited state.

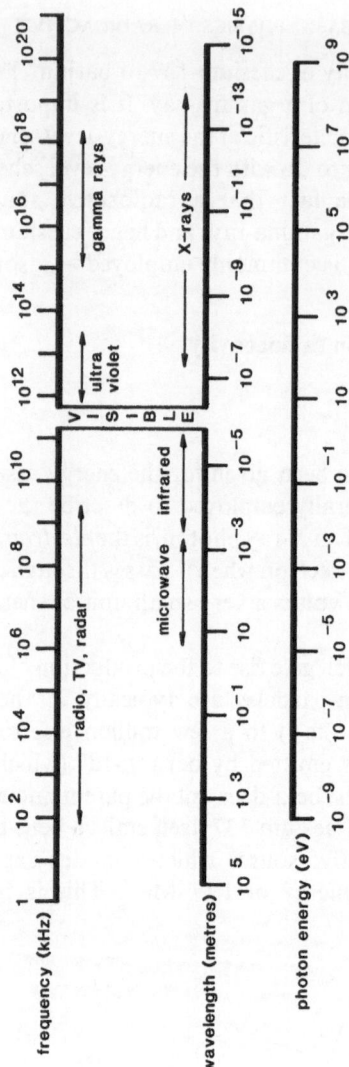

Figure 2.9. The electromagnetic spectrum.

caesium-137 nuclei in the source actually do decay straight down to the ground state of barium-137. When that happens, there is no energy left in the daughter nucleus to be shed as a gamma-ray. The decay of caesium-137 is illustrated in Figure 2.8.)

The energy of a material particle like an alpha-particle is fairly simple to understand. The greater the energy of the particle, the faster it is travelling. Energetic particles are often referred to as 'fast' whereas particles which have little energy, which are in thermal equilibrium with their environment or close to it, are often called 'slow'. Photons, on the other hand, are essentially wavepackets of electromagnetic radiation which, like light or radio waves, travel through space at 3×10^8 ms^{-1}, the so-called 'velocity of light', whatever their energy. Photons do have the property of momentum, however. An energetic photon has a large momentum. (Photons have no rest mass and when they stop moving, they cease to exist.) The momentum, p, and energy, E, of a photon are related to its wavelength by the equations

$$p = E/c \tag{2.1}$$

$$E = h.v \tag{2.2}$$

$$c = \lambda.v \tag{2.3}$$

where h is a constant (Planck's constant), c is the velocity of light, λ is the wavelength and v the frequency of an electromagnetic wave. Figure 2.9 shows the electromagnetic spectrum displayed on scales of wavelength, frequency and energy, all meaning the same thing, these scales being related by equations (2.1)–(2.3).

It is worth noting at this point that the distributions of X-rays and gamma-rays shown in Figure 2.9 overlap in energy. An observer would not be able to distinguish a 1-MeV X-ray in any way from a 1-MeV gamma-ray. Their properties would be the same. The distinction between X-rays and gamma-rays arises solely because of their different points of origin. This useful fact is worth remembering.

We have referred to 'slow' and 'fast' in the context of energy. It is common to hear of 'fast neutrons', for instance, 'hard' X-rays (meaning energetic ones), and 'soft' gamma-rays (meaning low-energy gamma-rays, say of about 100 keV and less). 'Soft' X-rays are sometimes referred to as 'mush' and are often disregarded.

2.5.2 Half-life

The amount of a particular radioisotope diminishes with time as it decays away. In other words, the radioactivity due to the decay of that isotope is reduced as the nuclei spontaneously change into some other form. This decay is expressed in terms of the *decay constant*, λ, not to be confused with the wavelength of electromagnetic radiation despite the fact that, for historical reasons, the same symbol λ is usually used.) If there are N nuclei, then

$$\frac{dN}{dt} = -\lambda N \tag{2.4}$$

which is another way of stating that the number of nuclei which decay in a given time interval is proportional to the number available to decay. The negative sign appears because the number of nuclei diminishes as time increases.

This equation can be transformed to express the number of nuclei which have not decayed by time t given an arbitrary starting point of N_0 nuclei at time zero. Thus

$$N = N_0 \exp(-\lambda t). \tag{2.5}$$

If we call dN/dt the 'activity' (or radioactivity) of the sample in question, expressed in disintegrations per second (dps) or counts per second (cps), then

$$A = A_0 \exp(-\lambda t). \tag{2.6}$$

The simple exponential relationships of equations (2.5) and (2.6) allow a *half-life* to be defined: that time at which half of the original atoms or activity will have decayed away, that is, when $N = 0.5N_0$. Putting this condition into equation (2.5) yields

$$\tau_{\frac{1}{2}} = (\ln 2)/\lambda. \tag{2.7}$$

The decay constant of a radioisotope is a statistical value expressing something about the instability of the nucleus in question. The more unstable, the larger the value of λ. In practice, however, the half-life, which is essentially an experimentally-derived quantity, is much more useful. It tells the observer just how the radioactivity of a specimen will diminish with time. If the observed radiation count rate at time zero is 100 000 per second and the half-life is known to be one hour, then it can be predicted confidently that in two hours a count rate of 25 000 per second will be observed, that is, after two half-lives have passed. The concept of half-life allows simple inferences to be made about the relationship between present and future levels of radioactivity without any thought for the underlying processes of radioactive decay. The concept of half-life only holds good because the statistics of very large numbers always applies to samples which contain radioactive materials, but if an observer could look at one nucleus alone, the half-life would not be able to tell him just when that nucleus will decay away, although during any given time interval, there is a certain *probability* that it will decay.

The probability of any one nucleus decaying, and hence the decay constant and the half-life for material containing the radioactive species, is not affected by chemical composition, temperature, pressure or any other normal outside agency of the everyday world. This is because the tendency to decay is associated exclusively with the structure of the nucleus which is isolated at the centre of the atom.

2.5.3 Units of radioactivity

Radioactivity is described in terms of the number of disintegrations that occur per unit time. The SI unit of radioactivity is the becquerel (Bq), 1 disintegration per second. The curie (Ci), 3.7×10^{10} disintegrations per second, is still in current usage. Units of radioactivity do not express anything about the types or energy of radiations emitted by the nucleus. They do not provide any information about the half-life of the nuclei, nor any connotation of relative hazard if different amounts of radioactivity of different radioisotopes are presented—one million becquerels of isotope x is not necessarily half as hazardous, useful or valuable as two million becquerels of isotope y. To make any kind of judgement about the value or nature of a radioisotope, all the different kinds of information must be considered, the number and type of radiations per disintegration, their energies, the rate of decay of the nuclei and so on.

2.6 Properties of radiations—interactions with matter

Different kinds of radiation have particular properties and interact with matter in different ways. This section outlines in a general way the main processes governing the interaction of radiations with simple materials, and compares and contrasts the way they deposit energy and penetrate matter.

2.6.1 Gamma-rays and X-rays

Gamma-rays and X-rays, both being electromagnetic radiations, interact with matter in much the same way—a gamma-ray of a given energy will be indistinguishable in its properties from an X-ray of the same energy. There are three main processes by which energetic photons interact with matter: photoelectric absorption, Compton scattering and 'pair production'.

Photoelectric absorption is important at fairly low energies. When a gamma-ray photon or X-ray photon interacts with an electron in an atom and imparts all of its energy to the electron, the electron takes up the energy of the electromagnetic photon (or *quantum* of energy) and travels away at high speed through the medium, leaving behind an ionized atom, while the photon ceases to exist. This phenomenon is known as the *photoelectric effect.*

Compton scattering occurs when a photon interacts with an electron and it is scattered with diminished energy, the electron taking up some of the energy of the photon. In effect, the photon collides with the electron, 'bouncing off' and knocking the electron forward. The products of this kind of interaction are an energetic electron, an ionized atom and a photon of less than its original energy. This photon is still able to interact with the medium through which it is passing, of course. The energetic electron, too, interacts very strongly with the other electrons in the medium and it is often capable of producing secondary ionization of atoms, generating tertiary electrons and so on, until the original energy of the photon is spread throughout the medium—that is, deposited in it.

Pair production occurs when a photon interacts with the electromagnetic field of an atomic nucleus to create an electron–positron pair of particles. The particles are created out of the energy that the photon originally possessed, and since the rest mass energy of the pair is 1.022 MeV this process cannot occur for photons with less than this minimum energy.

All of these processes involve energy being deposited in the medium through which the photons are travelling. The photons cause ionization of the medium and they are scattered and ultimately absorbed.

Over the energy range 0.5 MeV to about 2.5 MeV, it is Compton scattering which is the dominant process whereby photons transfer their energy to matter. The photons are interacting with the electrons in matter, leaving behind ionized atoms, regardless of the form of the matter. The most important factor is the electron density.

For a *narrow* beam of monoenergetic gamma-rays of intensity I_0 travelling through a medium of density d, the residual intensity after traversing a distance x is given by

$$I = I_0 \exp(-\mu \cdot d \cdot x) \tag{2.8}$$

where μ is a constant, called the *mass absorption coefficient*, for the particular energy of the beam.

Equation (2.8) is a simple exponential function, and just as a half-life could be defined for the exponential decay of a radioisotope, so a *half-thickness* or *half-range* can be defined for gamma-rays in this energy band. For example, the half-thickness for the gamma-rays from cobalt-60, which have an energy of about 1.3 MeV, is about $15 \, \text{g cm}^{-2}$, while for the gamma-rays from caesium-137, which have an energy of 0.66 MeV, it is about $9 \, \text{g cm}^{-2}$. (The dimensions of this term, g cm^{-2}, arise because it is given by the product of density and distance.) As already mentioned, in the energy range 0.5 MeV to about 2.5 MeV the physical form of the medium through which the gamma-rays are passing is not very important. It is the electron density, which is related to the bulk density of the medium, which is the important factor. This is equivalent to saying that for cobalt-60 gamma-rays 15 cm (about 6 inches) of material of density $1 \, \text{g cm}^{-3}$ (say, water) will reduce the intensity of a beam of gamma-rays to half its original intensity; or, by simple proportion, 5 cm of material of density $3 \, \text{cm}^{-3}$ will do the same thing. If the half-thickness is known for a particular energy of gamma-ray, which is another way of saying that μ is known, equation (2.8) can be used to determine density provided distance is known, and vice versa. (The practical application of this equation is discussed in Chapter 13.)

It should be understood that the equation only strictly applies to narrow beams of radiation, which are rather difficult to guarantee in practice. For

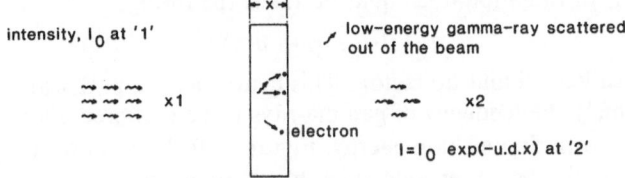

(a) Narrow beam conditions (monoenergetic beam both sides)

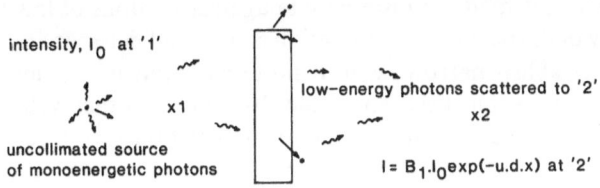

(b) Broad beam conditions

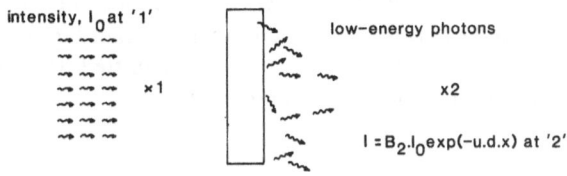

(c) Broad beam conditions

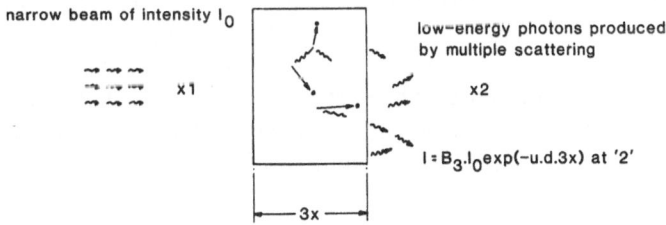

(d) Developing broad beam conditions

Figure 2.10. Absorption of gamma-rays and build-up. (a) Narrow-beam conditions (monoenergetic beam both sides); (b) broad-beam conditions; (c) broad-beam conditions; (d) developing broad-beam conditions.

broad beams of radiation, the equation takes the form,

$$I = B \cdot I_0 \exp (\mu \cdot d \cdot x) \qquad (2.9)$$

where B is called a 'build-up factor'. This factor takes a practical detail into account namely the tendency of gamma-rays to be scattered about through any medium with diminishing energy. Figure 2.10 shows how at a point of measurement on the other side of a block of absorber (some shielding material), gamma-rays which should not have reached the radiation detector at that point *do* reach it and are recorded. This phenomenon is called 'build-up' and refers to the component of low-energy photons at a point of measurement. Unless some effort is made to discriminate against photons of less than the original energy of the beam, or the radiation source and detector are carefully collimated to establish narrow-beam conditions, build up is inevitable. B usually takes some value between 1 and 10, but the exact value is very dependent on the geometry of the measurement, that is the precise relationship of radiation source and absorbers around it, and it cannot be predicted readily—it must be assigned from measurements on site. Usually, in real situations, B is ignored and a value of half-thickness determined on site. This is really an *effective* half-thickness for the geometry of the measurement. Another way of putting it is to say that a value μ_{eff}, for the mass absorption coefficient, is established empirically. As an example, the effective half-thickness for the gamma-rays from an uncollimated cobalt-60 source is about $20 \, \text{g cm}^{-2}$ when measurements of intensity transmitted through typical process vessels on chemical plant are conducted.

2.6.2 *Beta-rays*

Beta-rays are energetic electrons or positrons. They interact very strongly with the electrons in matter because of the long-range electrostatic forces due to their charges. Negative electrons, or negatrons, even if they are called beta-particles because of their point of origin, behave exactly as do other electrons in nature. They cause ionization of the media through which they travel and they are scattered and deflected by the forces they encounter, gradually giving up their energy and slowing down. As they slow down, they emit photons of bremsstrahlung as discussed in section 2.4.1. The electron, being a light particle, is deflected and bounced around in solid media very severely. Since it interacts very strongly, its range is relatively short—its penetrating power is low, although its exact range depends on its initial energy. The more energy it originally has, the more penetrating it will be. A 1.5 MeV electron will be stopped by a millimetre or two of aluminium. It will cause fairly intense 'hot spots' of ionization at discontinuous points along its erratic path inside a solid material. The beta-particle will behave exactly like a secondary electron generated by a gamma-ray interaction as described in section 2.6.1. The major processes, therefore, by which gamma-rays and beta-rays deposit energy inside materials have many common points. Although gamma-rays are much more penetrating than beta-rays, it is as if they actually generate a population of beta-particles inside an object exposed to them. The secondary processes are exactly similar.

Positrons are different in one important respect. Like negative electrons they interact strongly with the electrons already in matter and they are strongly ionizing, but unlike electrons, when they slow down they combine with an electron and the pair 'annihilate', creating two photons of energy 0.511 MeV each. This occurs because the positron is the so-called 'anti-particle' of the electron and it cannot survive free in the universe. (This photon creation is the exact opposite, the counterpart, of pair production discussed in section 2.6.1. Note that the energy involved, 2×0.511 MeV $= 1.022$ MeV, is the rest mass of the electron–positron pair. The reason why two photons are created with equal energies instead of one with 1.022 MeV is to conserve momentum as well as energy. The two photons leave the site of annihilation travelling in opposite directions, so there is no net momentum in the system.) It follows that positron emission is always accompanied by the emission of 0.511 MeV photons. Usually a radioactive source which emits positrons is also regarded as a source of 0.511 MeV gamma-rays since many of the positrons are annihilated while still inside the source, and even those that escape do not travel far. The photons are called gamma-rays because the process which initiated them was a nuclear decay, even though an intermediate step involved electrons. (It may occur to some that the status of bremsstrahlung as X-radiation is not firmly fixed. This may be why the special name has stuck.)

2.6.3 Alpha-rays

Alpha-particles interact very strongly with matter. They are highly charged, since they contain two protons, and cause intense ionization along fairly well defined, reasonably straight paths. In solids, they leave behind them a trail of damage which in certain materials is even visible through an electron microscope.

An alpha-particle is a large, heavy object compared with a beta-particle. If a beta-particle was about the size and weight of a small ball, say about 100 g, an alpha-particle would weigh about a three-quarters of a tonne, about the same weight as a small car. For a given energy, the heavier alpha-particle would be comparatively sluggish but it would transfer energy to the medium through which it was travelling very dramatically. Imagine that solid matter, which is a regular lattice-like arrangement of atoms, looks rather like a car park, with vehicles neatly parked in regular positions. Fire into this car park a beta-particle, a very fast tennis ball say, and it will bounce around quite erratically, being deflected and scattered until it rolls to a halt. Not much damage will have been done.

If an alpha-particle were fired in, however (as it were a small car moving at high speed), there would be intense local damage where the moving car crashed into the parked cars. At the point of impact, cars would be displaced from their neatly parked rows; they would be wrecked and bits would be broken off and littered around, though on the far side of the car park, little damage would have been done.

Put less picturesquely, alpha-particles are not very penetrating but they are intensely ionizing. An alpha-particle of 1.5 MeV will be stopped by the thickness of a sheet of paper. It will not penetrate human skin, but if the particle starts its journey on the inside of a person, as might happen if some radioactive material were swallowed, then it would do intense damage to the tissue inside.

2.6.4 Neutrons

Some mention should be made of neutrons as particulate radiation. Intense neutron fluxes are found inside nuclear reactors. Some radioisotopes also spontaneously emit neutrons. More likely to be encountered in industry, however, are small composite isotope sources which emit neutrons. The neutrons are actually produced as a result of a nuclear reaction when an alpha-particle is captured by a beryllium nucleus. For example, the alpha-particles from americium-241 in an americium/beryllium source yield neutrons by the following reaction:

$$_{95}Am^{241} \rightarrow _{93}Np^{237} + _{2}He^{4}$$
$$_{2}He^{4} + _{4}Be^{9} \rightarrow _{6}C^{12} + \text{neutron}.$$

These neutrons have high energy and they are referred to as 'fast neutrons'. Generally neutrons from this type of source have energies in the range 3–14 MeV.

Neutrons are not charged particles and so they do not interact with the electric fields of electrons or nuclei. They can only be stopped by direct collisions with nuclei, so fast neutrons are very penetrating, because matter is mostly empty space at the nuclear level. Interestingly, the best materials for slowing down fast neutrons and increasing their chances of absorption are ones which contain light nuclei, preferably hydrogen nuclei, that is protons. This is because the very light neutrons exchange momentum and energy best with particles of about the same size, and protons are the only nuclear particles which are the same size as neutrons. (The situation is the same as a billiard ball finds itself in when it bounces off the side of a billiard table. It does not exchange momentum very well with the massive table and so is reflected with much the same energy as it had when it struck the table. So it is with neutrons which tend to bounce off large nuclei with fairly undiminished energy. A billiard ball exchanges energy very well with another billiard ball, however, and it is not uncommon to see one ball come to a complete standstill, giving up all its energy to a target ball. And so it is with neutrons.) The best materials for shielding sources of fast neutrons therefore contain a lot of hydrogen per unit volume: water, paraffin wax or polyethylene.

Eventually fast neutrons lose most of their energy and come to thermal equilibrium with their environment. They are then called 'slow' or *thermal* neutrons. The process of slowing down is called *moderation* (see Chapter 15). The material causing the neutrons to be moderated is called the *moderator*. Thermal neutrons generally become absorbed in nuclei into which they drift by simple random, thermal motion. Interestingly, neutrons are unstable outside the nucleus and will decay by beta-decay to a proton and an electron with a half-life of about eleven minutes.

Select bibliography

The subject matter of this chapter can be dealt with at many levels of increasing sophistication and complexity. There is no shortage of material in print at the sophisticated end of the spectrum, where professional research physicists and university teachers spend their time. There are, however, few simple, direct texts which are accessible to the general reader or graduate in another discipline. The short list for further reading provided here are older books which give a more detailed treatment of the material included in this chapter in a reasonably readable but nevertheless rigorous way.

Modern Physics, Robert L. Sproull (2nd edn.), Wiley, New York. (1963)—Chapter 1, parts of Chapter 4 and Chapters 13, 14.

This book was originally written for engineers and it is strong on practical detail and experimental justification. The mathematics is not too daunting but can, in any case, be skipped over safely since the text is detailed.

Fundamental Atomic Physics, D. H. Tomlin, Blackie, Glasgow and London (1966)—pp. 155–165; pp. 185–192.

Although not as accessible to the general reader as Sproull's book, there are some good diagrams.

Quantum Physics (Vol. 4—Berkeley Physics Course), Eyvind H. Wichmann, McGraw-Hill, New York (1967)—Chapters 1–4.

An excellent, lucid account of the field developed in a convincing, logical and attractive way.

Applications of Nuclear Physics, J. H. Fremlin, English Universities Press Ltd. [Hodder and Stoughton, London] (1964).

This book is very readable and cogent. In places it is very entertaining and it is always pertinent. It is particularly good in its presentation of numbers and the sizes of things.

The Structure of Matter, R. M. Turnbull, Blackie, Glasgow and London (1979).

3 Radiation detection

K. JAMES

3.1 Introduction

It is usual to describe visible light in terms of its colour and its brightness. In a broader sense all radiations, whether particulate or electromagnetic, can be characterized by establishing (a) the energy and (b) the intensity of the radiation. Various detectors relevant to radioisotope techniques can measure one or both of these parameters. In this chapter we will review the most common types of detector and will see that virtually all detection techniques rely on the ability of the radiation to cause ionization. We will then describe in some detail the two most useful types of detector for industrial applications of radioisotopes and will conclude with a summary of the form of the electronic apparatus which takes the output from a detector and performs the necessary operations to accomplish the measurements.

3.2 Methods of detection

3.2.1 Gas-filled detectors

Several of the oldest and still most commonly used types of detector rely on the effects produced when ionizing radiation passes through a gas. The basic construction of a gas-filled detector is shown in Figure 3.1. In essence, the detector consists of a gas-filled chamber with a central electrode electrically insulated from the chamber walls. We apply a voltage between the wall and the central electrode through the resistance, R. When radiation enters the detector it causes ionization in the gas. The positive ions move towards the chamber wall and the negatively charged electrons move towards the central electrode under the influence of the applied field. Figure 3.2 shows a graph of the charge, Q, appearing on the capacitor as a function of the applied voltage, V. Two curves are shown: (a) for a weakly ionizing radiation; (b) for a strongly ionizing radiation. We can conveniently divide these curves into four regions as indicated.

Region I. The voltage applied is small. Electrons and positive ions take an appreciable time to travel to their appropriate electrodes and recombination of the ions can take place. Thus, only a small number of ions are collected. As V

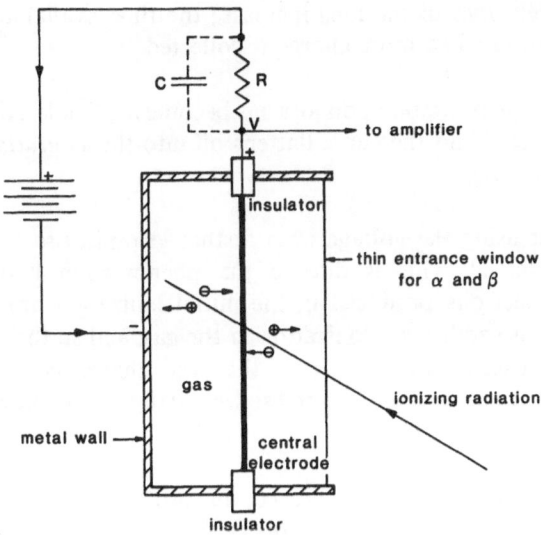

Figure 3.1. Gas-filled radiation detector: cross-section.

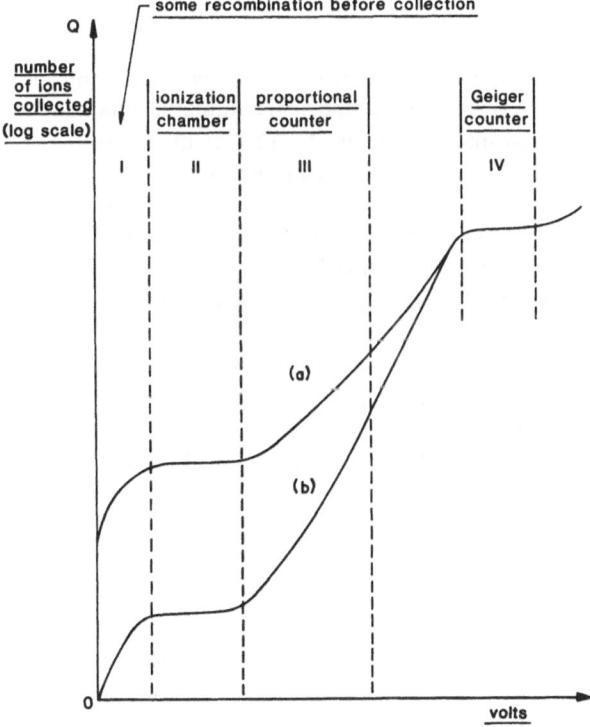

Figure 3.2. Output pulse height *v*. applied voltage (gas-filled radiation detector).

increases, the velocities of the ions increase; the time available for recombination decreases and so more charge is collected.

Region II. Here the recombination loss has become negligible. All of the ions created are collected and the curve flattens off into the *saturation region* or *ionization chamber region.*

Region III. Increasing the voltage even further, a rapid rise in the charge collected is observed. This is due to the phenomenon known as *gas multiplication.* Electrons produced in the initial ionization are accelerated sufficiently to cause additional ionization in the gas, and so the total charge collected is increased. Note, however that the charge collected is still proportional to the initial ionization and so we call this region the *proportional counter region.*

Region IV. If the voltage is again increased, this proportionality breaks down until eventually the number of ions collected is independent of the initial ionization and a saturation level determined by the characteristics of the counter is reached. This is known as the *Geiger–Müller region.*

Corresponding to regions II, III and IV of this graph, three types of detector are in use.

(i) *The ionization chamber.* These instruments can take many forms, but a typical construction consists of a cylindrical conducting chamber containing a central electrode and insulated from it—just as in Figure 3.1. The voltage applied between the wall and the central electrode is such that the counter operates in Region II of Figure 3.2. The gas filling can be practically anything, but dry air and the rare gases are popular. Ionization chambers can be used for the detection of all types of radiation but they are particularly useful for the detection of charged particles—alphas and betas. These instruments can be operated in either of two modes. (a) In the *pulsed* mode we detect the voltage pulses due to the ionization produced by the individual particles. This is used mainly for alpha-particles and always with low intensities of radiation. (b) In *mean-level operation* the current is measured and this is, of course, proportional to the *rate of detection* of the radiation. This mode is used for high intensities of radiation (alphas, betas and gammas) and is very commonly used in radiation monitors for health physics purposes.

(ii) *The proportional counter.* Here the applied voltage is such that the detector is operated in Region III of Figure 3.2. The gas multiplication factor can be of the order of 10^4. The counter is usually built to a cylindrical design. The gas filling is usually Ar, Kr or Xe, quite often at atmospheric pressure, and the counters are nearly always operated in the pulse mode. The size of the output is *proportional* to the primary ionization and this, in turn, is proportional to the energy of the radiation. The *number* of pulses recorded in a given time is a measure of the intensity of the radiation. Proportional counters are widely used for betas, alphas and *low-energy* gamma radiation. The counters cannot be used very successfully with gamma-rays of energy greater than about 60 keV because, at this energy, most of the gamma-rays simply pass straight through the filling gas without interacting.

(iii) *Geiger counters.* Geiger counters operate in Region IV of Figure 3.2. In this region the size of the output pulse is *independent of the primary ionization.* This means that the Geiger counter cannot give any information whatsoever about the energy of the radiation. Such counters are used only to measure the intensity. The Geiger counter is probably the most widely used of all detection devices, and is discussed in more detail in section 3.3. Because of the large number of secondary ions produced, these counters are much more sensitive than either the proportional counter or the ionization chamber.

3.2.2 *The scintillation counter*

This type of counter again depends upon the ionizing property of radiation. There are two distinct parts: a *phosphor* and a *photomultiplier tube.* The interaction of the radiation with the detector takes place in the phosphor. The initial ionization process releases electrons from the various electronic bands of the solid, leaving vacancies. Some of these vacancies are filled by electrons from luminescent centres which are present in the phosphor. When this happens, energy is released as a flash of light. The greater the initial ionization, the larger will be the resulting light flash. This light pulse then passes into the photomultiplier tube which in turn produces a current pulse at its output. The size of this pulse is proportional to the energy of the initial radiation, and this type of detector can therefore be used to measure the energy of the radiation in addition to its intensity.

Scintillation detectors are widely used in industrial problem-solving applications and are described more fully in section 3.4.

3.2.3 *Photographic techniques*

Historically, the first known method of radiation detection was to note the blackening produced by the radiation on a photographic plate. Methods akin to this are still widely used today. A common example is the X-ray film used in clinical radiography; similar films are used to detect gamma-rays in industrial applications. Once more, it is the ionizing property of the radiation which we make use of. Radiation, just like ordinary light, liberates electrons from the bromide ion and these electrons are subsequently captured to liberate free silver. This causes the blackening of the film.

The photographic method is widely utilized in personal monitor badges. These badges contain a strip of film and are worn throughout the time that a worker is exposed to radiation. The subsequent developing of the film allows us to estimate how much radiation exposure the worker has received. In the UK, the film badge is becoming less and less common and is being replaced, in general by the TLD badge (see below).

Photographic techniques can be applied to the detection of all types of radiation.

3.2.4 *Thermoluminescent detectors*

Thermoluminescent detectors (or TLDs) are also used for personal monitoring. They are, in a way, similar to scintillation counters. They consist of inorganic crystals (lithium fluoride being the best known), and the initial interaction of the radiation is to liberate electrons from some of the electronic bands. This, however, is where the similarity ends because, whereas in scintillation counters the vacancy created is immediately filled, in these crystals the vacancy remains empty. As more and more radiation falls upon the crystal, more and more vacancies are created, so that the number of vacancies is a measure of the total radiation exposure. After the exposure is over, the crystal is heated by placing in a special oven. This causes the vacancies to be filled and light to be produced, and if we then measure the intensity of the light, this is a measure of the exposure.

Such detectors are, in many establishments, used in parallel with the film badges to keep a check on the radiation dose received by personnel.

3.2.5 *Neutron detection methods*

Methods of neutron detection are simple modifications of the technique used for other types of radiation. Neutrons themselves do not produce ionization since, being uncharged, they do not interact with the orbital electrons. They do, however, interact with nuclei. In particular, they can interact with certain specific nuclei such as ^{10}B or ^{6}Li to produce alpha particles. If we then detect these alpha particles, then the number we measure is clearly related to the initial number of neutrons. To detect the alphas, we use any of the methods previously discussed.

Perhaps the most popular type of detector is the BF_3 proportional counter. This is just like an ordinary proportional counter except that the gas filling is BF_3. The alphas are produced in the body of the counter from the neutron–boron interaction and are detected in the usual way. A more recent neutron detector is a proportional counter filled with helium-3 gas, which produces ionizing particles via the reaction

$$\,^{3}_{2}\text{He} + \,^{1}_{0}\text{n} \rightarrow \,^{3}_{1}\text{H} + \,^{1}_{1}\text{p} + 765\,\text{keV}$$

Another useful detector is the Li-glass scintillator. This consists of a scintillation crystal impregnated with ^{6}Li. Alphas produced in the neutron–lithium reaction are detected by the scintillation technique.

3.2.6 *Semiconductor detectors*

(i) *Silicon or germanium–lithium drifted (GeLi).* This detector is normally only used in special analysis instruments. It consists of a semiconducting crystal, usually of germanium, in which is formed a relatively large volume 'intrinsic' region by a lithium-drifting process, which also forms a pn junction. The detector is operated at the temperature of liquid nitrogen to reduce generated noise. The output signal pulses caused by ionization are of the order of tens of millivolts and are usually amplified with an integral preamplifier. The detector exhibits very good resolution and produces an output which is linear with the energy of the ionizing radiation.

(ii) *Cadmium telluride.* An expensive semiconducting crystal which is operated at room temperatures, made from grown crystals up to 1 cm diameter, its main use is in applications where a physically small detector is required, e.g. medical work.

3.3 The Geiger counter in more detail

The Geiger counter is a simple, rugged, low-cost device which exhibits high sensitivity compared to the other gas-filled detectors and which produces a remarkably stable output under changing voltage and temperature conditions. Despite its age (having been introduced by Geiger and Müller in 1928), these features make the Geiger counter one of the most popular detectors for use in an industrial environment, particularly in applications which involve permanently installed nucleonic instrumentation i.e. level gauges, density gauges etc.

3.3.1 *Construction and operation*

A typical Geiger counter consists of a metal tubular case (cathode) containing a fill gas (for example the noble gases, particularly helium and argon) and a central rod-like anode which may be supported at one or both ends. A thin mica window positioned at one end of the tube will allow the counter to detect beta- or energetic alpha-particles.

When charged particulate radiation enters the counter, a number of ion pairs are created within the gas along the radiation track. Each original electron produced during this initial ionization is accelerated towards the

C

central anode, colliding with other gas molecules and liberating secondary electrons. The secondary electrons produce tertiary electrons, and ultimately an avalanche of electrons is collected at the anode.

In a proportional counter, each original electron leads to the formation of a single avalanche and because all avalanches are nominally identical, the charge collected at the anode remains proportional to the number of original electrons, i.e. proportional to the energy of the ionizing particle. However, in a Geiger counter a somewhat different set of conditions apply. Here the strength of the electric field around the anode is much higher and the number of electrons involved in an avalanche (typically 10^{10}) is very much greater than in a proportional counter. Similarly, the number of electron collisions which do not lead to further ionization but create 'excited' gas molecules is very much greater in Geiger avalanches than in proportional-counter avalanches. These excited gas molecules can return to their ground state by the emission of photons in the visible or ultraviolet region, and the interaction of one of these 'de-excitation' photons with a gas molecule elsewhere in the detector can liberate an electron through the photoelectric effect and initiate another avalanche. While the fill gas is relatively transparent to visible and ultraviolet light, so many de-excitation photons are generated during a Geiger avalanche that photoelectric absorption somewhere within the gas is highly probable. One Geiger avalanche will therefore trigger a whole chain of further, virtually simultaneous avalanches at random positions throughout the tube and the Geiger discharge will grow to envelop the whole anode (Figure 3.3.).

The Geiger discharge is terminated only because of the relative immobility of the positive ion which is created along with each electron in an avalanche. As the discharge proceeds, the concentration of positive ions builds up and begins to reduce the magnitude of the electric field at the anode to a level where no further gas multiplication is possible. For a fixed applied voltage, a particular density of positive ions will be needed to reduce the electric field below the minimum value required for further multiplication. Consequently, each Geiger discharge is terminated after developing the same total charge regardless of the number of original ion pairs created by the incident radiation. All Geiger output pulses are therefore of the same size, and their amplitude can provide no information about the energy of the incident radiation.

It is interesting to note that the average energy reached by a free electron between collisions with gas molecules depends on the ratio E/P, where E is the electric field strength and P is the gas pressure. Below a critical value of this ratio it is impossible to create a full Geiger discharge. Commercially available Geiger tubes have rugged anodes and tend to be operated at moderate voltages (typically 500 V between anode and cathode). Consequently, their fill gases are normally introduced at low pressure, typically a few tenths of an atmosphere, in order to maintain a high E/P ratio. At this low pressure, a gamma-ray of moderate energy will pass through the gas without causing ionization. Indeed, the detection of gamma-rays by Geiger counters depends

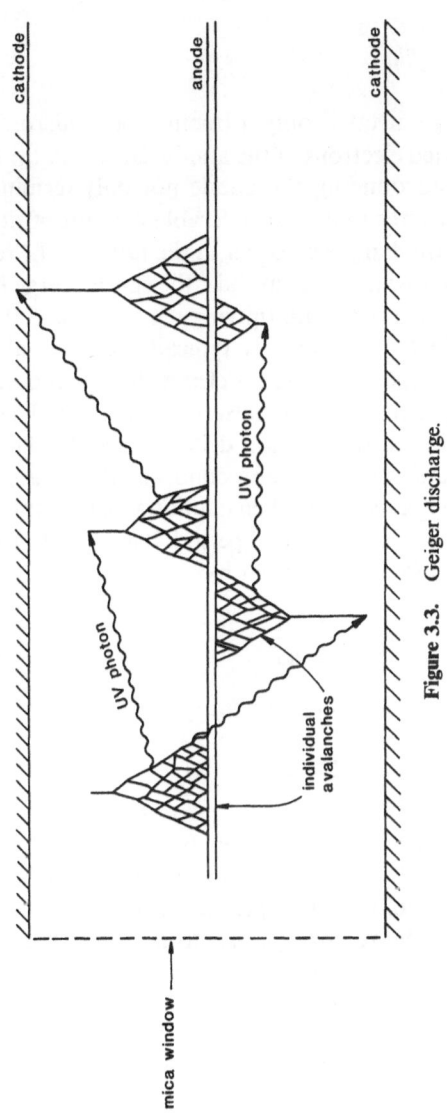

Figure 3.3.   Geiger discharge.

entirely on the ability of the incident gamma photon to interact within the wall of the tube (cathode) and eject an electron into the gas. Only a single ejected electron need enter the gas to produce a full Geiger discharge as described above.

3.3.2 Dead time and pulse profile

Following a Geiger discharge, it takes only a fraction of a microsecond to collect the avalanche-generated electrons at the anode. However, the cloud of positive ions which is left surrounding the anode not only terminates the Geiger discharge but also ensures that a considerable amount of time must pass before a second Geiger discharge can occur in the tube. As the relatively immobile positive ions move towards the cathode, the space charge becomes more diffuse and the electric field in the multiplying region close to the anode starts to return to its original high value. It typically requires about 100 microseconds of positive ion dispersal before the electric field near the anode is sufficiently high to allow a further Geiger pulse to take place. The tube is effectively dead for this period of positive ion drift—hence the term 'dead-time'. If a Geiger tube produces n pulses in one second, and the tube is dead for a period of T seconds following each pulse, then it follows that the tube is dead for a total of nT seconds within the one-second period. If no dead time existed, the tube would produce a countrate N_0 given by

$$N_0 = \frac{n}{1 - nT}.$$

Using the above equation we can define a quantity

$$\frac{N_0 - n}{N_0} = nT$$

which is the fractional reduction in countrate due to dead time losses. For a dead time of 100 microseconds it can be seen that this quantity is significant ($= 0.1$) at an observed countrate of 10^3 pulses per second, becoming unity when $n = 10^4$ pulses per second. For this reason, it is usually necessary to restrict the use of Geiger tubes to applications in which the observed countrate is no more than a few hundred pulses per second.

Dead-time loss is not the only factor which affects the countrate measured by a Geiger-based instrument in a fixed field of radiation. Some care must be exercised in the design of the counting circuit, otherwise some of the pulses coming from the Geiger tube will fail to be registered.

An equivalent counting circuit for a Geiger tube is shown in Figure 3.4. The voltage V across the load resistance R is the basic electrical signal. In the absence of any ionizing radiation this signal voltage is zero and all the applied voltage V_0 appears across the tube. In the presence of ionizing radiation, a

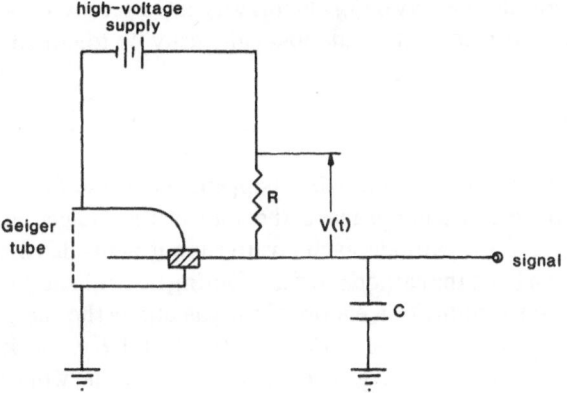

Figure 3.4. Equivalent counting circuit for Geiger tube.

Geiger discharge will take place, charge will be collected at the tube electrodes and the voltage across the tube will be reduced from its equilibrium value V_0. Simultaneously, a signal voltage V (equal to the amount by which the tube voltage has dropped) appears across the load resistance. This output pulse consists of two components: an initial fast slope corresponding to the collection of electrons and a subsequent more gradual rise corresponding to the collection of positive ions. If a signal which accurately reflects the charge of both the electrons and the ions is to be generated (i.e. a signal of maximum possible amplitude) the collection circuit time constant RC must be long compared with the time required to collect the positive ions, i.e. much larger than 100 microseconds. Under these conditions the Geiger tube must be operated at very low pulse-rates to avoid excessive pulse pile-up. If we wish to register all of the discharges taking place within the tube, we must employ short time constants (RC of the order of a few microseconds) which effectively eliminate the slow-rising portion of the pulse and leave only the fast leading edge, as shown in Figure 3.5. While a significant fraction of the potential

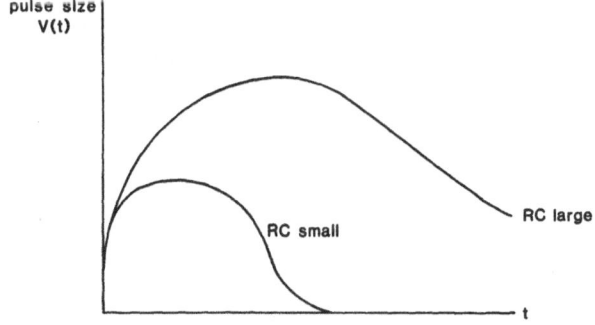

Figure 3.5. Effect of time constant RC on Geiger pulse shape.

output pulse amplitude may be lost in this way, a Geiger discharge produces so large a pulse that some amplitude loss can easily be tolerated.

3.3.3 Quenching

So far we have not considered what happens when the large positive ions generated during a discharge reach the wall of the Geiger tube (cathode). When it arrives at the cathode, each positive ion is neutralized by combining with an electron from the cathode surface. During neutralization an amount of energy equal to the ionization energy of the gas minus the energy required to remove an electron from the surface of the metal (i.e. work function) is liberated. If this liberated energy exceeds the value of the work function, it is possible that another free electron will be ejected from the cathode and enter the gas. In turn this would lead to a second Geiger discharge which, during positive ion collection, could lead to another electron being liberated, another discharge and so on. Under these conditions a Geiger counter, once initially triggered, would produce a continuous output of multiple pulses even if the source of radiation were to be removed.

In order to overcome this problem, Geiger tube manufacturers add a 'quench gas' to the fill gas. The quench gas always has a lower ionization potential than the primary gas component and may be organic (e.g. ethyl alcohol, ethyl formate) or, in more recent designs, a halogen such as chlorine or bromine. The positive ions formed in a discharge are mostly of the primary gas. These ions collide with many other gas molecules during their drift towards the cathode and some of these collisions will be with molecules of the quench gas. Since the quench gas has a relatively low ionization energy, there is a tendency for positive charge to be transferred from the fill gas ion to the quench gas molecule. If the concentration of the quench gas is sufficiently high, all of the ions which reach the cathode will be of the quench gas. When they are neutralized, the energy liberated goes into dissociation of the quench gas molecule in preference to ejecting a free electron from the cathode surface. Consequently, no unwanted additional avalanches are produced within the tube and the situation where the Geiger produces a continuous output of multiple pulses is avoided. Organically quenched tubes have a limited useful life because the dissociation of the quench gas is permanent. On the other hand, subsequent recombination of dissociated halogen molecules can take place and halogen-quenched tubes, therefore, have very long lives (typically 5×10^{10} counts).

3.3.4 Plateau characteristics

If a Geiger tube is held in a fixed radiation field, the observed countrate varies with applied voltage as shown in Figure 3.6. On increasing the applied voltage

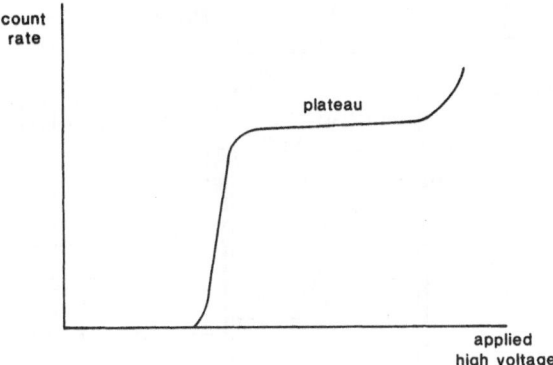

Figure 3.6. Counting plateau for Geiger tube.

from zero, no pulses are recorded until the starting voltage is reached. At this point on the curve the electric field around the anode is just sufficient to establish electron avalanches. With further increases to the high-voltage supply, the curve passes through a transition region (its 'knee') and then shows a substantially flat plateau for the next 100 volts or more. If the voltage is raised sufficiently high the plateau ends abruptly due to the onset of continuous discharge mechanisms in the tube, e.g. corona discharges from sharp irregularities on the anode.

Within the plateau region the observed countrate is virtually independent of applied voltage. Changes to the applied voltage in this region will alter the size of the voltage pulses produced but will essentially cause no change to their rate of production. The large voltage range of the plateau and its near flatness enable accurate measurements of radiation intensity to be made without the need for expensive highly-stabilized h.t. supplies—an attractive feature of detectors based on Geiger counters.

3.4 The scintillation counter in more detail

We have seen in the previous section that Geiger detectors can provide no information regarding the energy of the incident radiation and that their use is limited to the measurement of relatively low countrates due to dead-time considerations. At low countrates it is difficult to resolve small changes in radiation intensity unless long counting periods are used.

Scintillation counters sufffer from neither of these limitations: energy information is retained and countrates as high as 10^5 s^{-1} can sensibly be employed. Scintillation counters, therefore, find wide use in the industrial application of radioisotope techniques. However, their relatively high cost, fragility and sensitivity to changing voltage and temperature conditions make scintillation detectors a less popular choice than Geiger-based devices for

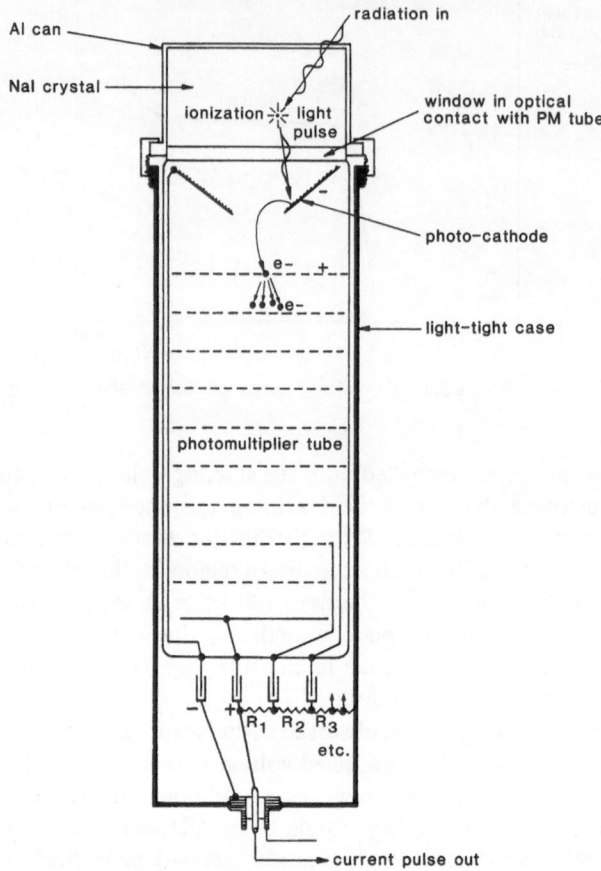

Figure 3.7. Scintillation counter (schematic diagram).

applications which require the equipment to provide continuous measurements over a long period of time (e.g. instrumentation for process control).

The construction of a typical scintillation counter is shown diagrammatically in Figure 3.7. Two distinct components are involved: a phosphor or scintillator where ionizing radiation produces light, and a photomultiplier tube which converts the light into a measureable electrical signal.

3.4.1 *The phosphor*

Phosphor material exists in a variety of forms—organic, inorganic, solid, liquid and plastic. Suitable phosphors for alphas, betas and gammas are available and for extremely low-energy particulate radiation the radioactive material can be dissolved in a liquid scintillator. The radiation is then produced in the body of the detector and can, therefore, be counted very efficiently. The most widely applied phosphor is sodium iodide doped with a little thallium, and we will use this material to illustrate the mechanism by which ionizing radiation creates scintillations.

In a pure sodium iodide crystal, the electrons are constrained to exist in discrete energy bands. The lower band, called the valence band, represents those electrons that are found at lattice sites, whereas the upper (conduction) band represents those electrons which have sufficient energy to be free to move throughout the crystal. Separating these two energy levels is the 'forbidden gap' in which electrons can never exist in the pure crystal. If sufficient energy is transmitted to an electron, it can be elevated from its normal position in the valence band across the gap into the conduction band, leaving a hole in the normally filled valence band. If the electron were to return to refill the hole a photon would be produced, but the gap width in the pure crystal is such that this photon would be of too high an energy to lie in the visible range and its subsequent detection would not be possible.

In order to enhance the probability of visible photon emission during de-excitation, about 10^{-3} mole fraction of thallium is added to the sodium iodide as an activator. The activator creates special sites in the lattice where the energy band structure is altered from that of the pure crystal. Energy states are created within the forbidden gap through which the electron can de-excite back to a lower energy condition. Since the energy is less than that of the full forbidden gap, de-excitation transitions now produce visible photons and therefore serve as the basis for the scintillation process.

Ionizing radiation passing through the phosphor creates a large number of electron–hole pairs by elevating electrons from the valence to the conduction band. The positive hole moves to the location of an activator site and ionizes it because the ionization energy of the 'impurity' atom is less than that of a normal lattice site. The electron migrates through the crystal until it comes across a positively charged activator. The electron combines with the charged activator to form a neutral impurity configuration which has its own set of excited energy states within the forbidden gap (see Figure 3.8). De-excitation to the activator ground-state energy quickly takes place and is accompanied by the emission of a visible light photon—410 nm wavelength for the most prominent transition in NaI (Tl).

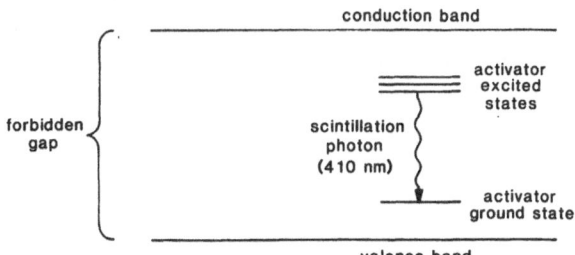

Figure 3.8. Schematic representation of energy band structure for NaI (Tl).

Thallium-doped sodium iodide exhibits excellent light yield with a light photon being produced for almost every electron–hole pair generated by the incident ionizing radiation. Moreover, the total light produced in a single scintillation pulse is linearly related to the energy of the incident radiation. Unfortunately, the NaI (Tl) crystal is rather fragile and deteriorates due to water absorption if exposed to the atmosphere. For this reason commercially-available crystals are often supplied in thin aluminium 'cans' which protect all surfaces of the crystal other than a glass-covered window which is brought into optical contact with the photomultiplier tube.

3.4.2 *The photomultiplier tube*

When ionizing radiation interacts within the phosphor, only a very weak flash of light is produced. While the number of photons created by an ionizing event varies linearly with the energy of the incident radiation, only about 10^4 photons are generated even for 1 MeV gamma rays. The purpose of the photomultiplier tube is to convert these weak light signals into measurable electrical signals.

The evacuated photomultiplier tube contains a photosensitive layer called the photocathode, and an electron multiplier structure (dynode chain). An external high-voltage source must be connected through a resistive voltage divider in such a way that the first dynode is held at a voltage which is positive with respect to the photocathode and each succeeding dynode is held at a positive voltage with respect to the preceding dynode. In a typical photo-multiplier with ten dynodes, it is usual to establish a potential difference of about 200V between the photocathode and the first dynode, and to maintain a voltage difference of about 100V between each of the following dynodes.

When visible light reaches the photocathode, a number of electrons are liberated via the well-known photoelectric effect. The number of liberated electrons is directly proportional to the number of photons generated in the scintillator and this in turn is linearly related to the energy of the ionizing radiation. The kinetic energy of an electron is very small when it first leaves the photocathode. However, each electron will be rapidly accelerated towards the first dynode and will typically reach a kinetic energy of 200 eV immediately before it strikes this plate. The energy required to liberate a secondary electron from the dynode surface is about 3 eV, so in theory it is possible for each electron striking the first dynode to liberate a further 60 or so secondary electrons. In practice, for conventional dynode materials such as BeO and MgO, the number of secondary electrons created for each electron striking the dynode is closer to 5 than 60. Nevertheless, if there are ten dynodes in the chain, each electron liberated at the photo-cathode will give rise to 5^{10}, i.e. about 10^7, electrons at the final charge collection plate (anode).

With gains of this magnitude it is easy to appreciate that measureable charge pulses will be produced by the photomultiplier when ionizing radiation interacts with the scintillator. The rate at which charge pulses are produced is a measure of the intensity of the radiation, and the amplitude of the pulses is directly related to its energy.

Since the various processes occurring in a scintillation counter are based on extremely short time-scales, it is convenient to ignore the effects of dead time when using this type of detector. However, there are two potential problems which, depending on the application, should be borne in mind when considering the use of scintillation counters. The first of these relates to thermionic emission from the photocathode. At room temperature there is a finite probability that an electron will be liberated from the photocathode due to thermal effects, giving rise to a small charge pulse at the anode. The probability of thermionic emission is dependent on temperature so, even in the absence of ionizing radiation, the photomultiplier will produce a background countrate (noise) which varies with temperature. Since this noise is comprised of small amplitude charge pulses, it is usual to establish a pulse amplitude threshold in the ancillary counting equipment so that most thermionic pulses fail to register. However, this introduces a second potential problem. The gain of the photomultiplier (i.e. overall electron multiplication factor) typically varies as V^6 to V^9, and small voltage variations will not only change the magnitude of the registered charge pulses but will also cause more or less thermally initiated pulses to exceed the threshold amplitude. Accurate measurements of radiation energy (pulse height) and intensity (pulse rate) therefore require a scintillation counter to be operated at a fairly constant temperature with a highly stable voltage supply.

3.5 Pulse processing equipment

Having described the various detector types, it is perhaps worthwhile to summarize the form of the ancillary equipment necessary to make meaningful radiation measurements.

A general detection system suitable for use with all detector types is shown schematically in Figure 3.9. Its basic components are as follows.

3.5.1 *EHT unit*

This provides the high voltage supply necessary to operate the counter. Depending upon which type of counter is being used, the unit must be capable of producing output voltages up to about 3 kV. Except in the case of Geiger counters, the EHT unit must provide a highly stable voltage output. Scintillation counters and proportional counters are sensitive to changes in the applied voltage.

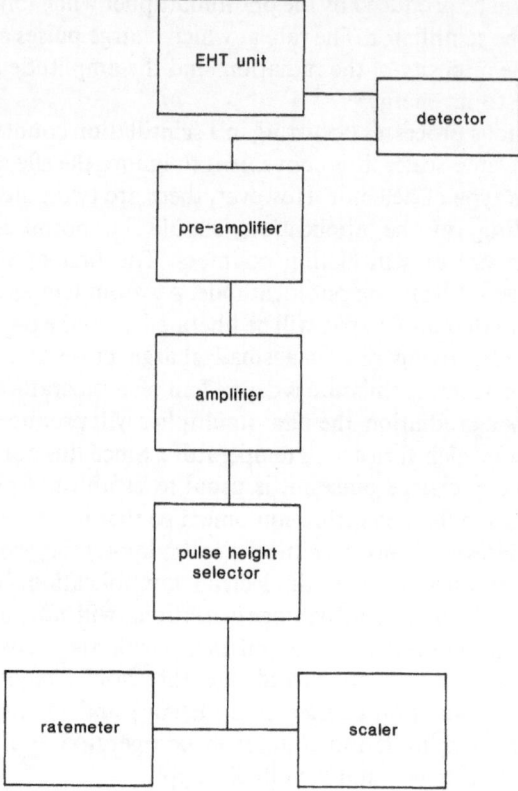

Figure 3.9. Typical radiation detection system.

3.5.2 *Preamplifier*

The fundamental output of all pulse-type radiation detectors is a burst of charge Q which is liberated by the incident radiation. For Geiger tubes, Q is large and the voltage pulse produced by integrating Q across the summed capacitance represented by the detector and the connecting cable is also relatively large. For other detectors, however, the output pulse from the detector is usually less than a millivolt in amplitude. The first element in the signal processing chain is, therefore, a pre-amplifier. This unit typically has a gain of about 10 and is usually placed physically close to the detector. It serves as an impedance matcher between the detector output and the cable leading to the main amplifier.

3.5.3 *Main amplifier*

The main amplifier can have a gain as high as 1000, usually adjustable over a wide range. After amplification the pulses emerge with amplitudes of several volts. For applications in which energy measurements are to be made it is important that the amplifier has a linear response to the incoming voltage pulses. In this way the proportionality between the amplitude of the output pulse and the energy of the radiation is preserved.

3.5.4 *Pulse-height analyser or selector*

As its name implies, the purpose of this device is to select pulses of a given amplitude for counting purposes. Often we are interested only in radiation of one particular energy. However, our detector will pick up all radiation present and in general, will produce a range of output pulses of corresponding amplitudes. The pulse-height analyser allows us to discriminate against unwanted energies by establishing upper and lower amplitude thresholds. Only pulses corresponding to energies which lie within the 'window' established by the threshold potentiometers will be recorded.

3.5.5 *Scaler/ratemeter*

After the pulses leave the analyser they may be fed directly into an electronic counting unit called a scaler. This counts each pulse separately and produces a digital display of the total number of counts recorded in a preset time interval.

Alternatively, (or additionally) the pulses may be passed into a ratemeter circuit. This unit effectively integrates the counts received over a time period characterized by the circuit time-constant and produces a continuous output voltage which is proportional to the countrate. The ratemeter output can be fed into a chart recorder should a permanent record of the countrate be required.

4 Radioactive sources

T. L. JONES

Radioactive sources are classified into two distinctly different categories: (a) sealed and (b) unsealed sources. There are several important differences between these categories, principally in the potential radiological hazards associated with them and the corresponding legislation which is reviewed in the next two chapters. There is also a basic difference in the way that they are used. Unsealed sources are usually added to a system with the objective of tracing a pathway or determining a distribution, i.e. they are used as tracers. Sealed sources are used as a source of radiation to investigate what can be described broadly as the interaction of a radiation with matter. Consequently the chemical and physical form in which the unsealed radioactive source is produced will depend upon the nature of the system under investigation. A sealed radioactive source is invariably produced as a solid, but the chemical form of the radioactive material is of little importance to the user who is concerned only with the quantity and type of radiation emitted and with the type of encapsulation.

4.1 Production of radioactive sources

Before the discovery of artificial radioactivity in the 1930s, the only radioisotopes available were the naturally-occurring parent isotopes uranium-235, uranium-238, thorium-232 and their decay products, known as daughters. The important work which was carried out at the turn of the century and which began to characterize radioactivity for the first time led to the isolation of several of these daughter products and introduced the names of Curie, Becquerel and Rutherford into the field of ionizing radiation.

After the 1930s a number of artificial radioisotope sources were produced by bombarding target elements with a whole range of particles produced in accelerators such as the cyclotron. It was not until the discovery of nuclear fission, however, with the subsequent availability of a high concentration of neutrons (usually referred to as a high flux) with which to bombard the target elements, that artificial radioactive materials became available in large numbers.

The higher energies which can be achieved in modern particle accelerators have widened the field to include other types of nuclear transformation, such as spallation, which is the breaking off of a series of light fragments from a heavy nucleus following collision with a high-energy (i.e. fast) particle.

At present there are approximately 180 radioisotopes in use, of which about 120 are obtained from neutron-irradiated targets, 30 by irradiation in particle accelerators, 10 from naturally-occurring materials and about 20 from fission products. Several radioisotopes can be produced by more than one method, and there has been a growing interest in the use of radioisotope generators to isolate short-half-life isotopes for specific applications.

4.2 Radioisotopes from natural sources

The radioisotopes in the uranium and thorium decay series have a wide range of radiation energies and decay schemes. Part of the decay scheme from uranium-238 is shown in Figure 4.1. Although the use of these radioisotopes as tracers is limited, they have been used extensively in the medical field, ^{226}Ra in radiotherapy for example, and their importance lies in the fact that they

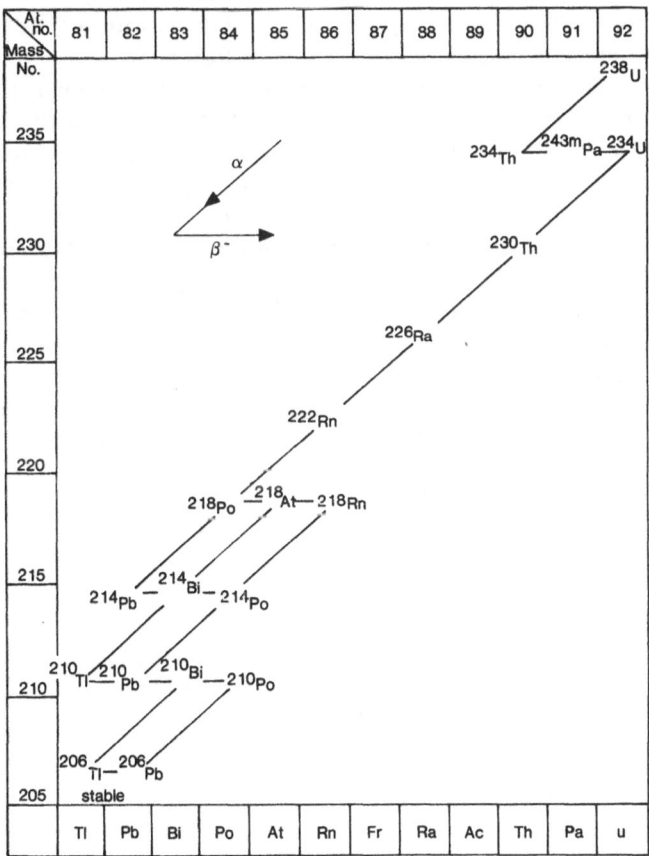

Figure 4.1. Uranium-238 decay series.

enabled the basic techniques of radioisotope tracer investigations to become established.

It might be assumed that the gradual, increasing development of techniques for producing artificial radioactive sources would end any interest in natural radioactive materials. In fact they continue to find uses in many areas up to the present time. In most cases, however, natural radioisotopes are more expensive to produce than artificial ones. The main reason for this is that, while artificial sources can be obtained, at least initially, in high concentrations, costly separation and enrichment techniques are required to isolate natural radioisotope sources at high specific activities. In fact the distinction between 'naturally-occurring' and 'artificial' radioisotopes is less than clear because some 'naturally-occurring' radioisotopes are best prepared artificially.

In addition to the daughter products of the uranium and thorium decay series there are a number of lower atomic number radioisotopes which occur naturally and some of these are listed in Table 4.1. Some of these isotopes are being produced continuously by nuclear processes in the upper atmosphere initiated by cosmic radiation, e.g. ^{3}H and ^{14}C. Others with very much longer half-lives were constituents of the earth at its formation.

Carbon-14 has been produced at a fairly constant rate for many thousands of years and has reached a steady-state concentration in the earth's biosphere over this period. As a result of this, living biological organisms through their metabolic processes have attained the same ^{14}C concentration in their carbon content as the biosphere. After death the ^{14}C in the material decays without replenishment by exchange. Thus, from a knowledge of the half-life of ^{14}C the age of the specimen can be determined. As a result of intensive study of biochemical processes and organic chemical reactions over the past few decades, carbon-14 has become the most widely used of all radiochemical tracers. It is now produced by neutron irradiation.

Of all the naturally-occurring radioisotopes the most abundant is potassium-40. This isotope constitutes 0.012% of terrestrial potassium and the

Table 4.1   Naturally-occurring radioisotopes

Element	Half-life (years)	Radiation
^{3}H	12.3	Beta
^{14}C	5.76×10^3	Beta
^{40}K	1.3×10^9	Beta, electron capture
^{50}V	4×10^{14}	Electron capture
^{87}Rb	5×10^{14}	Beta
^{115}In	6×10^{14}	Beta
^{138}La	1×10^{11}	Beta, EC, gamma
^{144}Nd	5×10^{15}	Alpha
^{147}Sm	1.3×10^{11}	Alpha
^{176}Lu	2.4×10^{10}	Beta, EC
^{187}Rh	5×10^{10}	Beta
^{190}Pt	5.9×10^{11}	Alpha

natural radiation (1.46 MeV gamma radiation) can be used for the continuous analyses of production processes involving potassium compounds. Without further concentration, however, it is of little value in other applications. Potassium-40 does also have a nuisance value in that the presence of potassium in radiation counting apparatus contributes to the instrument noise (background) level, and low potassium-content glass has to be used for the manufacture of counting vials, etc.

4.3 Fission products

The structure of the nucleus was described in Chapter 2 and Figure 2.2 illustrates graphically that with increasing size the nucleus contains an increasing excess of neutrons over protons in order to achieve stability. As the atomic number approaches 100 the balance between the repulsive forces of the protons and the surface forces holding the nucleus together reaches a critical stage. At this point it requires only a moderate input of energy to produce fission, which is defined as the splitting of a nucleus into two or more parts. The energy can be in the form of electromagnetic radiation, i.e. gamma-rays, or in the form of particles. In most cases, however, it is associated with neutrons and the phenomenon was first observed following the bombardment of uranium with neutrons.

It is impossible for the fission fragments to be stable because they will contain an excess of neutrons. Stability can be achieved by conversion of a neutron into a proton and emission of an electron by the process known as beta-decay. Following this transition the nucleus is in an excited state, i.e. not at its lowest energy level. The change to the lower (ground) state is accompanied by the emission of gamma-radiation. Also, because the fission products have a smaller total mass than that of the target nucleus, the process is always associated with the release of large amounts of energy. This energy can take the form of heat, electromagnetic radiation and kinetic energy of the particles released in the fission. This is the basis of nuclear power production and of some nuclear weapons. The fission products are an inevitable consequence of nuclear reactor operation and are separated when the fuel elements are processed to recover uranium-235 and plutonium-239. The fission yield from uranium-235 is shown in Figure 4.2.

Accumulated fission products consist essentially of a mixture of some 37 elements with atomic numbers ranging from around 30 to around 70 (the rare earth elements). This process followed by physical or chemical separation provides an abundant supply of radioisotopes such as krypton-85 and caesium-137. The fission process is also used for the preparation of some shorter-lived radioisotopes such as iodine-131 and xenon-133 which cannot easily be produced by other means, but are obtained by irradiation of a uranium target specifically for this purpose.

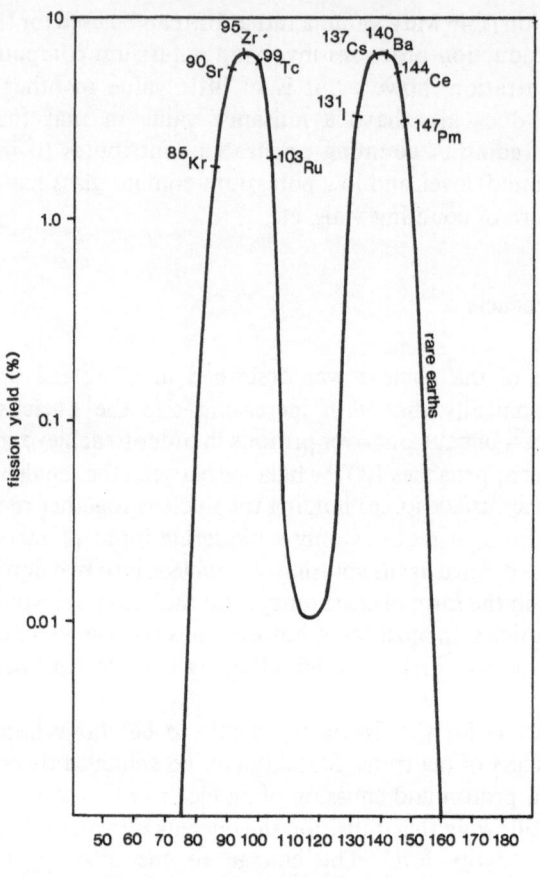

Figure 4.2. Uranium-235 fission yield.

4.4 Neutron activation

The radioisotopes produced by neutron bombardment of inactive targets are by far the largest group, and this process has been responsible for the very extensive developments in the use of radioactive materials in research, medicine and industry over the last 40 years. This expansion was initiated by the building of a number of nuclear reactors which continue to be the most useful source of neutrons. The reactors are capable of producing a high flux of both fast and thermalized, or slowed-down, neutrons. Thermal neutrons have a greater probability of colliding with a target, and consequently most neutron processes require thermal neutrons. A typical neutron flux in a reactor is of the order of 10^{12} neutrons $cm^{-2} s^{-1}$.

It is useful at this stage to introduce two concepts. These are the *nuclear equation* and the *effective cross-sectional area.*

Nuclear changes can be represented by equations in which the atomic numbers and the mass numbers balance on both sides. For example, the capture of a neutron by a sodium atom, yielding an isotope of sodium accompanied by the emission of gamma radiation, is written

$$_{11}^{23}\text{Na} + {}_0^1\text{N} \rightarrow {}_{11}^{24}\text{Na} + \gamma.$$

The subscript is dropped by convention and the reaction is commonly shortened to $^{23}\text{Na}(n,\gamma)^{24}\text{Na}$ and is referred to as an (n,γ) reaction. This process, known as neutron capture, is by far the most common form of interaction between neutrons and atoms. As a consequence it is used for the production of more than half of the radioisotopes which are in general use. The product of this particular type of reaction is an isotope of the target element, so chemical separation of target and product is not possible and thus the specific activity of the radioisotope is limited. For most industrial applications, however, this is not a problem and there are positive advantages in having targets and radioisotopes which are the same chemical element.

The effective cross-sectional area of the target element is a measure of the probability that a given nuclear reaction will occur upon collision of the incident particle with the target and is expressed as an area. It is usually referred to as the 'cross-section' but has no real, physical significance as a measurement of area. It is an extremely small unit, of the same order as the cross-sectional area of the nucleus, and, being regarded by nuclear physicists as big as the side of a barn, is called a 'barn' (1 barn = 10^{-28} m^2).

The sequence of events in many nuclear reactions is that the incident particle enters the nucleus and forms an excited 'compound nucleus'.

$$_y^x\text{A} + {}_0^1\text{n} \rightarrow {}_y^{x+1}\text{A}^*$$

This nucleus may then decay in one or more different ways:

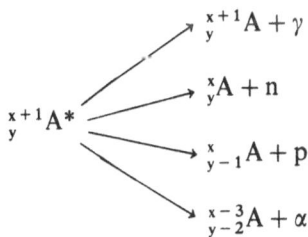

Each possible mode of decay has its own probability (cross-section) and this figure is used to calculate the yield of the nuclear reaction.

The yield of a reaction is controlled principally by three variables, the concentration of neutrons available (flux), the cross-section of the target element and the half-life of the radioactive product. There are standard, complex mathematical relationships by which the yield of a particular reaction can be calculated but it is not an unreasonable analogy to compare neutron irradiation of a target with the transfer of water from a reservoir to a leaking bucket by means of a ladle. The reservoir is the target element (relatively large volume), the ladle is the cross-section (range of sizes), the frequency of ladling is the flux and the leak rate is the half-life, i.e. a small leak is equivalent to a long half-life. The leak rate also increases with increased volume of water in the bucket.

The highest yield will be obtained by frequent transfers with a large ladle into a bucket with a small leak. This is a somewhat unusual state of affairs, and a more frequent result will be that an equilibrium is established when the rate of transfer to the bucket is equal to the leak rate from it. This situation is known as the saturation specific activity. At the opposite end of the scale would be the system using infrequent transfers with a small ladle into a bucket with a large leak, in which case it is doubtful if any net transfer will occur and the net yield will be zero.

In addition to the (n, γ) reactions referred to earlier, the alternative decay schemes following neutron capture also fulfil an important role in radioisotope production. Of the remaining schemes the (n, p) reactions are most significant.

$$\text{e.g.} \quad {}^{35}_{17}\text{Cl}(n, p){}^{35}_{16}\text{S},$$
$$ {}^{32}_{16}\text{S}(n, p){}^{32}_{15}\text{P}$$

The cross-sections for these reactions are usually small, and in general fast neutrons are required for the reaction to proceed. It is possible to obtain high-specific-activity materials from these reactions because the reaction product differs chemically from the target element and a chemical separation can be made.

A similar situation occurs with (n, α) reactions:

$$ {}^{27}_{13}\text{Al}(n, \alpha){}^{24}_{11}\text{Na}$$
$$ {}^{6}_{3}\text{Li}(n, \alpha){}^{3}_{1}\text{H}$$

There are also a number of multi-stage processes where the primary product decays with a short half-life to produce a radioactive daughter.

$$ {}^{130}_{52}\text{Te}(n, \gamma){}^{131}_{52}\text{Te} \xrightarrow[25 \text{ min}]{\beta} {}^{131}_{53}\text{I}$$

Further interactions can also take place by further bombardment of the primary product, giving rise to the following reaction:

$$ A(n, \gamma)B(n, \gamma)C \dashrightarrow D$$

A good example of this is the production of americium-241.

$$ {}^{239}_{94}\text{Pu}(n, \gamma){}^{240}_{94}\text{Pu}(n, \gamma){}^{241}_{94}\text{Pu} \xrightarrow[13.2 \text{ y}]{\beta} {}^{241}_{95}\text{Am}$$

The majority of the transuranic elements are produced by this type of process.

The fission process which was described earlier can also be described by this

Table 4.2

Target	Process		
	n, γ	n, p	n, α
^{23}Na	^{24}Na	^{23}Ne	^{20}F
^{35}Cl	^{36}Cl	^{35}S	^{32}P
^{37}Cl	^{38}Cl	^{37}S	^{34}P

type of notation:

$$U(n,f)^{137}Cs$$

One final point which should be stressed in reviewing neutron activation processes is that irradiation of a simple target such as sodium chloride gives rise to nine products as shown in Table 4.2.

However, in this process only sodium-24, chlorine-36, phosphorus-32 and sulphur-35 have significant half-lives and the cross-sections of ^{32}P and ^{35}S are small.

4.5 Cyclotrons and accelerators

Almost by definition the radioactive sources which are produced by neutron bombardment and by fission will contain excess neutrons in the nucleus. It would follow that nuclides which are deficient in neutrons cannot, in general, be produced in nuclear reactors. They can, however, be obtained by irradiating suitable target materials with protons, deuterons or heavy ions in a particle accelerator.

There are several types of accelerating machine, each one named according to the particle accelerated or the method used to produce the acceleration. The principle involved here is that a beam of charged particles, usually positively-charged, although electrons are also used, is injected into the machine and, by subjecting it to electric and magnetic forces, is accelerated either in a straight line (linear accelerator) or, more commonly, in a spiral (cyclotron). The high-energy particle thus created is then directed on to a target transferring sufficient energy to cause a nuclear reaction.

Examples of irradiation with protons, deutrons and α-particles are given below.

$$^{12}C(p, 3p, 3n)^{7}Be$$

(i.e. 3 protons and 3 neutrons are ejected from the carbon-12 nucleus and a beryllium-7 nucleus is formed.)

$$^{24}Mg(\alpha, \alpha)^{22}Na$$

(α-particles are emitted and sodium-22 nuclei are formed.)

$$^{52}Cr(\alpha, 4n)^{52}Fe$$

(4 neutrons are ejected and iron-52 nuclei are produced.)

Because the products from particle accelerators are neutron-deficient, they will strive to attain greater stability by conversion of protons to neutrons. In order to do this they must either capture an electron or emit a positron. The choice will depend upon the energy change of the transformation and some gamma-radiation is invariably produced during the decay process. Following electron capture, characteristic X-rays of the daughter element are produced as an electronic rearrangement occurs after the loss of an inner electron. From positron emission, 0.51 MeV gamma rays are produced from annihilation radiation following collision of the positron with an electron.

Several general points can be made about particle accelerator irradiation.

(1) The target and radioisotope product are chemically different, so that separation can be carried out to produce high-specific-activity material if this is required.
(2) The process produces a great deal of energy in the form of heat, and this can restrict the choice of target.
(3) The accelerated beam is small, so that the amount of material which can be irradiated at one time is limited, which in turn means that production costs are high.
(4) As in neutron activation, the product yield will depend upon the particle flux, the activation cross-section for the particular reaction, and the product half-life.

By accelerating particles to higher energies it is possible to transfer sufficient energy to the target to cause a series of fragments to be broken off from the nucleus. This process is known as 'spallation'. An example of this is the production of magnesium-28 from chlorine-37 by the reaction

$$^{37}Cl(\rho, 6\rho, 4n)^{28}Mg.$$

One specific radioisotope production process using accelerated particles which is of increasing interest in wear and corrosion investigation is the technique of thin-layer activation.[1] In this technique test pieces or plant component surfaces are irradiated with high-energy ion beams to produce low concentrations of radionuclides within surface layers, usually between 25 micrometers and 300 micrometers deep, depending on the energy of the ion beam, for example in iron by the reactions $^{56}Fe(\alpha, 2n)^{56}Co$ or $^{56}Fe(p,n)^{56}Co$.

Loss of metal can then be detected on-line by loss of radioactivity, corrected for natural decay.

4.6 Radionuclide generators

There has been increasing interest over the last 20 years in the use of very short half-life radionuclides, particularly in the medical field. There is also a growing interest in the use of these materials in industrial applications.

In tracer investigations many of the problems associated with residual

Table 4.3 Radionuclide generators

Parent		Daughter	
Radionuclide	Half-life	Radionuclide	Half-life
^{44}Ti	48 y	^{44}Sc	3.92 h
^{52}Fe	8.2 h	^{52m}Mn	21 min
^{68}Ge	287 d	^{68}Ga	68.3 min
^{81}Rb	4.7 h	^{81m}Kr	13 s
^{82}Sr	25 d	^{82}Rb	1.3 min
^{87}Y	80 h	^{87m}Sr	2.8 h
^{90}Sr	28 y	^{90m}Y	3.2 h
^{99}Mo	66.7 h	^{99m}Tc	6.02 h
^{113}Sn	115 d	^{113m}In	99.5 min
^{118}Te	6 d	^{118}Sb	3.5 min
^{122}Xe	20 h	^{122}I	3.6 min
^{132}Te	78 h	^{132}I	2.29 h
^{137}Cs	30.23 h	^{137m}Ba	2.55 min
^{178}W	21.5 d	^{178}Ta	9.4 min
^{191}Os	13 h	^{191m}Ir	4.9 s
^{195m}Hg	40 h	^{195m}Au	31 s

activity, waste disposal, etc., can be overcome by the use of short half-life materials. The main difficulty lies in providing a supply of short half-life material at a location situated some distance from a reactor or an accelerator, and it is to meet this requirement that interest has turned to the development of radionuclide generators[2].

These generators are based on a decay process involving a parent with a relatively long half-life and a daughter element with a short half-life. Chemical separation of the parent and daughter is usually effected by means of an ion-exchange system. The parent element is retained on the ion exchange resin and the daughter element is 'milked' from the system as and when it is required. For obvious reasons the system is frequently referred to as a 'gamma cow'. Some of the potential systems are listed in Table 4.3, and many of these are in commercial use.

In considering the suitability of a particular radioisotope for a specific application, whether it is obtained from a radionuclide generator system or from any of the production processes, several factors must be considered. The factors influencing the choice of radioisotope both for sealed source applications and for tracer studies are discussed in detail in subsequent chapters.

References

1. Conlon, T. W. (1974) *Wear* **69**, 29.
2. *Radionuclide Generators.* ACS Symp. Ser. **241**, 1984.

5 Biological effects of radiation

J. S. CHARLTON

5.1 Introduction

Because work with radioactive materials inevitably leads to a certain degree of exposure to ionizing radiations, it is important that we should consider and discuss methods by which the exposure, and the resulting damage to the body's tissues, can be kept as low as possible. Practical measures for protection (radiological protection) are described in Chapter 6. To understand the reasons for the precautions adopted it is necessary first to provide some background information on the effects of ionizing radiations on the body. It is also valuable to examine the risks involved in work involving radioactive materials and to put them into perspective by comparing them with hazards associated with occupations of a more familiar nature.

Our objectives, then are threefold: firstly, to describe how our knowledge of the harmful effects of ionizing radiations has developed and to describe the basis of current systems of protection; secondly, to outline the present state of our knowledge; thirdly, to demonstrate that with proper precautions the risks involved can be made negligibly small.

5.2 Ionizing radiations

We are concerned with alpha-particles, beta-particles, gamma-rays, neutrons and X-rays of energies ranging from a few keV up to a few MeV. These energies are several orders of magnitude higher than the binding energies of electrons in atoms (typically several eV, or tens of eV). The radiations can thus knock electrons out of their atomic orbits to create an ion pair. In fact, ionization is the principal mode of absorption of the above-mentioned radiations in matter. Clearly, because of their ability to disrupt atomic and molecular structure, the so-called ionizing radiations can produce damage in the cells which make up human tissue.

5.3 Harmful effects of radiation: historical perspective

That unrestricted exposure can cause biological damage was recognized shortly after the discovery of radioactivity and ionizing radiations. Some of the better-known examples are:

(a) Roentgen (the discoverer of X-rays) and his co-workers. X-rays were discovered in 1895 and by 1902 it had become recognized that large and repeated exposures (particularly to the hands) caused skin cancers[1].

(b) The Curies (who, as we know isolated radium from pitchblende). Shortly after the isolation of radium it was realized that exposure of the skin to the element could lead to damage. Indeed, it is recorded that Pierre Curie himself 'voluntarily exposed his arm to the action of radium during several hours. This resulted in a lesion resembling a burn that developed progressively and required several months to heal'[2].

(c) Painters of luminous dials. The luminizing industry gained momentum in World War I. Instruments were hand-painted (usually by teams of women) using a paint made up from a phosphor and radium. Before adequate controls were introduced it appears to have been the practice of the painters to 'point' the fine tips of their brushes by mouth. The radioactive isotope radium-226 is a bone-seeker. A higher than average incidence of bone cancers was noted some forty years ago[3].

(d) Uranium and (non-coal) miners. One of the early observations concerned the Joachimstal mines in central Europe. These mines were rich in copper, nickel and silver. Unfortunately, they were also rich in pitchblende which is now known to contain uranium. The miners throughout their working-lives were inhaling the gas radon-222—an alpha-emitting daughter of uranium-238. The incidence of lung cancer in these workers was observed to be higher than the norm.

(e) The 'A-bomb' survivors of Hiroshima and Nagasaki. Because of the severity and wide range of magnitude of the exposures, those exposed exhibited an equally wide range of ill effects. Indeed, much of the data that we possess on acute doses of radiation (see Table 5.2) derives from Hiroshima and Nagasaki. A longer-term but no less fatal effect was leukaemia[4] subsequently found to occur at higher-than-normal levels among the exposed population.

(f) The Marshall Islanders, inhabitants of Pacific islands inadvertently exposed to the fallout (fission products) from a test explosion in 1954. A significantly high incidence of thyroid cancers was noted[5].

5.4 Radiological protection: historical perspective

As the harmful effects of radiation became recognized so the need for protection became apparent. It is possible to summarize landmarks in the development of Radiological Protection (sometimes referred to as Health Physics).

1921: British X-Ray and Radium Committee formed
1922: USA Roentgen-Ray Society formed

1925: First International Congress of Radiology (to initiate international agreement on protection)
1926: Second International Congress.

This last event was of importance since at the Congress an international organization, 'the International X-Ray and Radium Commission, came into being'[6]. Its first recommendations were issued in 1929. This Commission in 1950 was re-named the International Commission on Radiological Protection (ICRP). The remit of this organization was to study the effects of radiation — both short- and long-term. ICRP is still very much in existence today and is the accepted world authority. Its recommendations form the basis of national legislations and codes of practice.

Before going on to ICRP's recommendations on radiological protection we shall discuss the units which express the exposure of humans to radiations.

5.5 Radiation dose

Damage produced in a given mass of human tissue is related to the number of ions produced, which in turn depends upon the energy dissipated in the tissue. It is not surprising, therefore, that the units of dose are those of energy deposited per unit mass. The *unit of absorbed dose* is the $GRAY(Gy) = 1$ joule kg^{-1}. This, of course, is the SI unit. Prior to its introduction, Absorbed Dose was expressed in terms of the *rad* $(= 100 \, ergs \, g^{-1})$.

As we have seen (Chapter 2) different types of radiation are absorbed in different ways. For example, alpha-particles are stopped (i.e. deposit their energy) over a very short path-length in tissue; gamma-rays of similar energy traverse much greater thicknesses and thus deposit their energy far more sparsely. If we wish to describe the damaging effects of the radiations, account must be taken of this. We can regularize the situation by introducing another unit, the dose equivalent.

The dose equivalent is equal to the absorbed dose multiplied by a *quality factor* which makes allowance for the way different radiations deposit their energy. Thus

$$\text{dose equivalent} = \text{absorbed dose} \times \text{quality factor}$$

The unit of dose equivalent is the *sievert* (Sv) which, like the gray, is 1 joule kg^{-1}.

For X-rays, gamma-rays and beta-particles, the Quality Factor is equal to 1: for alpha-particles it is equal to 20. For completeness, it should be noted that the pre-SI unit of dose equivalent was called the *rem* and it is worth remembering the relationships:

$$1 \text{ gray} = 100 \text{ rads}$$
$$1 \text{ sievert} = 100 \text{ rems}$$

since the rad and rem are still in common usage.

Table 5.1 Risk weighting factors

Organ	Factor
Testes and ovaries	0.25
Breast	0.15
Lung	0.12
Red bone marrow	0.12
Thyroid	0.03
Bone surfaces	0.03
Remainder	0.30
(up to 5 organs	0.06 each)
TOTAL (whole body)	1.00

In deriving the concept of dose equivalent, we have arrived at a quantity which allows us to compare directly the dose to any body organ from exposure to any type of radiation. However, in considering the total risk to any individual exposed to ionizing radiations we must go one step further. The problem is that different organs have different sensitivities to radiation. Consider for example, fatal malignancies: the risk from exposure of the red bone marrow is much greater than exposure of (for example) the thyroid gland. Similarly, if one considers a further type of hazard, hereditary damage, it is clear that a dose to the testes or ovaries is much more significant than a dose to the lung.

It is, in fact, possible to compile a table of *risk weighting factors* for the various body organs[7] (Table 5.1).

If any *single* tissue receives a given dose equivalent we can, in effect, convert this to a dose equivalent to the whole body by weighting it with the appropriate risk weighting factor. Thus, a dose equivalent of 20 sieverts received by the lung would carry with it a risk of fatal malignancy and hereditary harm equal to $20 \times 0.12 = 2.4$ sieverts received uniformly throughout the whole body.

Similar calculations may be performed if several body organs receive non-uniform irradiation—as is frequently the case when radioactive material is ingested. In this latter case, if the appropriate weighting factors are applied to each irradiated organ, the sum of the products (dose equivalent × weighting factor) is the dose equivalent which, received by the whole body, would yield the same risk overall. This sum of weighted dose equivalents is called the *effective dose equivalent*.

5.6 The hazards of ionizing radiations

Having defined units, it is now possible to discuss, in a systematic way, the various hazards associated with excessive exposure to ionizing radiations.

5.6.1 *Early somatic effects*

The early somatic (occurring in the irradiated body) effects are observed within
a few hours to a few weeks following acute whole-body exposure (i.e. a large
dose received in a short period of time). This is what is usually referred to as
'radiation sickness': the effects are summarized in Table 5.2.

A second early effect which occurs when the skin is exposed to a high dose of
weakly penetrating radiation (e.g. beta-particles) is radiation erythema
(burning). Approximately 3 Gy of low-energy radiation is necessary to induce
this effect. Larger doses give rise to radiation burns, which differ from heat
burns in one important respect—they take longer to heal.

These effects are called 'non-stochastic', i.e. the severity of the effect varies
with the size of the radiation dose. There is also a threshold effect, i.e. a dose
beneath which none of the above effects are observed.

*We should stress that these effects only occur as a result of doses received over
a very short period.* They are due, in the main, to the destruction of cells in
rapidly dividing stem cell populations. The classic symptoms of radiation
sickness—bleeding, increased susceptibility to disease, loss of hair, reduced
blood count—are temporary and, because of the presence of surviving stem
cells in the bone marrow, disappear in a matter of weeks or months after
irradiation. It is this ability of bone marrow cells to survive and re-populate
the bone marrow which is the main factor responsible for recovery from doses
of a few sieverts. Of course, at progressively higher doses, there is less and less
chance that there will be surviving cells, hence the increased risk of death at
high doses (Table 5.2).

Table 5.2 Expected effects of acute body radiation

Acute dose (sieverts)	Probable effects
0 to 0.5	No obvious effect, except minor blood changes.
0.5 to 1.2	Vomiting and nausea for about 1 day in 5–10% of exposed personnel. Fatigue but no serious disability.
1.3 to 1.7	Approximately one-quarter of those exposed will experience vomiting and nausea for about 1 day, followed by symptoms of radiation sickness. No deaths anticipated.
1.8 to 2.2	Vomiting and nausea in about 50% of personnel for about 1 day, followed by other symptoms of radiation sickness. No deaths anticipated.
2.7 to 3.3	As above except that at this level of dose there would be 20% deaths within 2 to 6 weeks after exposure; survivors convalescent for about 3 months.
4 to 5	As above but with 50% mortality.
5.5 to 7.5	Vomiting and nausea in all personnel within 4 hours from exposure, followed by other symptoms of radiation sickness. Up to 100% deaths; few survivors convalescent for about 6 months.
10	Vomiting and nausea in all personnel within 1 to 2 hours. Probably no survivors from radiation sickness.
50	Incapacitation almost immediately. All personnel will be fatalities within 1 week.

The other point to note is that because of the body's recovery mechanisms, large doses accumulated over *long* periods of time do not give rise to acute somatic effects.

5.6.2 *Malignant diseases*

The principal 'late' effect of radiation is cancer. Cancer is an over-proliferation of cells in an organ of the body and it is thought that it results from irradiation of a single cell, causing damage to the control mechanisms. The cell divides more rapidly than a normal cell, and since this defect is duplicated in the daughter cells the population of abnormal cells builds up to the detriment of the normal cells in the organ. The estimation of increased cancer risk is complicated by the fact that cancers have long (and variable) latent periods and by the fact that radiation-induced cancers are not distinguishable from those which arise spontaneously or are caused by other carcinogenic agents.

Cancers are a stochastic effect of radiation, i.e. the statistical risk of cancer is proportional to the radiation dose; the severity of the effect is not.

5.6.3 *Genetic effects of radiation*

These effects result from damage to reproductive cells and take the form of mutation in the hereditary material of the cell (the genes). This, also, is a stochastic effect. It is interesting to note that *no* direct evidence for hereditary defects arising from radiation *has ever* been found in humans—even in the children and grandchildren of survivors of the 'A' bomb attacks on Hiroshima and Nagasaki. However, laboratory tests on mice populations indicate that the risk of serious hereditary defects in the first two generations following irradiation would be 1 in 250 per sievert received.

5.7 Dose and risk

We have stated the risk of hereditary defects resulting from ionizing radiations and have further stated that the risk is entirely derived from tests on animals. What can we say about the risk of malignant disease—and more particularly, fatal malignant disease? Here we do have human data, of a sort. However, this information derives entirely from high doses delivered over a short period of time (e.g. A-bomb survivors, persons subjected to high-intensity radiation therapy). In the normal course of events, this type of situation does not arise: instead persons are usually exposed to low doses of radiation over a long period of time. The problem, therefore, is how to relate the risk, as derived from the high-dose data, to the normal situation. The problem is shown diagramatically in Figure 5.1.

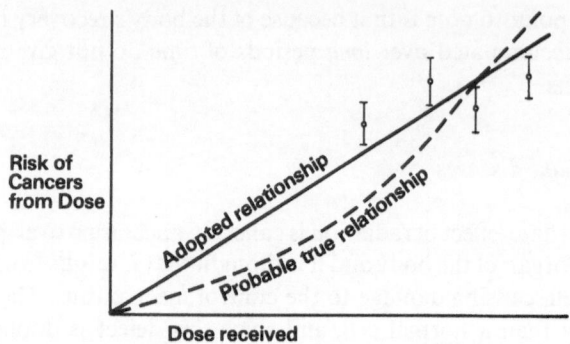

Figure 5.1.   Extrapolation of risk at high dose rates to low dose rates.

Experimental data derived from four studies are shown. These are all at high dose rates. To obtain a risk estimate at low doses, extrapolation is required. Two possibilities are shown. The dotted curve probably represents the true situation. It is derived from experimental and theoretical radiobiology. Unfortunately its exact shape cannot be predicted for cancer in humans and so we adopt a straight-line relationship between dose and risk. This probably leads to an overestimate of the risk at low doses, but at least this is erring on the side of caution. The straight-line relationship has other implications. Thus the risk factor (risk per unit dose) for cancers does not decrease with decreasing dose or doserate and all doses, whenever received, are additive in their effects. Thus the risk associated with 1 microsievert is assessed as being one-millionth of the risk associated with 1 sievert. Also, it does not matter whether a dose is received instantaneously or over several years—the effect (i.e. risk) is the same. Yet a further implication is that any dose, no matter how small, creates a finite risk of cancer.

Genetic effects are assumed to follow the same rules. However, the risk factor for cancer estimated in the manner described is 1 in 80 per sievert received, i.e. it dominates the genetic risk.

5.8 ICRP recommendations

The International Commission on Radiological Protection has been mentioned earlier. It is (and has been for over 50 years) the central authority on Radiological Protection. It receives a vast amount of information from laboratories all over the world. At appropriate times, when it perceives a need, a new volume of recommendations is published. In earlier years, the frequency of publications was much higher than now, presumably as the various sub-committees got to grips with their particular subjects. In recent years the number of fundamentally new publications has been small—rather, the effort has gone into refining existing recommendations. This reflects the increased confidence ICRP has in its basic pronouncements on the acceptable dose for man, and, more particularly, the exposure particular organs may have.

The present system of Radiological Protection is based upon three main requirements:

(a) No practice shall be adopted unless the risks associated with it are outweighed by the benefit
(b) All exposured shall be kept *as low as reasonably achievable*, economic and social factors being taken into account (the so-called 'ALARA' principle)
(c) Dose equivalents to individuals shall not exceed limits specified by the Commission.

There can be little argument that doses should be kept as low as possible but it is clear that there is considerable scope for debate about the acceptability of certain practices *vis-à-vis* the associated benefits and hazards.

We will attempt to put this into perspective later. For the present let us confine ourselves to the question of dose limits.

5.8.1 Dose limits

The recommended levels aim to prevent:

(a) Acute somatic effects (burns)
(b) Long-term somatic effects (cancers)
(c) Genetic effects.

Since, as we have seen, the hazards associated with malignancies dominate the genetic hazard, it is the prevention of cancer which is the issue at the core of the dose-limitation system.

The dose limits are based on the assumptions (already discussed in section 5.7), that

(a) No threshold dose exists and any dose received involves a proportionate risk of induction of cancers.
(b) Doses delivered on separate occasions act cumulatively and linearly in the induction of cancer.

It is worth repeating that, if anything, these assumptions err on the side of safety.

The dose limit is set at 50 mSv (5 rems) per year for an occupationally exposed worker (Radiation Worker). The dose limit for a member of the public is 5 mSv (0.5 rems) per year.

The rationale behind the differences in these limits is that throughout his working life a radiation worker is subject to routine medical checks (to ensure that he has not developed any form of ill health due to non-radiation effects which might mask the effect of radiation) and is also subject to regular

Table 5.3   Dose limits for radiation workers and members of the public (mSv per year)

	Worker	Public
Effective Dose Equivalent	50	5
Dose equivalent to a single organ or tissue	500	50
Dose equivalent to eye	150	15

personal monitoring, such as the wearing of a dosemeter. It is simply impractical to monitor members of the public in this way. Also, of course, in the case of the public there is a chance that children will be irradiated. In addition, a member of the public might be exposed for 70 years or more as opposed to the 45 years of occupational exposure.

It cannot be too firmly stressed that these figures represent limits which should not be exceeded. Overriding this, however, there is the ALARA requirement. Any practice which regularly allowed a worker to approach the dose limit would *not* be deemed acceptable. Dose limits recommended by ICRP form the basis of the national legislation of most countries. The recommended limits are listed in Table 5.3.

The dose limit to a single organ is set so as to avoid non-stochastic effects. It is recognized that with organs of relatively low sensitivity the dose required to produce an effective dose equivalent equal to 50 mSv (the limit relating to protection against cancer and hereditary effects) might be so high as to produce other somatic effects. Similarly, the dose limit to the eye (recently reduced from 300 mSv to 150 mSv)[8] is set with a view to protection against cataract induction.

Women of reproductive capacity (radiation workers) have the same annual dose limit of 50 mSv, but it must be received uniformly at not more than 13 mSv per calendar quarter. Pregnant women have a limit imposed on the annual effective dose equivalent of 15 mSv per year until the end of pregnancy, i.e. not more than 10 mSv after diagnosis.

5.8.2 *Comparison of risks*

Table 5.4 compares the average annual risk of death in the UK from accidents in various industries with those from cancers potentially induced among radiation workers. The annual average effective dose equivalent to workers in the UK is 4 mSv. Remembering that the risk factor for cancers is 1 in 80 per sievert, the average risk of fatal cancer to radiation workers is 1 in 20 000. This is seen to be closely comparable with the general risk of death in employment, thus giving credibility to the protection procedures adopted.

Turning from the radiation worker to the general public, the average effective dose equivalent from artificial sources to a member of the public is

Table 5.4 Average annual risk of accidental death in various occupations and from cancer induced in radiation workers

Industry	Risk of death per year
Textiles	1 in 40 000
Food and drink	1 in 30 000
All employment	1 in 20 000
Radiation workers	1 in 20 000
Metal production	1 in 7 000
Construction	1 in 5 000
Coal mining	1 in 4 000
Deep-sea fishing	1 in 400

Table 5.5 Average annual risk of death from several causes and from cancers induced by radiation of artificial origin

	Risk of death per year
Radiation exposure (0.5 mSv)	1 in 160 000
Accidents at work	1 in 20 000
Accidents at home	1 in 10 000
Road accidents	1 in 5 000
Smoking 20 cigarettes a day	1 in 200

0.5 mSv. This implies an annual risk of death of 1 in 160 000. This is compared with other everyday risks in Table 5.5. As in the case of occupational exposure, the risk is of comparatively low magnitude.

5.9 Doses in perspective

Having discussed large doses (section 5.6) which can give rise to sickness, death or the 'long-term' somatic effects, and having also looked at the system of dose limitation, it is useful, to preserve a sense of balance, to briefly discuss radiation to which we all are exposed from birth—so-called 'natural radiation'. This arises from several sources:

(a) *Cosmic radiation.* The Earth is constantly bombarded by radiation from outer space. This is very high-energy radiation—typically thousands of MeV, much of which is absorbed in the atmosphere. Some radiation makes its way to ground level, however, and therefore contributes to the effective dose equivalent which we receive. The typical annual effective dose equivalent is approximately 0.31 mSv.

Table 5.6 Average annual effective dose equivalents from radiation of artificial origin

Source	mSv
Medical procedures	0.51
Weapons fall-out	0.01
Discharges of artificial radioactivity	0.003
Miscellaneous	0.017

(b) *Radioactivity ingested.* One of the results of cosmic-ray interaction with the nitrogen in the atmosphere is the production of the radioisotope carbon-14. This enters the food chain and becomes incorporated in human tissue. Other radionuclides, from radioactive isotopes present in the Earth's crust, potassium-40, lead-210 and polonium-210, are also absorbed. The typical annual effect dose equivalent is about 0.37 mSv.

(c) *Gamma-radiations of terrestrial origin.* These arise from radioactive materials in the Earth's crust, uranium and thorium (and their daughter isotopes) and potassium-40. Many of these materials are extracted for use in building materials and hence, in varying degrees, almost all these materials are slightly radioactive. The typical annual effective dose equivalent is about 0.38 mSv.

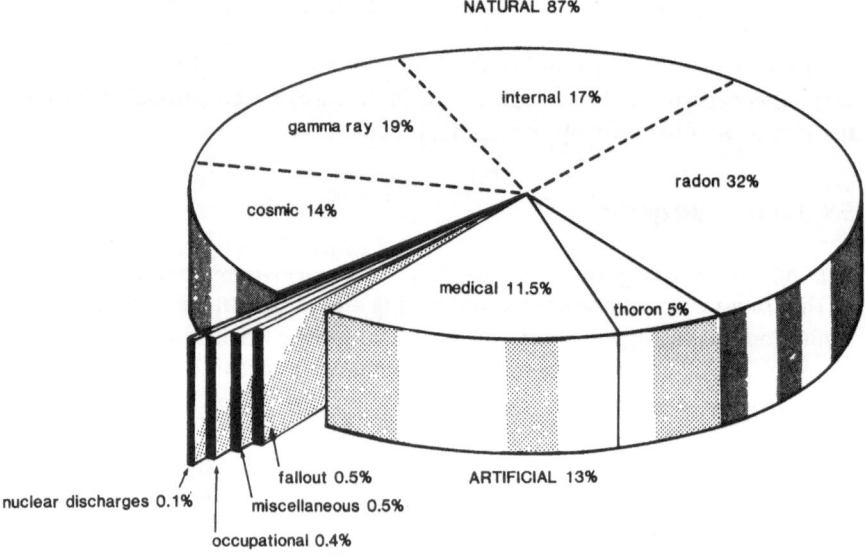

Figure 5.2. Average annual radiation dose to the population in the UK. (Courtesy National Radiological Protection Board.)

(d) *Radon decay products.* Radon gas is a daughter isotope of naturally occurring uranium-238. The decay products of radon itself are solids which can attach themselves to dust particles and be ingested into the lung. The typical annual effective dose-equivalent is about 0.8 mSv, but there are very pronounced variation about this mean figure. In total the annual effective dose equivalent from all natural sources is 1.86 mSv.

In addition to natural radiation, the population is also subjected routinely to radiation of artificial origin. By far the major contributor is medical procedures, X-rays and radiation therapy, but there also small contributions from other sources.

Table 5.6 shows the annual effective dose equivalent to an average member of the population arising from artificial sources. It can be seen from the table that the average annual effective dose equivalent to members of the population is typically 2.4 mSv.

The various contributions to the average annual dose of the population of the United Kingdom are illustrated in Figure 5.2. It will be noted that the natural background accounts for approximately 80% of the dose received.

5.10 Conclusions

In spite of numerous studies carried out on different groups of exposed personnel in all parts of the world, no harmful effects have been observed in human populations at the low dose corresponding to the ICRP dose limits. Indeed, in the case of hereditary damage no effect has *ever* been observed in humans, even at doses orders of magnitude higher.

These facts must lend credibility to the present system of dose limitation. Nevertheless, complacency cannot be afforded—nor is it. Data continue to be collected and the central issue in radiological protection—the estimate of risk at low doses—continues to be keenly debated and researched.

In this context mention should be made of a major study which has been commenced by the United Kingdom National Radiological Protection Board[7]. A National Registry for Radiation Workers has been established in which the lifetime dose and cause of death of individual radiation workers are recorded. Analysis of these data will, ultimately, determine the relationship between mortality and dose. In particular, evidence of excess cancers in radiation workers will be sought. The problems are formidable: approximately 1 person in 5 in the United Kingdom dies of cancer. Against this background the difficulties in identifying excess cancers which can be positively correlated with radiation dose are immense. It is felt that within *a decade* it should be possible to positively demonstrate whether or not the present risk factors are correct to *within an order of magnitude* and as time progresses the bounds of uncertainty will further diminish. This is for the future. Our present state of understanding may perhaps be best summed up by the statement: 'We know that the effects (of low levels of radiation) are small, what we don't know is *how* small they are'[10].

References

1. Pochin, E. (1981) The development of radiation protection. *J. Soc. Radiol. Protection* **1**, 17.
2. Glasstone, S. (1958) *Sourcebook on Atomic Energy.* Van Nostrand, New York, 588.
3. Pochin, E. (1981) Estimation of Radiation Risk in Man, Supplement to *Radiological Protection Bulletin* No 42, National Radiological Protection Board, Chilton, 7.
4. Folley, J. H. *et al.* (1952) Incidence of leukaemia in survivors of the atomic bombs in Hiroshima and Nagasaki. *Amer. J. Med.* **13**, 311.
5. *The Effects of Nuclear Weapons*, US Government Printing Office, Washington, DC, 1957.
6. Sievert, R. M. (1958) The Work of the International Commission on Radiological Protection, *Proc. Second Int. Conf. on the Peaceful Uses of Atomic Energy*, Vol. 21, United Nations, Geneva, 3.
7. Recommendations of the International Commission on Radiological Protection, ICRP Publication 26, *Ann. ICRP.* **1** (1977).
8. Recommendations of the International Commission on Radiological Protection, *Br. J. Radiol.* **53** (1980) 816.
9. *Living with Radiation*, National Radiological Protection Board (1981) 23.
10. Agard, T. E. (1982) Public education on radiation effects. *Health Physics* **43**, 85.

6 Radiological protection

G. REED

The fundamental objective of Radiological Protection is to reduce the hazard arising from the use of radioisotopes to as low a level as is reasonably achievable, social and economic factors being taken into account. Requirements are contained in the various items of legislation concerned with the protection of radiation workers, the general workforce, the population at large and the environment, the assumption being that if humans are protected then other living species are unlikely to be harmed. These legislative requirements should not be considered as working levels but rather as maximum levels, never to be exceeded except possibly under accident or planned emergency conditions. Radiation doserate records appear to indicate that this is the view taken by the vast majority of employers, since mean yearly doserates for classified workers average about one-tenth of the maximum permitted levels laid down in the legislation, with less than 1% of workers exceeding the maximum levels. Even so, it is these radiation workers who are

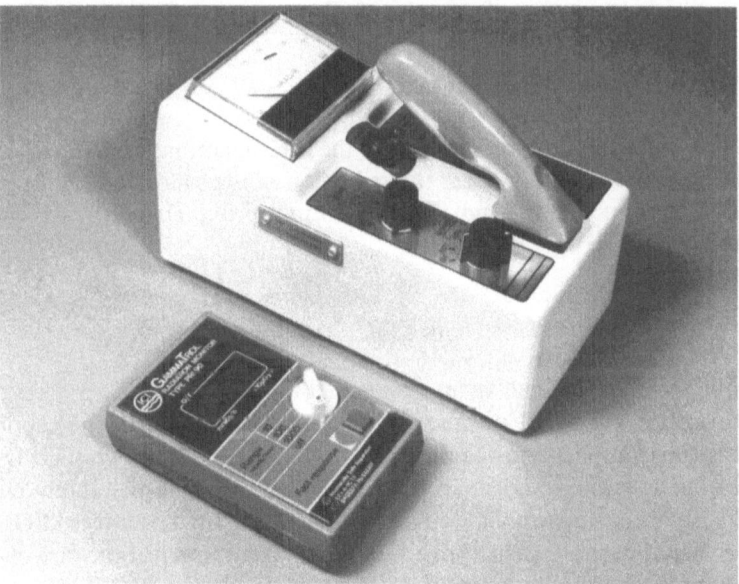

Figure 6.1. Gamma dose-rate meters manufactured by Physics and Radioisotopes Group of ICI PLC.

most at risk since they are in close contact with the radioactive material and most effort must be expended on their protection.

Process personnel on the plant where work is being carried out must also be considered but it is usually a fairly simple matter to exclude them from radiation areas (defined using a monitor of the type illustrated in Figure 6.1) where sealed sources are exposed and concentrated unsealed materials are in use. Once materials are injected into plant streams, tremendous dilutions usually occur; pulses of tracer travel at considerable velocities in the plant fluids, and exposure times and doses are correspondingly low. Similarly, members of the general public are at no risk from the use of radioactive materials on process plants. They do not come into contact with sealed sources, and concentrations of unsealed radioactive materials discharged to the environment may be monitored and controlled so that a negligible risk to the general population results.

Protection against hazards arising from the use of ionizing radiations is conveniently considered under two classifications, those arising from materials external to the body and those arising from material which may be taken into the body. The external hazard may arise from sealed or unsealed sources, the internal hazard only from unsealed sources, at least on process plant.

6.1 Protection against external radiation

There are three methods used to reduce the external radiation hazard, each of which we will consider in turn: the reduction of exposure period, the use of distance to reduce dose, and the use of appropriate shielding around the source.

6.1.1 Reduction of exposure time

The quantities of radioactive materials used to carry out many process plant investigations are small compared with the most commonly encountered radioisotope application—industrial radiography. Radiographic applications apart, sealed and unsealed gamma-emitting sources rarely exceed 8 GBq (of, for example, cobalt-60 or bromine-82) neutron sources 80 GBq of ^{241}Am/Be, and noble gas isotopes 400 GBq. Most applications are carried out with much smaller quantities. At these levels the classified workers involved are most unlikely to suffer somatic effects; their risks are stochastic in character. The total dose received during an operation is, therefore, the important criterion, rather than the doserate being received. Work may be carried out in quite high radiation doserate areas, but because it is carried out quickly and efficiently, only small radiation doses to the operator result. However, any new type of handling of radioactive materials in significant quantities should always be preplanned and practised to ensure proficiency.

Trainees should always receive training in the manipulation of handling tongs and clamps used in the dispensing of unsealed material and handling of sealed sources. Solutions of dye and dummy sources may be substituted for the real thing until a sufficient standard of proficiency is achieved.

Probably the largest quantity of unsealed radioactive material ever to be handled by Physics and Radioisotopes Services Group was a batch of 22 terabecquerels of bromine- 82. The objective of the work was to investigate the fate of effluent discharge from a large petrochemicals site into a local river; how much was swept out to sea on the falling tide, how far and in what concentration the effluent progressed upstream, layering in the river etc. Injection into the effluent was to take place over seven days and the large quantity of tracer was necessary because of the high dilution anticipated and the sensitivity required. Considerable effort was expended in preparing the injection vessel and erecting 25 cm of lead and 1.5 m of sand shielding around it for the protection of personnel involved in the injection. However, the greatest hazard existed during the transfer of the bromine-82 from the shielded transport container to the injection vessel in a doserate area of approximately 2.5 sieverts per hour, the physical weight of the material precluding the use of handling tongs of greater than 2.0 m in length. Two experienced operators practised the transfer at length, using dummy material, and at the critical moment achieved a successful transfer. Neither received more than 0.8 mSv body doserate during the whole operation, illustrating the value of practice in reducing exposure time.

Short-lived tracers must be used wherever possible, and invariably where a liquid physically-compatible tracer rather than a chemically-compatible tracer is adequate. Their use results in a reduction of the exposure time of the process worker and general population, and here recent advances in the development of isotope generators or 'cows' for producing short-lived daughter isotopes on the plant have been particularly effective (Chapter 4). Caesium-137 cows producing barium-137 m with a 2.6-minute half-life are now regularly used for pulse velocity flowrate measurements. Similarly argon-41 (half-life 110 minutes) is now used in preference to krypton-85 (half-life 10.6 years) even at distant locations, an additional bonus being that it is possible to use less than one-hundredth of the quantity because of the higher percentage gamma-emission from argon-41 (100%) than from krypton-85 (0.7%).

6.1.2 Use of distance to reduce exposure

Distance is a particularly valuable factor in radiation protection since the doserate due to a point source is inversely proportional to the square of the distance from the source. We may write the inverse square law as

$$D \propto \frac{1}{r^2} \quad \text{or} \quad D = \frac{k}{r^2}$$

Table 6.1 Doserates from 370 GBq, Cs-137 at various distances

Distance	Dose per hour
1 m	31 mGy
10 m	310 μGy
100 m	3.1 μGy

Table 6.2 Doserates from 370 MBq, Cs-137 at various distances

Distance	Dose per hour
1 m	31 μGy
1 cm	310 mGy
1 mm	31 Gy

Therefore $Dr^2 = k$ where k is a constant for a particular radioactive source.

Therefore
$$D_1 r_1^2 = D_2 r_2^2$$

where D_1 is the doserate at a distance of r_1 from the source and D_2 is the doserate at distance r_2 from the source.

It is, of course, fairly well appreciated by people using radioactivity that large sources give high radiation doserates. Table 6.1 shows the doserate at various distances from a 370-GBq source of caesium-137.

Paradoxically, it is less commonly appreciated that on approaching a source of radiation the doserate rises very rapidly. Table 6.2 shows doserates from a source of only one-thousandth of the size of that described in Table 6.1, i.e. a 370 MBq, Cs-137 source.

This last distance (1 mm) is about equivalent to direct handling of the source and at a doserate of 31 Gy per hour it becomes very obvious why even the smallest radioactive source should never be handled with the bare hands, but must be manipulated with suitable tools. For the smaller source above, ordinary pliers or crucible tongs would be adequate for a few seconds' work; for the larger source, long handling tongs would be appropriate.

It may be found convenient to permanently locate a sealed source which is in regular use in a source holder which ensures a sensible separation between the source and the hands of the operator. The source need never leave the holder (which is stamped with information relevant to the source) except for the statutory periodic leak tests. Storage at base may be in a loose-fitting lead shield, and the source may be transported in a specially designed Type 'A' container which accepts the screw fitting on the holder. The source can be used on plant in a holder which again accepts the screw fitting. Figure 6.2 shows

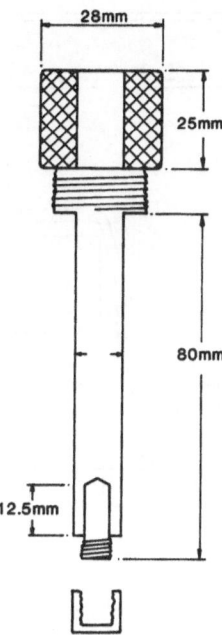

Figure 6.2. Pencil-type source holder in brass—for storage, transport and use of sealed gamma sources.

such a holder and Figure 6.3 a transport container, which in itself meets the requirements for Type 'A' containers but is more often located in an outer drum during transport. Such an arrangement ensures a separation of some 10 cm between the hands and the source, but more importantly, facilitates transfer of the source between storage, transport and use containers, each transfer taking only one or two seconds to complete.

Undoubtedly the highest doses on plant will be experienced by the radioisotope practitioners when a source is exposed, particularly in, for example, the cramped conditions of distillation column platforms. It should be stressed that these operators must always keep as far away as possible from the source. Opening of liquid transport containers and dispensing of liquid radioactive materials should always be carried out at a distance using remote-handling tongs and other suitable equipment.

6.1.3 Use of shielding to reduce exposure

6.1.3.1 *Gamma-radiation shielding.* Unlike alpha- and beta-radiations, which have a fixed range in any given materials, gamma-radiation is exponentially attenuated by matter but (theoretically) never stopped, and gamma-emitting isotopes are most often used in process plant investigations. The doserate, D_t,

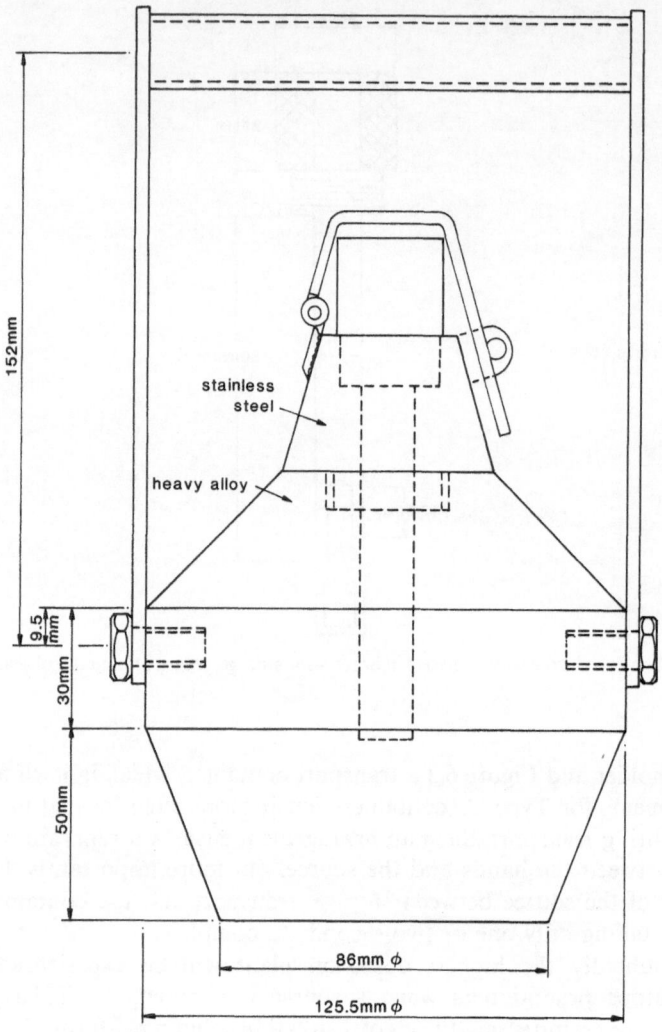

Figure 6.3. Heavy alloy sealed source transport container.

due to gamma-radiation emerging from a shield of thickness t, can be written as

$$D_t = D_0 e^{-\mu t}$$

where D_0 is the doserate incident on the shield; μ is termed the linear absorption coefficient and is a function of the density and elemental composition of the shield and also the energy of the incident radiation. It has dimensions of (length)$^{-1}$ and is usually expressed in m^{-1} or mm^{-1}.

It is convenient to separate the effects of density (ρ) and composition by rewriting the equation as

$$D_t = D_0 e^{-\mu_m \rho t}$$

where μ_m is the mass absorption coefficient with dimensions (length)^2 (mass)$^{-1}$.

In this form it becomes apparent that dense shields are most effective for reducing the intensity of gamma-radiation beams and the most cost-effective material for shielding small-volume sources is lead (density $11.3 \, g \, cm^{-3}$). Even so the volume of lead required to reduce radiation doserates to an acceptable level under the Transport Regulations or Sealed Source regulations is often found to be such that there is a serious risk of physical injury in attempting to lift a container, and more dense materials are necessary to provide a portable, more handleable shield. The two in common use are depleted uranium (density $18.9 \, g \, cm^{-3}$) and sintered alloys of tungsten and copper (density range $17-18 \, g \, cm^{-3}$).

Figure 6.4 illustrates the reduction in transmitted radiation from cobalt-60 achieved by various materials, and it can be seen that to achieve the same reduction more than twice the thickness of lead is required than of uranium. If a 5-centimetre radius cylindrical shield of uranium was necessary to produce a given reduction in doserate, this would weigh 1.49 kg per centimetre height. The 10.6 centimetre radius lead shield to produce the same reduction would weigh 3.99 kilograms per centimetre height; in addition, the cylinder would need to be taller to reduce the doserates at the top and bottom to the required level.

When the mass of shielding required is considered in this light, it can be seen that the volumes of radioactive material handled must be kept as low as possible. For unsealed materials this means as concentrated as possible, or at least of sufficient concentration to allow them to be handled and transported in a portable container. If large volumes are required on site then the final dilution should be carried out on site and local shielding erected *in situ*. Suitable shielding is often difficult to achieve, and more often than not it becomes sensible for the operator to carry out his tasks in the vicinity of the material as quickly as possible and then move to a safe distance, viewing the operation through field glasses if necessary.

Where dilution of the tracer material is not necessary on site, the material for injection, gas or liquid, can conveniently be left in the transport container which then provides shielding for the operator. The use of angled preformed plastic bottles as the primary containment for liquid tracers, which fit snugly into and are held rigidly by the shielded transport container, eliminates the use of one pair of handling tongs when they are being opened and the operation can easily be carried out by one operative rather than two.

6.1.3.2 *Neutron radiation shielding.* Whereas gamma radiation is most effectively shielded by heavy (high-density) materials, neutron shielding is

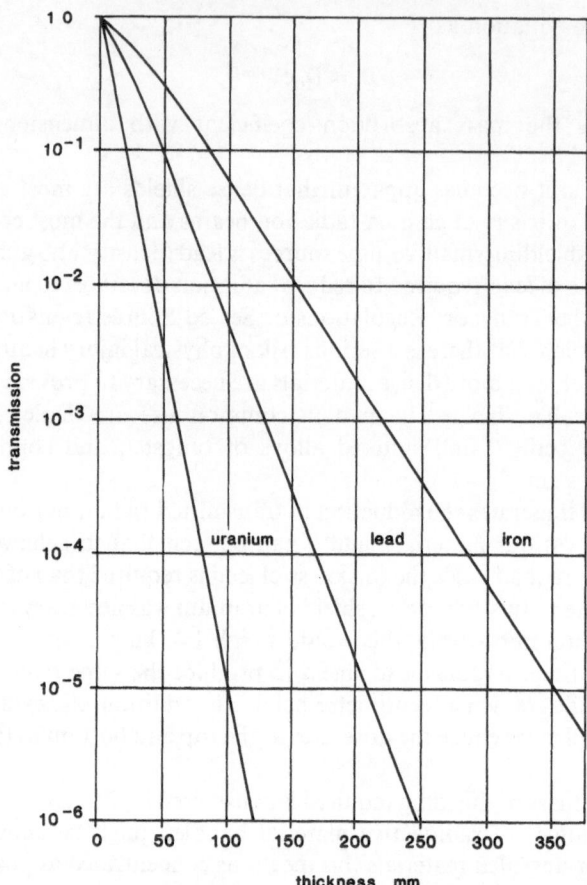

Figure 6.4. Transmission of gamma rays from ^{60}Co through barriers of various materials.

complicated by two facts. The first is that the dose is very dependent on neutron energy, for example at a neutron energy of 5 KeV a flux of 570 neutron cm^{-2} s^{-1} will give a doserate of 25 μSv h^{-1} while at a neutron energy of 1 MeV only 18 neutrons cm^{-2} s^{-1} will give the same doserate. The second complicating fact is that there are no efficient absorbers for neutrons of greater energy than a few eV. Fast or energetic neutrons have to be reduced (moderated) in energy to a few eV when they can be readily absorbed by elements such as boron and cadmium.

This moderation is most easily achieved by materials containing appreciable amounts of hydrogen, such as water or paraffin wax, by an elastic scattering mechanism. A shield for fast neutrons would, therefore, consist of a hydrogenous layer to reduce energy, a thermal neutron absorber of cadmium or boron and finally a heavy gamma absorber since thermal neutrons almost always produce gamma-radiations when absorbed by matter. Composite shielding materials of polythene, cadmium and lead are available or they can

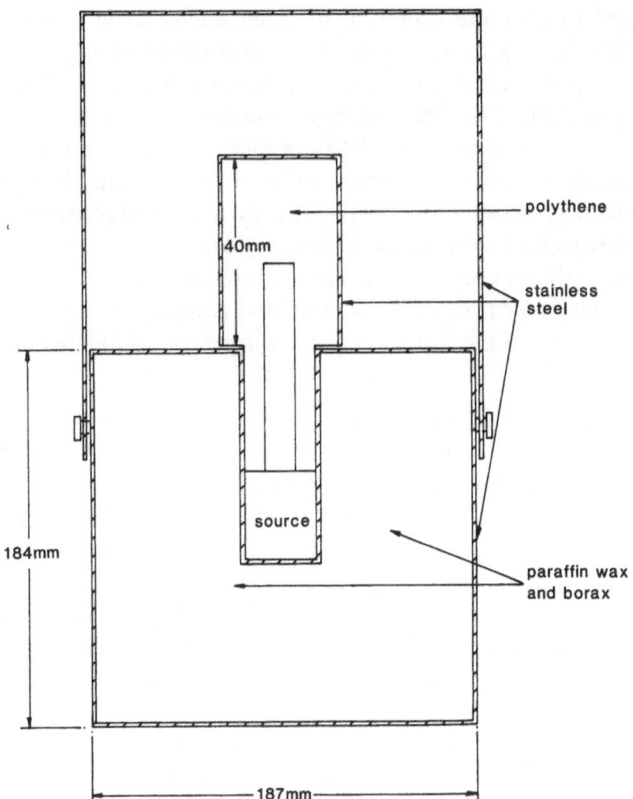

polythene

40mm

stainless
steel

source

184mm

paraffin wax
and borax

187mm

Figure 6.5. Neutron source container.

be produced in house to a specific design. Figure 6.5 shows a transport container produced from paraffin wax laced with borax which is adequate for americium-241/beryllium neutron sources of up to 70GBq.

6.2 Protection against internal radiation

There are two principal ways in which radioactive materials can enter the body: by direct inhalation of airborne materials, and by ingestion through the mouth. A further subsidiary way is by entry through the skin by absorption or through a contaminated wound. This minor route will not be considered in any detail; suffice to say that a wound should be allowed to bleed for a while to assist in flushing out the contamination and expert medical help sought. Petroleum bandages may be used to reabsorb organic materials from the skin surface with some success.

Once radioactive materials are in the body little can be done to remove them effectively. Chelating agents such as ethylene diamine tetracetic acid may be administered to promote excretion of some highly toxic isotopes such as ^{239}Pu. Antacids may be taken to reduce absorption of ingested materials through the gastrointestinal tract, and tritiated water if swallowed can successfully be diluted by drinking large volumes of water. However, there is no really effective treatment for intake of many isotopes and the aim of the health physicist must be to prevent or reduce to a minimum the possibility of entry in whatever manner. To do this it is necessary to consider:

(i) The provision of suitable work facilities, laboratories, equipment etc.
(ii) Regular radiation and contamination monitoring.
(iii) The provision of protective clothing and equipment.
(iv) The provision of laboratory and plant operational safety rules.

6.2.1 *Laboratory and workplace design*

Because of the difficulty of controlling environmental conditions of plant or arranging suitable facilities, it is essential that as much dispensing and handling of the tracer material is carried out in the laboratory as is possible. As a general rule tracer should never be carried in the form of a solid to be dissolved on plant because of the increased risk of dispersion by the wind. Even though transport containers for liquids and gases need to be tested to much more rigorous standards and consequently tend to be correspondingly more expensive, the extra cost is justified. Similarly, tracer gases should preferably be dispensed into steel cylinders (from which injection aliquots may be drawn) in the laboratory rather than transporting them in the glass phials in which they are frequently purchased.

For the quantities of radiotracers commonly used for plant investigations, a laboratory meeting the International Atomic Energy Agency's standard for Class Two Laboratories is adequate. The primary requirements are an efficient and reliable ventilation system, easily decontaminated floor, wall and bench surfaces and personal decontamination facilities. Dispensing of solid, liquid and gas tracers should be carried out in a high-efficiency fume cupboard to prevent dispersion into the laboratory atmosphere. Solids and liquid operations should be carried out in trays lined with sufficient absorbent material to easily absorb any spillage. Lead walls or other suitable shielding should be erected to reduce the external radiation hazard to the practitioner, and remote-handling tools used to carry out manipulation of the tracers. On plant, dispensing of aliquots or other operations with liquid tracers must also be carried out on trays lined with absorbent material and remote-handling tools used.

Injection systems must be pressure-tested prior to introducing radiotracer into them and thoroughly flushed with inactive material at the end of the work. Equipment contaminated with tracer and radioactive waste materials must be carefully sealed into strong plastic bags or steel drums for subsequent disposal by an authorized route, or allowed to decay prior to normal disposal.

6.2.2 *Monitoring procedures*

6.2.2.1 *Personnel monitoring.* All persons leaving an area in which contamination may exist should be monitored to ensure that they are not contaminated with radioactive material. With the quantities involved in plant investigations a portable contamination meter of the type illustrated in Figure 6.6 is adequate to examine the hands, feet and clothing.

If contamination is found which cannot be readily removed by washing, the fact must be reported to the person authorized to carry out decontamination procedures. This personal contamination may indicate that there is unknown surface contamination in the active area; this should be investigated and the contamination removed if found. All personnel must, of course, wear a statutory personal dosemeter at all times when working with radioactive materials.

6.2.2.2 *Area monitoring.* Surveys of surfaces of areas which may be contaminated should be carried out at frequent intervals to ensure that there is no build-up of contamination, although with short-lived tracers this may not be a serious problem. If contamination is found it may indicate poor work procedures or inadequate control.

In the absence of a gamma-background, these surveys may be carried out using a portable contamination monitor. If a high gamma-background exists which will interfere with the monitor reading, areas to be surveyed are wiped with moistened absorbent material and the activity of the wipe measured in an area remote from the high-background area. Air sampling using personal air samplers or static air samplers should be carried out while dispensing operations are being performed to confirm the efficacy of the fume cupboard and room ventilation.

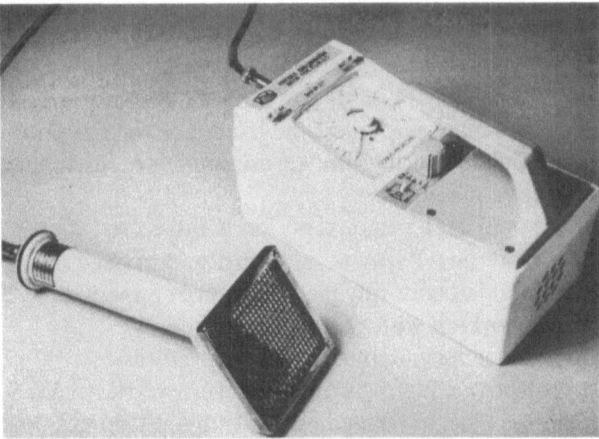

Figure 6.6. PCM5 monitor and DP2R probe by Nuclear Enterprises Ltd. The combination is capable of measuring and differentiating between alpha-,beta- and gamma-contaminants.

Those areas of plant which may have been contaminated by tracer materials, such as the dispensing area and injection point, should also be monitored for residual contamination. If contamination is found it should be reduced to a safe level by washing if possible. Where washing proves ineffective the area should be isolated or covered with absorbing material until the contamination is reduced by natural decay.

6.2.2.3 *Biological monitoring.* Biological monitoring is rarely necessary in plant investigation work unless large quantities of tritium are being used. Urine assay will show intake of tritium into the body and indicate that work procedures need to be improved.

6.2.3 *Protective clothing*

In the laboratory, a laboratory coat, disposable rubber gloves and possibly overshoes should be sufficient to afford protection. If dust masks or force-fed face masks are necessary then the fume cupboard or room ventilation is inadequate. Similar protective equipment will suffice for the plant work, coveralls usually being substituted for the laboratory coat. Often the type of coverall used will be dictated by the need to keep the operator warm and dry rather than to prevent contamination.

6.2.4 *Safety rules and instructions*

It is not possible to list a complete set of safety rules and those below should be regarded as a minimum, being added to as a particular work situation demands. The first five are statutory requirements in most countries.

1. All cuts or breaks in the skin should be covered with a waterproof dressing before starting work with radioactive materials.
2. No person should make use of any sanitary convenience while working with radioactive materials unless the hands are washed prior to doing so.
3. No person should eat, drink, smoke, use snuff or cosmetics while working with radioactive materials.
4. No person should use a personal pocket handkerchief while working with radioactive materials. Paper tissues must be provided and used.
5. No person should operate any piece of plant or apparatus by means of the mouth while working with radioactive materials.
6. Personal doserate meters must be worn at all times.
7. Suitable protective equipment must be worn when required.
8. Radiation areas must be defined using a radiation monitor and marked by erecting suitable barriers at which warning notices must be displayed.
9. No direct handling of radioactive material is permitted, suitable remote-handling tools must be used.

10. Manipulations must be carried out as speedily as possible (consistent with safety) to minimize exposure.
11. Manipulations involving solid or liquid tracers must be carried out in trays lined with absorbent material.
12. Shielding must be used wherever possible to reduce exposure to radiation.
13. Radioactive gas transfers must take place only in closed systems.
14. Contamination surveys and decontamination procedures must be carried out as appropriate.
15. Transport of radioactive materials is only permitted in containers which meet the appropriate transport regulations.

6.3 Conclusions

Our experience shows that with a proper understanding of the process, knowledge of the ultimate fate of tracer materials, choice of suitable short-lived isotopes, sensible restriction of quantities used and the use of experienced operators, plant investigations may be carried out in an extremely safe manner. Of the several hundred such investigations proposed and carried out per year, no more than one or two have to be rejected on safety grounds.

7 Radioactive tracer applications

T. L. JONES

The basic principle of a tracer investigation is to label a substance, an object or a phase and then to follow it through a system or to carry out a quantitative assay of the tracer after it has left the system. Looking at this principle from an alternative (problem-solving) point of view, if problems of fluid transport can be described in terms of 'When?', 'Where to?' and 'How much?', then they can probably be solved by means of tracer techniques. The techniques are basically of two types—chemical labelling and phase tracing.

Tracer methods using coloured dyes and chemical salts have been in use for some considerable time. With the introduction of radioactive materials as tracers, there was an initial tendency for most emphasis to be placed on chemical labelling in biochemistry and biological/medical research for example, and this tended to overshadow the development of physical tracers. Although chemical labelling, and carbon-14 studies in particular, continues to dominate the field, there has been considerable development in industrial tracer techniques over the last 20 years. Most industrial process investigation work involves physical or phase tracing applications, and so the main emphasis here will be on phase tracers.

The basic requirements of a tracer are as follows: it should behave in the same way as the material under investigation (this is not usually a problem unless there is a change of phase in the system); it should be easily detectable at low concentrations; detection should be unambiguous; injection, detection and/or sampling should be performed without disturbing the system; the residual tracer concentration in the product from the system should be minimal.

All of these criteria can be met by the use of radioisotope tracers and by careful selection of the most appropriate tracer for a particular application. Frequently more than one radioisotope can be chosen, and the factors which are important in the selection of the tracer are *half-life, specific activity, type of radiation, energy of radiation* and *physical and chemical form*. In most instances the freedom of choice is not unrestricted and specification of two or three of these factors will reduce the choice in the others.

7.1 Half-life

The choice of half-life is, like many of the others, a compromise. The half-life must be long enough to allow time to transfer the tracer from the nuclear reactor, etc., to the work site, prepare the tracer for use and complete the measurement. In order to reduce the level of residual tracer in the exit streams, however, a short half-life tracer is desirable.

There are alternative means of reducing the level of residual activity, of course. A smaller quantity of starting material could be used and/or the tracer could be extracted from the product stream.

However, reducing the amount of tracer which is used could affect the accuracy of the measurement, and extraction processes in order to be effective are likely to be time-consuming and expensive, thereby nullifying the advantages of the radioisotope method.

Apart from the radiological safety considerations, the advantages of using a short half-life tracer become far more significant if a series of replicate measurements is to be carried out on a closed, recycling system. The increasing level of activity in the system will require increasing amounts of tracer to be added to achieve the same statistical accuracy of measurements unless a short half-life tracer is used.

The compromise between the problem of transferring the active material to the site and the requirement of a short half-life necessitates the use of radioisotopes with half-lives of approximately 2–40 hours. A practical solution to this problem lies in radionuclide generators. A number of these systems were listed in Table 4.3. As they allow radioisotopes with half-lives of a few minutes (or even seconds) to be produced on site, there has been a steady increase in their use in both medical and industrial applications.

7.2 Specific activity

For any tracer investigation it is clearly imperative that the total amount of activity which is added to the system is such that, allowing for dilution and possible splitting of the process stream, sufficient remains for detection or measurement downstream. The amount of activity added and the specific activity will also be dependent upon whether the activity is to be detected *in situ* or analysed. In the case of analysis, the samples or the sample stream should contain sufficient activity for the analyses to be performed with an acceptable statistical error.

Having established the minimum level of activity, radiological safety considerations will determine the maximum limits. It may be, of course, that the latter will be smaller than the former, in which case an alternative solution will have to be sought.

From a knowledge of the total activity which has to be added to the system and the type of measurement to be performed, the specific activity of the radioisotope tracer can be determined. However, most industrial applications require dilution of the tracer with a carrier, and measurements involving high specific activity are rare. An advantage in transporting radioactive material at a higher specific activity than is required for the measurement is that the shielding requirements are more manageable for a smaller volume.

The tracer is diluted with an inactive carrier to ensure that the amount of radioactive material added to the system is not so small that its behaviour pattern is uncertain, i.e. to reduce adsorption or partial solubility effects. In making measurements of 'fluid' transport of solids or powders when the number of actual active particles will be finite, it is clearly desirable to add a large number of low-specific-activity particles in order to achieve a statistical distribution within a large system, particularly if samples are to be taken.

7.3 Type of radiation

Although there are obvious advantages is using gamma emitters in industrial applications, the use of this group is by no means exclusive, and radioisotopes which decay by alternative processes are frequently used, either because of the particular circumstances of a measurement or because other solution criteria take precedence and no gamma-emitting isotope is available.

Alpha-emitters are generally avoided, for two reasons: alpha-emission is more difficult to detect than other forms of emission unless the isotope or the daughter products also give rise to beta- or gamma-radiation; secondly and more fundamentally, unsealed alpha emitters are biologically very hazardous. The Annual Limit of Intake (ALI) of alpha-emitters is low, and the difficulty in containing tracer material within the prescribed limits, together with the measurement difficulties, restricts their general use outside the laboratory.

By comparison with both alpha- and gamma-emitters, beta-emitters offer a number of advantages, particularly in the relative ease with which they can be handled. They do not have the same degree of toxicity as alpha-emitters, and shielding requirements even for large amounts of activity are small.

The use of a pure beta-emitter, i.e. one with no associated gamma-radiation, does have the constraint that detection of the tracer in a system will involve intrusive measurement or sampling followed by radiation measurement. In this case intrusive measurement will probably be by means of a thin-wall Geiger–Müller tube, and the 'off line' measurements will usually be by liquid scintillation counting after some sample preparation.

In dual-isotope techniques, differentiation between two beta-emitters is possible using their energy differences. However, the difference in energy (E_{max}) has to be about an order of magnitude to be effective.

From the Table 7.1 it can be seen that it is possible to determine ^{3}H and ^{14}C or ^{32}P and ^{35}S in the presence of each other, but not ^{14}C and ^{35}S or ^{35}S and ^{45}Ca.

The advantages of using gamma-emitters for industrial process investigation work are well known. There are two specific properties of gamma-radiation which are significant. The first of these is the ability of electromagnetic radiation to penetrate dense materials such as steel pipe or vessel

Table 7.1

Isotope	Energy (MeV)
^3H	0.018
^{14}C	0.159
^{32}P	1.71
^{35}S	0.167
^{45}Ca	0.254
^{90}Y	2.27

walls. The obvious advantage is that many measurements such as flow rates, leak detection, residence times etc., can be carried out simply by attaching radiation detectors to pipe or vessel surfaces and monitoring the tracer as it passes.

The second important property of gamma-radiation is that the radiation from a specific radioisotope is characterized by a unique energy spectrum, thus making identification and measurement of the radioisotope relatively simple. Consequently, dual-tracer experiments have greater scope and are much more effective when gamma-emitters are used.

Shielding to reduce the external radiation to an acceptable level has to be given due consideration when handling gamma-emitters. The shielding, usually in the form of lead, is necessary during the transportation/initial handling stage of a measurement and, as previously stated, a high specific activity of the tracer at this stage reduces the isotope volume and also the mass of lead required for shielding.

The type of radiation which is used will have a direct bearing on the total amount of activity which can be accommodated safely within a given system. After injection, self-adsorption by dilution with the process stream and the additional shielding from vessel walls will reduce the external radiation to levels which must, of course, be within the legal limits. The extent to which one can allow for this in the calculation of the total activity required will depend on whether a beta- or gamma-tracer is used.

The lower-energy electromagnetic radiation from X-ray sources has some of the advantages of gamma-emitters, but these radioisotopes are usually chosen for specific properties associated with absorption or scattering, and have limited use as tracers.

The distinction between the various types of radiation is not always clearcut. Most radioisotopes decay by complex processes and it is usually the predominant rather than the exclusive type of emission which is considered in the selection process.

7.4 Energy of radiation

The choice of the energy of the radiation of the tracer will depend upon the measurement system which can be used and also upon shielding considerations. (As alpha-emitters are rarely used as tracers, there is little point in discussing the criteria for energy selection, in this case.)

The higher the energy of a beta-emitter the easier it is to detect and measure, particularly when it is used as a tracer in an industrial environment, i.e. in process fluids. The analysis of low-energy beta-emitters in industrial samples is affected by chemical or colour 'quenching' of the liquid scintillation process, i.e. a loss of efficiency in the measurement of activity. 'Cleaning' the sample is not always possible or practical. The choice of energy is also of importance in dual tracing techniques (see Table 7.1).

Autoradiography is occasionally used as a detection/measurement technique in industrial applications such as the mixing of constituents in resins and the distribution of material held in or on membranes. In these cases, unless very thin samples are involved, high-energy beta-emitters will generate a high 'background' effect from scattered radiation from below the surface, and will mask the radiation from specific particles on the surface. In the circumstances the best results are obtained by using a tracer with the lowest energy available.

Similar criteria are used to select the appropriate energy of a gamma-emitter. The actual energy is not critical, but clearly for an *in-situ* measurement the energy must be high enough to penetrate the vessel walls without having to use large quantities of material. This situation can be illustrated by examining the factors which influence the choice of an inert gas tracer. The relevant properties of the three available tracers are shown in Table 7.2.

The absorption half-thickness is the amount of material required to reduce the transmitted radiation by a half. Thus for xenon-133 the energy, and therefore the absorption half-thickness, is sufficient to penetrate only thin steel-walled vessels. On the other hand argon-41 has a high energy with a corresponding high absorption half-thickness and also a high percentage of transformations at this energy. What this means in practice is that relatively small quantities of argon-41 tracer will be required for efficient detection, even through very thick-walled vessels.

The gamma-energy of krypton-85 is suitable for detection through all but the thickest-walled vessels, but as only 0.7% of its transformations are by gamma-emission (the remainder are by beta-emission), a very large quantity of material will have to be injected into a system in order to produce the same response at the detector as argon-41. The factor is of the order of 10^2–10^3.

Table 7.2

Radioisotope	Half-life	Principal gamma energy (MeV)	Absorption half-thickness in lead ($g\,cm^{-2}$)	% of transforma- tions
^{41}Ar	1.83 h	1.29	13	99.1
^{85}Kr	10.6 y	0.51	4.5	0.7
^{133}Xe	5.3 d	0.081	0.1	35.5
		0.16	0.4	0.5

In making these comparisons, however, it could be that, because the investigation site is remote from a reactor, half-life considerations will take precedence.

Although a high-energy gamma-emitting radioisotope may be desirable for the reasons given above, it is likely to be less so when considering shielding requirements for transportation and dispensing. Also there are several applications, for example on-line leak detection and residence-time studies, when effective shielding of the downstream detector from the bulk of the tracer material is difficult or impossible and a high-energy tracer would give rise to high background 'pick-up' which would mask or distort the measured radiation. In these cases a low–medium energy gamma emitter would be the optimum choice.

7.5 Physical and chemical behaviour

Most industrial radioisotope tracer applications are based on phase tracing rather than chemical labelling; consequently, except in a few special circumstances, the physical form of the tracer is of greater importance when selecting tracer material. In assessing the behaviour of the tracer in a system, however, and particularly with regard to its ultimate fate, the chemical form should also be selected with care.

In carrying out investigations into the movement of particulate material one of the most important factors in choosing a tracer is to ensure that the tracer will behave in the same way as the process material. Attention must be paid to the tribology of the two materials as well as to the particle mass and size distributions. The ideal tracer in these circumstances is undoubtedly the irradiated process material itself.

An example of this is the determination of the residence-time distribution of sodium carbonate in rotary dryers. Irradiation of the process material in a nuclear reactor produces sodium-24 by an (n, γ) reaction. The half-life is short enough (15 h) for the material to decay during processing, and there are no complicating radioactive by-products.

Clearly this technique has limitations, and an alternative procedure is to adsorb a radioisotope tracer on to the surface of the particulate process material. Labelled material prepared by soaking a sample of a solid in a solution of the tracer has been used very successfully for many application such as catalyst recycle rate measurements, and there are many more cited in the literature[1].

For transport studies on fluids the main consideration is the solubility of the radioactive material in the process stream. In practice, material which has high chemical activity is avoided because of the possibility of chemical interaction and loss of the tracer, for instance by absorption on to vessel walls.

Attention should also be paid to the pH, or potential changes in pH, of the process stream. It is possible, for example, to lose radioactive anions by precipitation in alkaline media and also to lose radioactive cations by reduction in acid media and transfer from liquid to gaseous phase.

In general the best tracers for gas- or vapour-phase investigations are the inert gases. There are problems associated with these tracers, as illustrated earlier in Table 7.2, but the overriding advantage of using a chemically inert tracer in an industrial process is that the tracer will be discharged from the process through vent stacks together with other process inerts.

In considering vapour-phase transport studies there is the additional constraint that isokinetic sampling of vapour and tracer is impossible to achieve unless the tracer is chemically identical with the process material. The simplest solution to this problem in process investigations is to carry out measurements on vapour streams using 'on-line' detection methods and to avoid sampling.

There is, however, the special case involving steam measurements. For flow rate and distribution studies, leak determination and 'carry-over' measurements, the most suitable tracer to use is tritiated water. Tritium cannot be detected 'on-line' because it is a low-energy beta-emitter but, because the tracer and the process stream are chemically identical, sampling presents no problems.

The final choice of a radioisotope tracer for a particular investigation will be made after consideration of all of the factors discussed, many of which may be mutually exclusive. Once the decision has been made, the next stage in the procedure is the planning and execution of the measurement.

7.6 Planning a radioisotope tracer investigation

Many of the aspects of planning a radioisotope tracer experiment have been covered in the earlier part of this chapter, particularly those factors associated with the choice of tracer material. At all stages of the planning process, consideration given to basic radiation safety is of paramount importance. The basic principles of radiological safety are embodied in a series of monographs published by the International Commission for Radiological Protection[2]. These ICRP publications provide the guidelines on the amount of radioactive substances which can be handled safely and the acceptable levels of activity in waste material.

The first step in any radioisotope tracer measurement is to define the objectives of the investigation and to select the appropriate technique. The objectives will determine the duration of the measurement, the number of repeat measurements and the required accuracy. These factors in turn will enable decisions to be made on

(i) The quantity of activity to be used
(ii) Sampling systems *v.* 'on-line' detection
(iii) Detection system and efficiency of detection

(iv) Dispersion of the tracer in the system and the ultimate fate of the tracer
(v) Safety precautions.

All of these points are interdependent and this interdependence could simplify or confuse the final choice. In the following paragraphs each will be discussed briefly.

7.6.1 *Quantity of activity*

The total amount of activity to be used in an investigation will be controlled by the same criteria which governed the choice of specific activity, and also by the number of repeat measurements which are to be made. Although the radioisotope selection may call for a relatively low specific activity, investigations on a very large system or a large number of repeat measurements are likely to involve manipulation of a large quantity of activity in the initial stages. This in itself need not be a major problem, but it will require consideration of how and where the dispensing should take place and whether transport in smaller quantities may be appropriate.

The final decision on the maximum quantity which will be used will depend upon the likely radiation dose rate which will be received by the operator after all shielding and remote operating precautions have been taken.

7.6.2 *Sampling v. 'on-line' detection*

In general, 'on-line' detection is used to monitor the response to an injected pulse stimulus and, consequently, while the injection system is relatively simple, sophisticated nucleonic monitoring/recording systems are required on site. Sampling systems, on the other hand, are usually used following a continuous, steady injection of tracer material into a process stream and, as a result, the injection systems tend to be more complex and the sampling systems relatively simple. There are several exceptions to this somewhat simplistic comparison but it does underline the basic differences in the two techniques. The primary advantages of 'on-line' detection systems are:

(1) A relatively simple injection system in which the precise amount which is injected is not critical
(2) An immediate result is obtained
(3) After injection no further handling of the tracer is required
(4) Many replicate measurements are possible, leading to good statistical accuracy
(5) Access to the process stream is required at only one point
(6) Very short half-life tracers can be used
(7) Handling of toxic or disagreeable process fluids is avoided.

The advantage of 'sampling' techniques are:

(1) Sampling systems do not require complicated equipment on site
(2) A lower limit of detection is possible because laboratory-based counting systems can be used
(3) Lower specific activity material is required and/or better precision can be obtained
(4) Beta- or gamma-emitters can be used
(5) Direct assay of tracer concentration gives the result as a true volume flow-rate without requiring a knowledge of the pipe cross-sectional area or temperature and pressure of the process stream.

For a team equipped with a mobile laboratory containing nucleonic monitoring equipment, the first choice may well be on-line detection but the ultimate choice will depend upon the type of investigation, the accuracy required and the nature of the process stream.

7.6.3 *Detection systems and efficiency of detection*

Radiation detectors have been discussed at some length in Chapter 3 of this book. Obviously the type and energy of the radiation of the tracer determines which detector system is used.

For on-line detection the radiation of the tracer is exclusively gamma and the choice of detector system is between a scintillation detector or a Geiger–Müller tube. The greater counting efficiency and shorter 'dead time' of the scintillation detector makes it an ideal choice for on-line pulse detection. For sampled systems there is a wider choice of tracer material and the physical form and specific activity of the process material will also influence the selection of the detection system.

For gamma emitters, liquids or solids, scintillation counting is the most effective technique and the choice is between counting 'internally' or 'externally' to the scintillation crystal or plastic scintillator. The most efficient system in terms of sample disintegrations detected is the internal or well-counting method. If sample volume is not a constraint, however, large volumes, ∼ 5 litres, can be counted externally and this is particularly useful for measuring low-specific-activity samples.

External counting can be carried out either by placing the sample container over the scintillator in a 'Marinelli' cell, or by 'dip counting' using a Geiger–Müller tube. The 'dip counting' technique can also be of use in beta-counting (by allowing for the self-absorption of beta-radiation in the medium) and is used effectively for counting tracer/gas mixtures. The gas sample is collected in a vessel which has a Geiger–Müller tube sealed into it. Although the counting efficiency of the Geiger–Müller tube is inferior to the scintillation counter, the greater stability of the Geiger–Müller tube allows many measurements to be carried out on a previously calibrated gas counting vessel with a high degree of accuracy.

There are several other possibilities for sample counting, including liquid scintillation counting, end-window Geiger counting, particularly for solids, and gas proportional counting, but the final choice will depend on the form and specific activity of the sample.

7.6.4 *Dispersion of tracer in a system and the ultimate fate of the tracer*

Whether an on-line detection system or a sampling system is to be used for detection, the dilution of the tracer in the system has to be determined at the planning stage. It frequently happens that the dilution factor is the parameter which is being measured but there is usually sufficient data available to enable the calculation to be made with a reasonable degree of accuracy.

It is important at this stage to consider all of the possible routes that the tracer could take—this is particularly important in leak detection measurements, for example. Usually the further downstream that the tracer travels, through holding tanks, stock tanks, etc., the greater will be the dilution. There are some circumstances, however, where it is possible to re-concentrate the tracer in a part of a downstream process, and further dilution calculations will be necessary after that process. An example of this is the determination of gas flow rates to a gas–liquid reactor, or hydrogen flow rates to burners. If a chemically inert tracer gas is used, then the disappearance of the process gas stream in the reactor/burner will concentrate the tracer gas in the off-gases. In normal circumstances further dilution will occur in scrubbing towers, vent stacks etc., but one must always be aware of the possibility of reconcentration.

When introducing a radioisotope tracer into a process there are three possibilities (or a combination of these possibilities) which must be considered relating to the fate of the tracer.

(1) The tracer will be removed from the process with inert gases or liquid waste, i.e. to vent stacks and thence to atmosphere, or to drain
(2) The tracer will remain in the system in a closed loop where the activity will be reduced by natural decay
(3) It will appear in the product from the system.

The first two possibilities present no major problems, as these mechanisms can be encouraged by careful technical planning and by using chemically inert and/or short half-life tracers. The level of activity at the point of discharge must be within the prescribed limits set by the Competent Authority and these will be based on the recommendations of the ICRP.

If the nature of the process is such that it is not possible to remove the tracer from the system by any means before it appears in the product, and if the product will become available to the general public, then additional care must be taken to ensure that the Derived Working Limit* for concentration of the tracer is not exceeded.

One way of achieving this is by using tracers of very short half-life (of the order of seconds or minutes) from a radionuclide generator (see Chapter 4). Alternatively, it may be possible to remove or 'hold' the product batch from the investigation to allow the tracer additional time to decay. In this case the calculation of final activity should always be confirmed by measurement.

If none of the above procedures is applicable then it may be possible to use an 'activable' rather than a radioactive tracer to obtain the required information. This will involve addition of a suitable trace element to the system, sampling and the analysis of the concentration of the element in the samples by neutron activation analysis or, indeed, by any suitable analytical technique. There are obvious limitations to this approach but it has been used successfully for many applications.

7.6.5 *Safety precautions*

In all operations involving radioisotopes, having established the technical feasibility of obtaining the required data, decided upon the technique and tracer to be used, and determined the ultimate fate of the tracer, the next step is to define the safety precautions which must be taken to limit radiation exposure to a minimum.

The radiation dose rates expected to be encountered can be estimated by carrying out a hazards study on the investigation process which will include a review of the operating and emergency procedures.

Guidelines for assessing the potential hazards of a particular application are embodied in the recommendations of ICRP publication No 9[2] and are underlined in an IAEA publication[3]. The basic principles of radiological safety which are included in these publications are as follows:

(i) Unnecessary exposure should be avoided
(ii) Operational control should be such that the resulting radiation doses comply with the ALARA principle (i.e. as low as reasonably achievable)
(iii) Compliance with the relevant dose limits should be assured
(iv) The resulting dose to the whole population should be much smaller than the corresponding limits.

When considering the radiation levels involved it is important to distinguish between radiation workers and the general public. In this context 'radiation

---

*$DWL_c = $ *Annual Limit of Intake*$/V$ where V is the annual intake of air or water whichever is appropriate.

Table 7.3 Some useful isotopes for tracer studies

Isotope	Half-life	Radiation	Energy (keV)	Chemical form	System suitability
^{3}H	12.3 y	β	19	H_2O or H_2	Water or steam
^{14}C	5730 y	β	155	Organic compounds	Biosystems or laboratory use
^{24}Na	15.02 h	γ	1370 2755	Na_2CO_3	Solid and aqueous systems
^{35}S	87.2 days	β	167	S or SO_4	Solid and aqueous systems
^{41}Ar	1.83 h	γ	1294	Gas	Gas and vapour systems
^{45}Ca	165 days	β	254	$CaCO_3$	Solid and aqueous systems
^{46}Sc	83.8 days	γ	320 889	Sc_2O_3	Solid and aqueous systems
^{68}Ga	68 min	γ	510 1080	Ga in solution from Ge 'cow'	Aqueous or organic systems
^{76}As	26.5 h	γ	560–2080	As_2O_3 or AsH_3	Aqueous or gaseous systems
^{82}Br	35.4 h	γ	550–1480	NH_4 Br $C_2H_4Br_2$	Aqueous systems Organic systems
^{85}Kr	10.7 y	β γ	840 510	Gas	Gas and vapour systems
^{133}Xe	5.25 days	β γ	346 81	Gas	Gas and vapour systems
^{140}La	40.3 h	γ	487 1596	La_2O_3	Solid and aqueous systems
^{198}Au	2.7 days	γ	412	Au	Aqueous systems
^{203}Hg	46.7 days	γ	279	Hg	Mercury weighing

workers' will be personnel involved in work inside the controlled area, i.e. in mixing, dispensing and injecting radioactive material. In industrial process application investigations this controlled area is the principal, and frequently the only, controlled area on the site.

In order to reduce radiation dose rates to radiation workers, the operating procedure will, when appropriate, require shielding and remote handling equipment. For measurements involving very large quantities of radioactive material it is sensible to rehearse the operations and to perform 'dummy runs'. A procedure should also be established to decontaminate equipment after use and to transport and dispose of radioactive waste material.

The emergency procedure should include methods for dispersing spillages of high-specific-activity material followed by contamination monitoring. The actual method to be used for dispersing any spillages will depend upon the actual location and on the chemical/physical form of the radioactive tracer.

Finally, the notification process to commence and complete the measurement, including the emergency procedure, must be established between the site management and the investigating team at the planning stage of every radioisotope tracer investigation.

References

1. *Int. J. Appl. Radiat. Isotopes* **889**, 28 (1977).
2. ICRP Publication No. 9.
3. *Safe Use of Radioactive Traces in Industrial Processes*. IAEA Safety Series No. 40, IAEA, Vienna.

8 Measurement of flow using radioactive tracers

P. JOHNSON

8.1 Introduction

Radiotracers form the basis of several methods[1-4] for the accurate and convenient measurement of flow in practical field situations. Gaseous, liquid and solid phases may be measured individually or, if required, together. By employing detection equipment of high efficiencey only small amounts of radioactive material are necessary for determinations of high accuracy. Radiological hazards to personnel therefore can be made extremely small, including those which could, in principle, arise from contamination of plant and of the general environment. Some radiotracers suited to the measurement of flow are listed in Table 8.1.

A knowledge of flow rate is of fundamental importance in the efficient operation of process plant. This is especially true in many of the sophisticated large-scale continuous chemical processes which are operated in advanced countries over the world today. Knowledge of flow allows the process efficiency to be considered in relation to the maximum theoretically obtainable and subsequent adjustments in operating conditions observed in terms of flow used to assess optimization.

In process plant, radiotracer methods are frequently used in the following circumstances:

(a) To measure rates of flow to a greater accuracy than that possible by means of the installed instrumentation, e.g. for special purposes such as the determination of process efficiency.

(b) In order to provide data on streams for which flow information is not normally needed and for which permanent instrumentation has not been provided.

(c) To check the calibration of conventional flow instrumentation, e.g. which is suspected of malfunction either due to an intrinsic fault or which has been affected by the process fluids or conditions of operation.

Use of a radiotracer method in such circumstances is favoured by the fact that none of the methods interferes with the process plant—often there is no requirement to interrupt production at all. Methods which call for the provision of injection or sampling points in the flow can frequently make use of

Table 8.1   Radioisotope tracers suitable for flow measurement

Isotope	Half-life	Emission used	Chemical form	Medium
^{24}Na	15 h	γ	Carbonate	Aqueous liquid
			Salicylate	Organic liquid
^{82}Br	36 h	γ	Ammonium bromide	Aqueous liquid
			Potassium bromide	Aqueous liquid
			p-Dibromobenzene	Organic liquid
			Methyl bromide	Gas
^{3}H	12 y	β	H_2O	Aqueous (steam)
^{41}Ar	1.83 h	γ	Gas	Gas
^{85}Kr	10 y	β or γ	Gas	Gas
^{133}Xe	5.27 days	γ	Gas	Gas

existing access to the line, including thermocouple pockets and pressure-measurement tappings.

8.2 Pulse velocity method

A suitable radiotracer is injected over a short period of time so that a 'pulse' of radioactivity moves with the stream to be measured. The passage of the pulse is timed between two detectors, positioned at a known distance (d) apart. The linear velocity of the stream is converted to a volume rate of flow, as the diameter of the line carrying the stream is known. The general arrangement of the equipment is shown in Figure 8.1, with typical resultant traces from a chart recorder from which the transit time is derived. Evidently it is desirable that

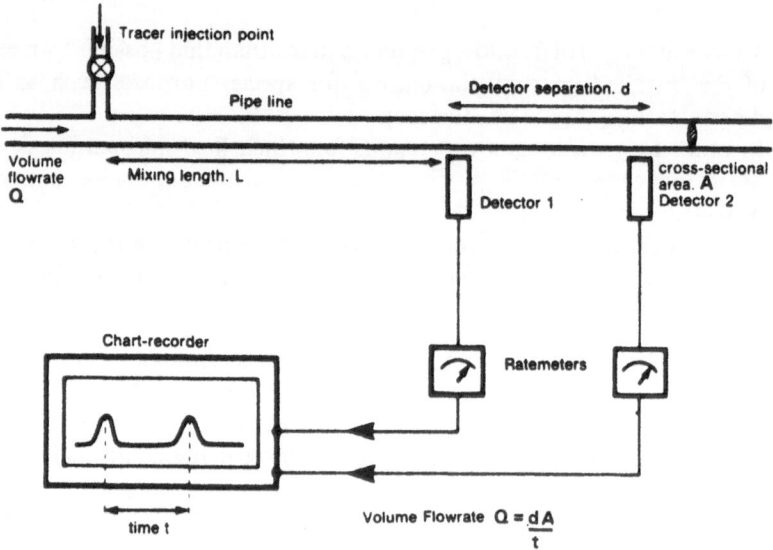

Figure 8.1.   General arrangement for pulse-velocity measurement of flow.

the traces should be as sharply defined as possible to minimize uncertainty in the determination of the time interval. The validity of the method rests on the flow being turbulent, with the first detector positioned at or beyond the distance (L) downstream at which complete lateral mixing has taken place.

For the liquid phase, several gamma-emitting radiotracers are available. The chemical form is chosen so that the tracer remains in solution in the stream to be measured and the radiation sufficiently penetrating to be detected with detectors situated outside the containment of the stream, a factor which simplifies experimental procedure considerably. In some circumstances (e.g. with gaseous flows) a suitable gamma tracer is not available, although it may be practicable to employ a thin-windowed cell in parallel with the main stream, through which beta-radiation from the chosen radioactive isotope can be transmitted.

For aqueous systems, compounds of sodium or of bromine are especially suitable, the radioisotopes having highly penetrating gamma-radiation with fairly short half-lives, 15 and 36 h respectively. A more comprehensive list of radioisotopes with appropriate chemical form suitable for the various media is given in Table 8.1.

In the pulse velocity method, the objective is to introduce the radiotracer into the stream to be measured as an instantaneous pulse. Any injection process takes a finite time and hence this represents an ideal situation. A practical system is designed to minimize the time over which the radiotracer is released into the stream and it must not, in the course of the operation, impose a disturbance on the stream in the sense of changing significantly its rate of flow. For a liquid stream, the small volume of radiotracer needed can be fairly readily introduced by use of a mechanical pump, whereby the single stroke is sufficient to overcome the internal pressure of the stream without interfering with the rate of flow at the section where the experimental observations are to be made. Liquid radiotracer may also be injected by means of gas pressure. The exit valve on a reservoir of gas (e.g. nitrogen) can be opened to the radiotracer charge, located initially in a small section of branch pipe having access to the flow through a final valve. Sudden operation of the final valve from the closed position to open and back again to closed allows a suitable 'pulse' to be introduced.

For the injection of gaseous radiotracer, mechanical systems are available such as that described by Clayton[1] in which the radiotracer gas contained in a cylindrical section moves under the action of a spring to replace the cylindrical section normally in place in the line carrying the flow. The last-mentioned flow is usually that in a branch connected to the stream to be measured and is at a somewhat higher pressure, sustained for sufficient time to sweep the tracer gas into the stream completely and with minimum disturbance.

A less sophisticated system relies on high-pressure gas applied to the radiotracer gas in a branch pipe. The opening of the final valve to the stream sweeps the tracer gas into the system to be measured. This method has to be

E

used with care as it is possible to induce shock waves in the gas flow being measured.

As it is essentially an on-site method in which gamma-emitting radiotracers are employed, the most desirable form of detector is the scintillation type, with high efficiency for gamma-radiation detection. To obtain sharply-defined responses on the passage of the tracer pulse, the scintillation detectors must be fitted with collimators made of lead or of some other material which effectively absorbs gamma-radiation.

The speed of passage of the tracer pulse carried by the stream requires attention to the integration time constant of the ratemeters employed and to the speed of response of chart recorders receiving the output of the complete detection system. When chart record of the faster transit times is required, it is necessary to use a non-contact ultraviolet or optical system. The transit time is measured from the distance on the chart between centres of area of the two responses, knowing the linear speed at which the recorder chart was operated.

For liquid flow, error in the distance between measurement points and that in the determination of transit time can both be made small; often in industrial pipelines, a larger source of error is associated with the internal diameter of the pipeline where only nominal bores are known. Frequently, this is the dominant source of error in the calculation of volume rates of flow in the pulse velocity method. Nevertheless, it is possible to make industrial measurements with a probable error of $\pm 2\%$ on a routine basis provided due attention is paid to the experimental procedure.

Gaseous flow frequently involves faster linear rates and therefore shorter transit times, yielding greater possibility for error in this part of the determination. In addition, it may be necessary, where a significant pressure drop occurs between the measurement points, to measure the actual temperatures and pressures so that the volume rate of flow may be related to appropriate reference conditions.

Experimentally the simplest of the radiotracer flow methods to carry out, all of the work is completed on the test site. The method finds application to the solution of many problems of flow measurement as well as to other industrial process problems capable of reduction to the measurement of flow—as illustrated in the practical examples given at the end of this chapter.

8.3 Dilution methods

8.3.1 Constant-rate injection method

In this method, radiotracer is injected at a constant measured rate (Q_1) in the stream to be measured. Samples are taken from a point sufficiently distant downstream to ensure complete mixing of the tracer with the stream. The specific counting rates of injected tracer (C_1) and of sampled material (C_2) are measured, the required volume rate of flow (Q_2) being computed from

$$Q_1 C_1 = (Q_2 + Q_1) C_2$$

or, as is usually the case, with $Q_2 \gg Q_1$

$$Q_2 = \frac{Q_1 C_1}{C_2}$$

An important advantage of the constant-rate injection method is that it is independent of cross-sectional area of the pipe in which it is carried. Indeed, provided the mixing requirement is satisfied, the method can be used for flow in irregular containment, for example, flow in open channels or rivers. Compared with the pulse velocity method, as there is no requirement for the diameter of the pipe to be known, it follows that it is not necessary for the flow to be full-bore.

Errors of measurement arise principally in the following factors:

(1) Incomplete mixing of the injected radiotracer
(2) Variation in and measurement of injection rate
(3) Measurement of specific counting rates of samples and of injected radiotracer (the latter including any error in dilution procedure).

(1) *Completeness of mixing.* In practical situations, the approach to complete mixing is assisted by using the longest length of pipe the test system offers, so that dispersion of the tracer is maximized by the turbulent flow regime. The use of bends, restrictions and pumps also assists in reducing errors from inadequate mixing as well as multiple (cf. single) injection points. Inadequacy of mixing is demonstrated by a wider scatter in counting rates of replicate samples from the stream than would be expected from statistical considerations.

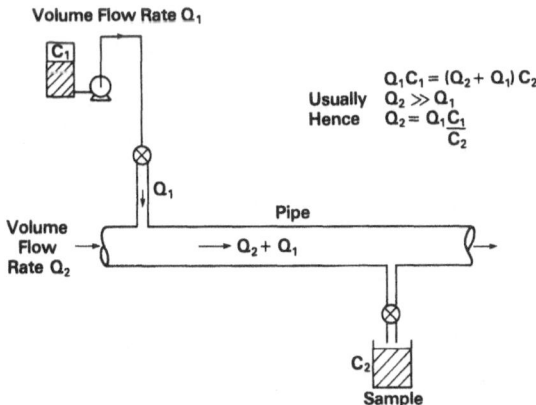

Figure 8.2.   General arrangement for constant-rate injection method.

It is possible in some cases to conduct test sampling across diameters of the test pipe carrying the flow to be measured. The consistency of measurements at various points of sampling will indicate the completeness of mixing. Evidently, such a procedure greatly complicates the measurement and may not be practicable. For the routine measurement, it is usual to adopt one or other of the means of improving mixing previously mentioned where doubt about dispersion exists.

(2) *Constancy and measurement of injection rate.* Constancy of injection rate requires the fabrication of an accurately machined injector, such as that described by Clayton[1]. In this system, a single-stroke piston expels liquid radiotracer from a precision cylinder, the former driven by a synchronous motor. The pump is calibrated by direct weighing of liquid expelled over a measured time interval in a separate test. Where accuracy of the flow measurement is required only to a few per cent, as in the case of many routine process plant measurements, a simpler type of reciprocating pump (frequently used in small-scale laboratory work) will be found to be satisfactory. An even less sophisticated system may be assembled in which radiotracer is fed by gravity into the stream to be measured from a vessel in which a constant head of liquid is maintained.

Up to the present, the injectors mentioned refer to the introduction of liquid radiotracer into liquid streams. In the case of gaseous systems, the experimental requirements are made more complex by the compressibility of the medium. Injector systems have to be designed with the particular conditions of the flowing medium in mind. Some examples of the method applied to the measurement of gaseous flow are discussed at the end of this chapter.

(3) *Measurement of counting rates—samples and injected radiotracer.* It is usual to count samples in a particular flow measurement in the same assay vessel, thereby ensuring that the same volume of sample and the same geometry (i.e. the same counting efficiency) is attained in successive measurements. It is convenient to retain a reference sample which may be set up in the counting assembly in a reproducible geometry in order to keep a running check on the stability of the counting equipment and, as necessary, to confirm the effective decay rate of the radiotracer species employed throughout the test period. If different counting vessels are employed, i.e. of the same nominal dimensions, it will be important to determine correction factors for the individual vessels to allow all counts to be reduced to a common basis.

Injection material (set aside from that radiotracer actually used in the site test) must be counted under the same conditions as the samples from the stream, using the same vessel as for stream samples. Usually, the injection radiotracer is several orders of magnitude higher in terms of activity per unit volume, and in order to avoid counting resolution errors, it is necessary to dilute the material by an accurately known factor. (This is likely to be necessary for safety reasons as well.) The dilution is carried out using ordinary radiotracer laboratory techniques, bearing in mind that any error in the dilution will show up as a proportionate error in the overall error of the flow measurement.

8.3.2 Total count method

The total count method has been considered in some depth by Hull[5]. If a detector is placed downstream of the injection point such that the usual requirement of complete lateral mixing is achieved, then as the pulse of radiotracer (which unlike the pulse velocity method, need not be a sharp pulse) passes, the number of counts recorded will be inversely proportional to the rate of flow of the stream.

If the activity injected, A, has specific activity C and the volume of solution injected is V,

$$A = CV \tag{8.1}$$

At the detector station, let the instantaneous concentration of tracer be C_t.
$dA = C_t dV$ and as $dV = Q dt$

$$A = Q \int C_t dt \tag{8.2}$$

Counts recorded by detector $dN \propto C_t dt$

$$N = F \int C_t dt \tag{8.3}$$

where F is a constant depending on the counting efficiency of the detector and the geometrical arrangement. From (8.2) and (8.3),

$$Q = AF/N \tag{8.4}$$

A static calibration is carried out using an experimental arrangement accurately simulating the system. A short length of pipe, plugged at its ends, and of such length that any increase in length does not cause any increase in counting rate in the procedure to be described, is filled with liquid of the same nature as in the measured stream. A small quantity of the radiotracer solution (specific activity C—as used in the flow to be measured) is diluted by a factor (f) and thoroughly mixed with the liquid in the calibration pipe. The same detector system as employed on the actual stream is arranged in identical geometry to the calibration pipe. Counts (n) are recorded in time t.

Equation (8.3) then applies, F being the same and the concentration C/f being constant:

$$n = FC/ft \tag{8.5}$$

From (8.5), using (8.1) and (8.4),

$$Q = \frac{nfV}{Nt} \qquad (8.6)$$

Evidently, by adopting this calibration procedure neither A nor F needs to be known explicitly, only the two counts, the dilution factor, the time for the 'calibration' count and the volume of the injected solution.

If a branch pipe can be attached to the main flow (at a point beyond the mixing distance) the flow fraction is equal to the activity fraction in the branch. Hence, when calibrated for a particular detector system and type of radiotracer the branch can be used for any flow.

The sample stream may be passed through any containing vessel—not necessarily a pipe. It is convenient, as in some of the practical examples on flow measurement given later, for the vessel to comprise the detection system itself. This is especially so where it is necessary to count beta-radiation from the tracer employed. The counting procedure must be continued for an interval that embraces the passage of the complete tracer wave. The modification using the sample stream is sometimes referred to as the 'continuous sample method'.

Another variation requires samples to be taken from the stream at regular intervals throughout the passage of the tracer wave (T). Individual sample counting rates are plotted throughout the period, the mean counting rate of the samples being determined. Alternatively, the samples may be mixed and the mean counting rate of a convenient part of the whole determined by assay directly. This method is sometimes called the 'discrete sample method'. It has the advantage that counting can be continued for a sufficiently long period to reduce statistical error to the desired level.

8.4 Application of radiotracer flow methods

8.4.1 Examples of pulse velocity method

(1) *Tube failure in waste-heat boiler.* Because of the experimental simplicity, the pulse velocity method is widely applied in the chemical process industry both to flow measurement *per se* and also to the study of working processes with the objective of identifying plant faults. One example involved a large waste-heat boiler[6] in which an abnormally high rate of tube failures was noted. Feed water was pumped to the boiler through a number of branches, (1) to (6) in the Figure 8.3. Pulses of ^{24}Na, as sodium carbonate solution, were injected just before the feed pump at I. Responses were observed for each branch in turn following successive injections, by detectors D_1, D_2, positioned in turn on particular branches. The volume rates of flow, determined from the measured velocities, showed that the feed supply was not uniformly divided between the branches. The two branches, (5) and (6), were found to be taking a much

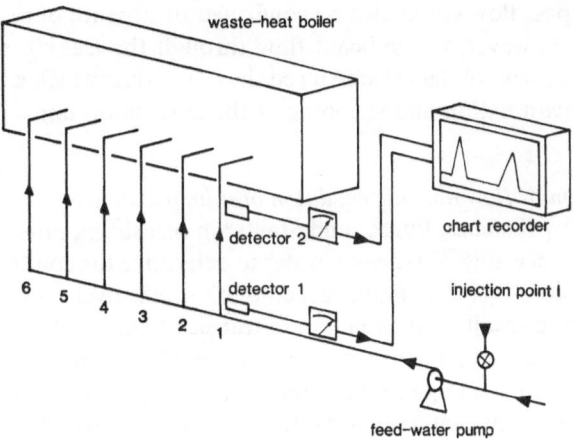

Figure 8.3. Distribution of feed-water to a waste-heat boiler.

greater proportion of the feed than the others, the maldistribution resulting in failure of the boiler tubes. By modification of the feed pipe system to equalize the flows the failure rate was greatly reduced.

(2) *Establishing nature of fault in a large-scale process.* Another example in which pulse velocity was used to establish the nature of malfunction in an adipic-acid plant is reported by Charlton and Polarski[7]. In this problem, an unusually high discharge of nitrogen oxides was observed, in spite of the bypass to the absorption section being isolated (Figure 8.4). Injection of ^{41}Ar at (I) allowed the flow to the absorption unit to be measured by means of D_1

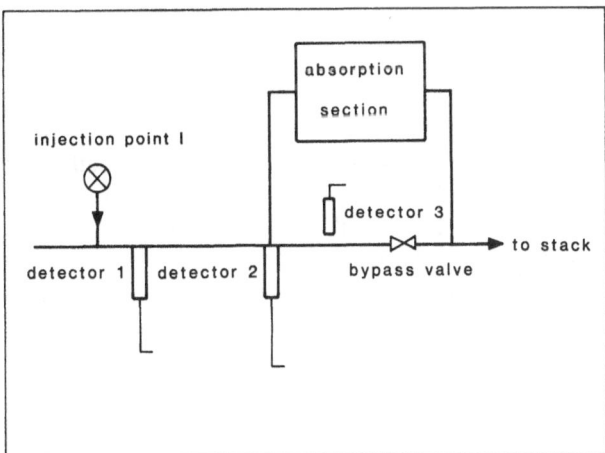

Figure 8.4. Detection of bypass flow in adipic-acid plant.

and D_2. The bypass flow could also be confirmed by absence of response on D_3. In this test however, a significant flow through the leaking valve was observed. Replacement of the valve reduced the down-time which would have resulted from investigation and stripping of the absorption unit.

(3) *Determination of flow and line condition over long distances.* [41]Ar has also been employed by Kniebes, Burket and Staats[8] in measuring rates of flow of natural gas over very long distances in order to determine the condition of the pipeline. Corrosion, dirt and liquid accumulation effectively decreases the bore of the line and results in greater cost of transfer. [41]Ar is sufficiently long-lived (provided access to a nuclear reactor is available) for the measurement, and yet decays rapidly enough to make residual activity hazards to personnel negligible. An energetic gamma-emitter, the tracer can be readily detected by equipment located external to the pipe line. Quantities of approximately 2mCi were injected into the line by means of a system operated by high-pressure nitrogen, such quantities giving accurate pulse flow measurements some 15 km downstream with a line pressure of 40 atmospheres.

(4) *Measurement of flow in an alkylation plant.* The use of the pulse velocity method for the measurement of the flow of sulphuric acid in the recirculation system of an alkylation plant[9] has been described, in which [198]Au was used as the tracer. The flow—approximately $1 m^3 s^{-1}$ in a pipe 0.5 m in diameter—was especially suited to the velocity method because of the rapid recirculation of acid in the process system. Although [198]Au has a relatively long half life of 2.7 days the initial concentration of activity in the pulse was rapidly diluted on circulation through the plant, which allowed repetition of the measurement, as required.

(5) *Study of flow in gas distribution networks.* Clayton et al.[10] describe a series of measurements on a gas distribution network in which the flow data, obtained by the pulse velocity method, were used to determine accurate values for pipe friction factors which, with pressure measurements, allowed reliable network analogues to be obtained capable of accurately predicting the behaviour of the system. Conventional methods of measuring gas flow (e.g. orifice plates) suffer from several deficiencies: pressure drop, the difficulty and expense of installation as well as large errors introduced by condensates. The pulse velocity method does not have these disadvantages; furthermore, flow measurements on several branches of the network can be made from a single injection. [85]Kr was employed as the radiotracer. As [85]Kr emits gamma-radiation in only about 0.5% of its disintegrations it was necessary in this case to employ beta-detectors so that small amounts of radiotracer would suffice. At each detector station, a sample of the gas was withdrawn continuously from

the line and passed through a cell comprising a plastic phosphor and scintillation counter. This arrangement was capable of high detection efficiency, the beta-radiation from ^{85}Kr having a maximum energy of 0.67 MeV with a yield of 99%. Existing tapping points on the gas main were utilized for the sample streams to the flow detectors. Separations between detectors varied between 300 and 900 feet. The time between responses was measured on integrating ratemeters, the actual interval being measured between half-amplitude values. Injection of the radiotracer was accomplished by means of a cylinder of small volume which could be presented to a propane stream—by operation of a trigger—and the ^{85}Kr pulse (contained in the cylindrical volume) swept into the gas flow to be measured. Gas temperature and pressure at each measuring station allowed the flow to be corrected to standard conditions. It is estimated that pulse velocity measurements at distances greater than ten miles downstream from the injection would be possible.

8.4.2 *Examples of constant-rate injection method*

(1) *Flow determination by constant-rate injection checked against direct weighing.* This example[11] is concerned with the use of the constant-rate injection method for use in the calibration of orifice plate meters for which it was desirable to have an independent assessment of the accuracy of measurement. For the purpose, the rate of flow of an organic liquid from a stock-tank was measured over the period in which the liquid was piped into a road tanker. The tanker was located on a weighbridge throughout the test. Hence, the weight of liquid transferred could be checked directly (Figure 8.5). The liquid was transferred by pump (*B*) augmented by the head in the stock tank. The line was of 3 inches nominal bore and a run of approximately 120 feet was available between the selected injection point (*I*), for the radiotracer and the sampling point *D*. Together with a number of bends (not shown in

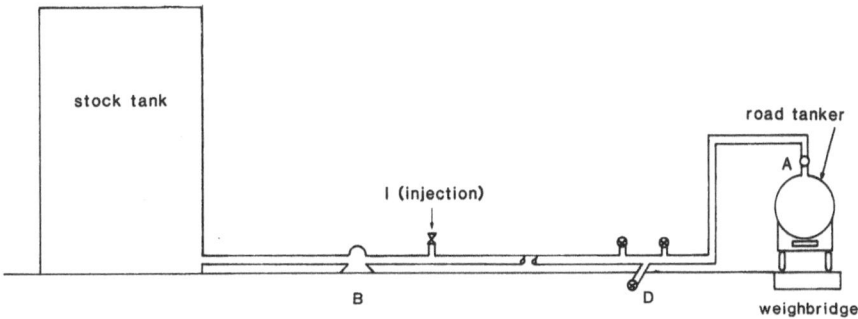

Figure 8.5. Comparison of flow by constant-rate injection and direct weighing.

diagram) thorough mixing of the radiotracer with the stock-tank liquid was assured.

[82]Br was selected as a suitable isotope in the form of para-dibromobenzene which was completely miscible with the liquid flow under test. Injection was carried out by means of a single stroke piston-in-cylinder type of pump (similar to that described by Clayton[1]).

Replicate tests of the injector confirmed the constancy of mass of material injected at a fixed temperature. Dilution of the injection solution retained from the test was carried out volumetrically, with the organic liquid at the reference temperature and all counting rates of samples from the line and of injection material were determined using 'identical' counting vessels (with minor corrections for differing counting efficiencies) containing the same volume of material at the reference temperature. The counting equipment—a sodium iodide scintillation counter and scaler—was checked for drift at intervals through the test by means of a standard source and all counting rates corrected for decay as necessary, using a common reference time. Results of measurements on samples of organic liquid are given in the table below. The sample time quoted is the time of origin after stable conditions of injection had been obtained.

The trend in the specific activity measured for the samples is real and corresponds to the decrease in flow rate (higher specific activity of samples) as the hydrostatic lead in the stock tank decreased progressively through the period of test. Calculation of the mass of liquid transferred from the constant-rate injection flow data (treating the small change in rate of flow as linear with time) yielded a value of 8.61 ± 0.06 tons compared with a value by direct weighing of 8.63 ± 0.01 tons.

For liquids, the constant rate injection method is regarded as a reliable means of calibrating orifice plates and similar conventional in-line equipment, provided that adequate attention is paid to experimental detail.

(2) *Effluent surveys on large chemical manufacturing sites.* The constant-rate injection method has been used on many occasions to conduct surveys on manufacturing sites. The need to control effluent discharges to the environ-

Table 8.2 Specific counting rate of samples.

Time (min)	$c \, sec^{-1} \, g^{-1}$	$(\times 10^{-1})$
4	3.77	
6	3.77	
8	3.80	
10	3.82	
12	3.81	
14	3.82	
16	3.84	
18	3.85	

ment has become increasingly important over the last decade or so, and hence the requirement to measure the rates of flow of aqueous streams carried in channels. A large chemical manufacturing complex may have several hundred plants contributing to the effluent network. A comprehensive survey, therefore, will call for a convenient, fairly accurate means of measuring large numbers of streams covering a wide range of flow rates which can be readily used where no provision for measurements has been made. The constant-rate injection method meets many of these requirements: basically the method is independent of the geometry of the channel in which the flow is carried and the flow does not have to fill the channel or containment. Effluent channels, being at atmospheric pressure, present no problem for injection of radiotracer which may be accomplished by means of a reciprocating pump readily available in the support laboratory. Sampling is also straightforward, and usually a simple sampling procedure, e.g. by bucket, will be satisfactory. In some establishments with a need for a fairly frequent survey, it may be desirable to employ sampling through a continuous flow cell, so that sampling counting rates are obtained on-site, avoiding transportation back to the laboratory for assay.

(3) *Measurement of gas flow rates in large-scale chemical processes.* The constant rate injection method is useful in the measurement of the flow gases, although it is to be expected that complications following from the compressibility of gases will arise. Consequently, special attention must be paid to the measurement and control of pressure and temperature in the procedure.

In this example[12] measurement of a mixed flow of hydrogen, carbon monoxide, carbon dioxide, together with smaller quantities of argon, was required in order to calibrate the orifice plate system installed in the main of 1.5 m diameter. The radiotracer selected was ^{133}Xe, because it was recognized that the element xenon lent itself to concentration by refrigeration and hence replicate measurements would be possible with relatively small quantities of radioactive material in spite of the large rates—in excess of $10^5 \, m^3 \, h^{-1}$ required to be measured.

Injection was carried out through a small precision orifice plate with flow regulation at $c.\ 10 \, M^3 \, h^{-1}$ in which the pressure drop could be accurately measured by a sensitive manometer. Temperature of the injected gas was measured to $0.2°C$ and was supplied from a reservoir containing radioactive and carrier xenon pressurized to 100 atmospheres with nitrogen.

Samples of main gas ($c.\ 5$ litres) were brought to equilibrium at known temperature and pressure and the CO_2 component removed by the addition of KOH solution. The sample was then circulated over the liquid nitrogen trap causing the xenon to solidify (vapour pressure $10^{-3} \, mm \, Hg$). The trap was isolated and the xenon transferred to a 5 ml counting vessel. Usually three circulation processes were necessary to recover sensibly all of the ^{133}Xe from the sample; on each circulation the partial pressure of xenon in the system was raised to approximately $0.2 \, mm \, Hg$ by the addition of inactive xenon. The small counting vessel was then placed in the well of a sodium iodide crystal for assay.

Samples of injection gas were treated in the same way, although here the determination of specific activity was simpler as there were no condensable gases (other than xenon) present and the volume circulated could be reduced because of the higher specific activity. Evaluation of the component errors in the measurement gave an estimated error of $\pm 1\%$ for flow rates of $10^5 \text{ m}^3 \text{ h}^{-1}$.

8.4.3 *Examples based on total count principle*

(1) *Application to flow measurement of rivers.* A considerable amount of literature exists on the measurement of liquid flow in open ditches and in rivers utilizing the total count method. Good accounts of this earlier work are given by Hull and Macomber[13]. Variations of the principle have been used in various manufacturing processes and they are especially useful where rapid recycle would interfere with the application of other radiotracer methods.

(2) *Process gas flow measurement.* Fries[14,15] has described the application of ^{85}Kr to the measurement of gas flow. A sample stream is taken from the test line and passed through a counting vessel in which a thin-window G. M. tube is mounted axially. The total count recorded during the complete passage of the tracer wave is corrected for background and the volume rate of flow deduced from $Q = AF/N$. The factor F, depending on the counting efficiency of the counting assembly, was obtained by injecting a known amount of ^{84}Kr into the counting vessel and determining the counting rate. Air was employed in the determination and a correction curve for gases of different densities (and hence different self-absorption) was established, from which correction could be made for particular streams under test. Considerable variation in counting efficiency was observed between G. M. tubes. A convenient way of dealing with this employed ^{204}Tl (which emits beta-radiation of similar energy to ^{85}Kr) in a subsidiary test. The ^{204}Tl was deposited on a filter paper and attached to the interior of the counting vessel. The factors for different G. M. tubes were then determined and their relative efficiencies used to normalize observations between tests.

The injection procedure involved the transfer of a measured amount (by counting the gamma-emission) of ^{85}Kr to a small glass vessel containing a steel ball. The vessel was inserted into a steel cylinder which was rocked vigorously to break the glass vessel. The cylinder could be connected to a high-pressure nitrogen cylinder when injection of the tracer pulse was required.

The total count method based on ^{85}Kr has been used for the measurement of process gas flow (for example, methane) and also for the investigation of leakage flows.

(3) *Steam flow by total sample method.* Fries[16] also has applied the total sample method to the measurement of steam flow. For the measurement of a medium which can be present in either gaseous or liquid phases it is essential that an isotopic tracer be employed which will follow the medium faithfully, regardless of phase. Consequently, tritiated water was chosen as the tracer. It was injected into the steam flow as water by means of high pressure gas (e.g. nitrogen). The method depends on the vaporization of the tritiated water and mixing with the steam before sampling is carried out from the main. Samples were taken through a side stream passed through a condenser, over a period covering passage of the complete tracer wave. A liquid scintillation counting system was employed for assay, necessary for the counting of the very low-energy beta-radiation from tritium.

References

1. Clayton, C. G. (1964) The measurement of flow of liquids and gases using radioactive isotopes. *J. Br. Nucl. Energy Soc.* **3**, 252.
2. Broda, E. Schonfeld, T. (1966) *The Technical Applications of Radioactivity*, Vol. 1. Pergamon, Oxford.
3. Webb, J. K. (1979) Radiotracer techniques for flow measurement and process investigation. *I Chem E Symp. Ser.* No. **60**, 73.
4. Ljunggren, K. (1967) Review of the use of radioactive tracers for evaluating parameters pertaining to the flow of material in plant and natural systems. *Proc. Symp., Radioisotope Tracers in Industry and Geophysics*, Prague, 1966, IAEA, Vienna, 303.
5. Hull, D. E. (1955) The total count technique: a new principle in flow measurement. *Int. J. Appl. Radiat. Isotopes* **4**, 1.
6. Charlton, J. S., Heslop, J. A. and Johnson, P. (1982) Radioisotope techniques for the investigation of process problems in the chemical industry. *Proc. Symp.: Industrial Applications of Radioisotopes and Radiation Technology*, Grenoble, 1981, IAEA, Vienna, 393.
7. Charlton, J. S. and Polarski, M. (1983) Radioisotope techniques solve CPI problems. *Chem. Eng.*, Feb., 21.
8. Kniebes, D. V., Burket, P. V. and Staats, W. R. (1960) Argon-41 measures natural gas flow. *Nucleonics* **18** (6), 142.
9. Hull, D. E., Fries, B. A. and Gilmore, J. T. (1965) Acid circulation volume, replacement and entrainment measured in an alkylation plant with radiotracer. *Int. J. Appl. Radiat. Isotopes* **16**, 19.
10. Clayton, C. G., Evans, G. V., Spackman, R. and Webb, J. W. (1969) A mobile system for measuring flow in a gas distribution network. *Atom* **151**, 128.
11. Whiston, J., Johnson, P. Internal Report.
12. Johnson, P. (1967) Application of the dilution principle to the measurement of gas flow rates in large-scale chemical processes. *Proc. Symp. Radioisotope Tracers in Industry and Geophysics*, Prague, 1966, IAEA, Vienna, 615.
13. Hull, D. E. and Macomber, M. (1958) *Proc. Second Int. Conf. on the Peaceful Uses of Atomic Energy*, Geneva, 1958, 19, United Nations, New York, 324.
14. Fries, B. A. (1962) Gas flow measurement by the total count method. *Int. J. Appl. Radiat. Isotopes* **13**, 277.
15. Fries, B. A. (1977) Krypton-85. A versatile tracer for industrial process applications. *Int. J. Appl. Radiat. Isotopes* **28**, 829.
16. Fries, B. A. (1965) Steam flow measurements by the total sample method. *Int. J. Appl. Radiat. Isotopes* **16**, 35.

9 Measurement of residence times and residence-time distributions

G. REED

9.1 Introduction

It is not the intention of this chapter to treat exhaustively the theory of residence-time measurements; there are many excellent textbooks on the subject[1,2]. A brief review of ideal and non-ideal models, together with an indication of how the more important parameters of the models may be calculated, is sufficient for our purposes. The effect of process malfunctions on residence times and residence-time distribution will be discussed and illustrated using relevant examples.

All process vessels are designed to carry out a specific function. When the vessel fails to perform this function there may be either a design fault or a malfunction of some sort, and residence-time distribution analysis is often used to gain an understanding of what is happening inside the vessel.

We will adopt the so-called 'control volume' approach. The system under investigation (it can vary from part of a particular item of equipment to a whole plant) is called the control volume. A known stimulus is fed into the control volume and monitored at the exit or some intermediate point. The system is then described in terms of the effects it has on this known stimulus as it passes through the control volume. Almost any type of stimulus which can be detected may be applied to the system. For example, a step increase in concentration of one of the reactants could be used, but we will confine ourselves to consideration of the injection of a tracer which does not disturb the flow pattern in the control volume.

Radioactive materials are particularly effective as tracers in process investigations of vessel systems because of the wide variety of isotopes available, chemical and fluid compatibility, the low concentrations required, and the ease with which sharp pulses of tracer may be injected into most systems. The ability to choose between external detection of tracer or sampling to determine tracer concentration is an additional advantage.

Figure 9.1. Ideal reactors. (a) Plug-flow; (b) stirred-tank reactor.

9.2 Flow through ideal reactors

Consider fluid flowing through the two reactors shown in Figure 9.1. The first type of reactor is known variously as the plug-flow, slug-flow, piston-flow or ideal tubular reactor. We will use the term 'plug-flow' reactor. If we assume that all elements of fluid move through the reactor at the same rate, with no mixing of the fluid elements either forwards or backwards (although mixing may occur in the lateral direction), then we have an ideal-flow situation. A sharp pulse of tracer injected at the inlet line will demonstrate the same tracer concentration distribution at the exit of the vessel (Figure 9.2a). The second type of reactor is called the ideal stirred-tank reactor, the well-mixed reactor, the backmix reactor or the constant-flow stirred-tank reactor. We shall use the term 'ideal stirred-tank' reactor. If the stirrer is perfectly efficient then as soon as a pulse of tracer is injected into the reactor it is equally distributed immediately throughout the total volume. The fluid leaving the reactor then has the same tracer concentration as the material within the reactor, and we have another ideal situation. As fresh fluid enters the vessel and is immediately distributed, then the tracer concentration at the exit will show an exponential drop with time, as shown in Figure 9.2b. In practice these ideal reactors are never encountered, although they may be closely approached and are, of course, very desirable from a design point of view.

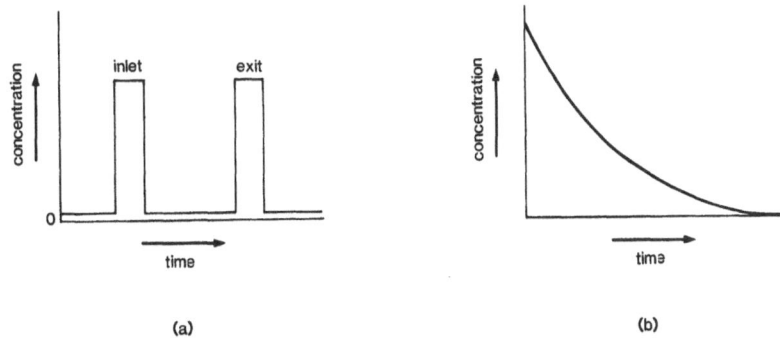

Figure 9.2. Tracer concentration distribution curves. (a) Ideal plug-flow reactor; (b) ideal stirred-tank reactor.

9.3 Flow through non-ideal reactors

If we now consider fluid flowing through a real reactor where back and forward mixing may occur or the stirrer is not 100% efficient, then different elements of the fluid may take different paths through the vessel and consequently take different lengths of time to pass through the reactor. The distribution of these times is called the exit age distribution (E) or the residence-time distribution (RTD). A typical E curve is shown in Figure 9.3. A tracer which is physically compatible with the fluid flowing in this reactor, injected into the inlet line, would distribute itself and follow the same flow patterns as the elements of fluid. If the tracer is, in fact, injected as a step input, of concentration C_0, into the fluid entering a closed vessel, then a time record of tracer in the exit stream from the vessel measured as C/C_0 is called the F curve. This is shown in Figure 9.4, and C/C_0 always rises to 1.

For a normalized E curve, that is, when the total area under the curve = 1, or

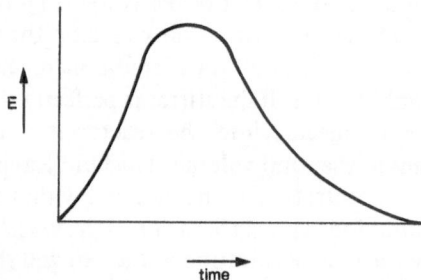

Figure 9.3. Exit age distribution (E) curve or residence-time distribution (RTD) curve—non-ideal reactor.

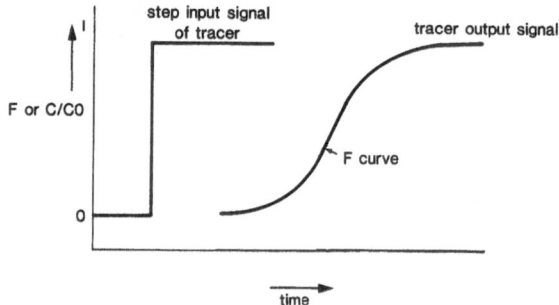

Figure 9.4. F curve—dimensionless tracer concentration v. time for step injection.

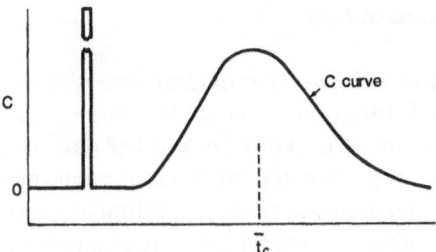

Figure 9.5. *C* curve—dimensionless tracer concentration *v*. time for instantaneous pulse (delta function) injection.

$$\int_0^\infty E\,dt = 1 \tag{9.1}$$

then
$$F = \int_0^t E\,dt$$

or differentiating,
$$\frac{dF}{dt} = E. \tag{9.2}$$

Alternatively if the tracer is injected as an instantaneous pulse (delta function) into the closed vessel, the normalized response, given by dividing the measured concentration, C', by Q (the area under the concentration–time curve) is called the *C* curve. Figure 9.5 shows a typical *C* curve. The normalization equation may be written:

$$\int_0^\infty C\,dt = \int_0^\infty \frac{C'}{Q}\,dt = 1, \text{ where } Q = \int_0^\infty C'\,dt \tag{9.3}$$

Since the *C* curve represents the residence-time distribution for that batch of entering fluid it must also be representative of any other batch of entering fluid, i.e. the *C* curve is the same as the *E* curve.

The mean time, $\bar{t}$, of fluid in the closed vessel at steady-flow conditions is, of course, given by

$$\bar{t} = \frac{\text{vessel volume}}{\text{flow rate}} = \frac{V}{U} \tag{9.4}$$

and it can be shown that

$$\bar{t} = \bar{t}_C = \bar{t}_E \tag{9.5}$$

where $\bar{t}_C$ and $\bar{t}_E$ are the mean times of the *C* and *E* curves respectively. $\bar{t}$ is generally termed the mean residence time (MRT). This relationship is true only for closed vessels; for open vessels where the *C* and *E* curves may differ it is not valid.

9.4 Models for non-ideal flow

So far we have considered two types of ideal-flow systems, the ideal plug-flow reactor and the ideal stirred-tank reactor, together with a non-ideal system in which a certain amount of backmixing or other deviation from ideal flow is occurring. We have seen that the latter case is intermediate in tracer exit concentration between the former two. Inspection of tracer distribution curves may well provide sufficient information for comparison with their design criteria, and show reactors to be near enough to or widely different from their expected performance. More often, a closer comparison is necessary and, because most real reactors differ significantly from the ideal, in order to describe these real systems it is necessary to propose some model. The normal type of model is based in measuring deviations from ideality.

9.4.1 Stirred-tank equivalents model

In this model it is assumed that the actual system under investigation can be represented by a series of equal-sized ideal stirred tanks, shown in Figure 9.6, the sum of the individual volumes being equal to the real volume of the system. The response of n stirred tanks in series to an impulse input can be shown[3] by using the convolution integral or Laplace transforms to be

$$C(\theta) = \frac{n^n \theta^{n-1} e^{-n\theta}}{(n-1)!} \qquad (9.6)$$

where n = number of stirred tanks (or stirred tank equivalents)
    θ = dimensionless time = time/mean residence time = $t/\bar{t}$
  $C(\theta)$ = dimensionless concentration = C/C_0 at time θ.

The family of curves of $C(\theta)$ v. θ for different values of n are shown in Figure 9.7. The stirred-tank equivalent, n, for a given reactor may be estimated by comparison. When $n = 1$ the model reduces to the ideal well-mixed reactor; when $n = \infty$ the model reduces to the ideal plug-flow reactor. Most process situations will lie somewhere between these two limits and it is recommended that this type of model is used for systems which approximate to the backmix case.

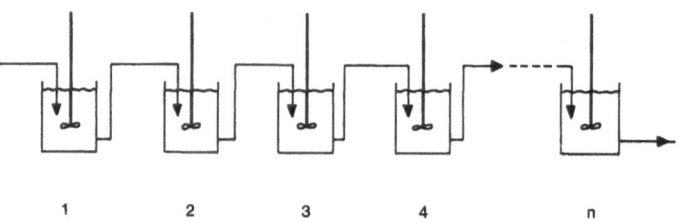

  1   2   3   4   n

Figure 9.6. Stirred-tanks equivalent model.

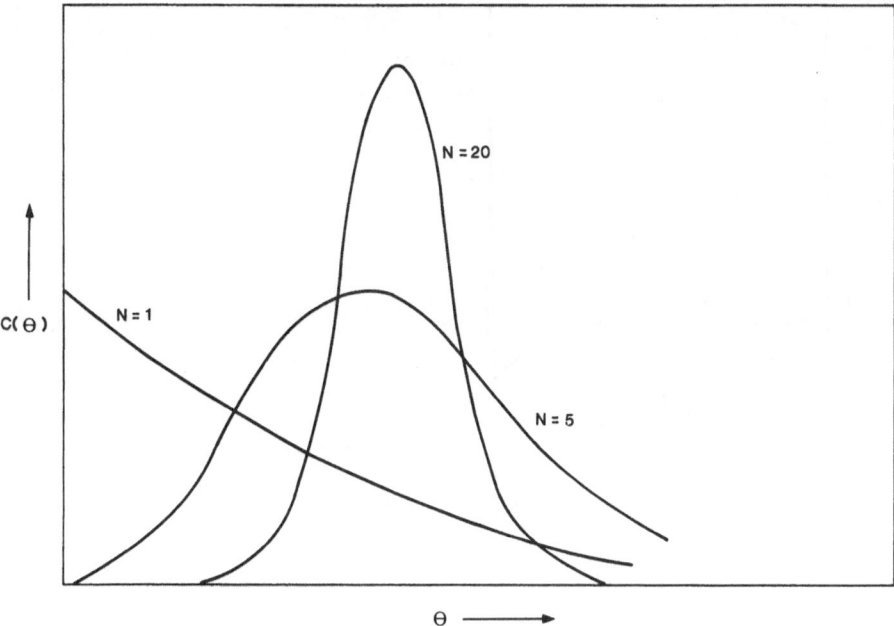

Figure 9.7. Examples of RTD curves for stirred-tank equivalent model.

9.4.2 *The dispersed plug-flow model*

In this model it is assumed that the flow can be represented by plug flow with a certain amount of longitudinal backmixing superimposed. The amount of backmixing is independent of position within the system. Assuming that this backmixing is similar to molecular diffusion we can apply Fick's Law of Diffusion and it can be shown[1] that the response to a system which can be described by plug flow and some dispersion is given by

$$C(\theta) = \frac{1}{2\sqrt{\pi D/\mu L}} \exp \left\{ \frac{-(1-\theta)^2}{4(D/\mu L)} \right\} \qquad (9.7)$$

where D is the longitudinal or axial dispersion coefficient, μ is the liquid velocity and L is the length of the reactor.

The group $D/\mu L$ is the Inverse Peclet Number. When $D/\mu L = 0$ the model reduces to the ideal plug-flow reactor. When $D/\mu L = \infty$ the model reduces to the ideal stirred-tank reactor, as shown in Figure 9.8. Most process situations will be somewhere between these two limits but it is recommended that this type of model is used for situations which approach the plug-flow case.

So far two relatively simple models have been proposed. One model measures deviations from ideality in terms of deviations from backmix flow, while the other model approaches the problem from the other extreme. In

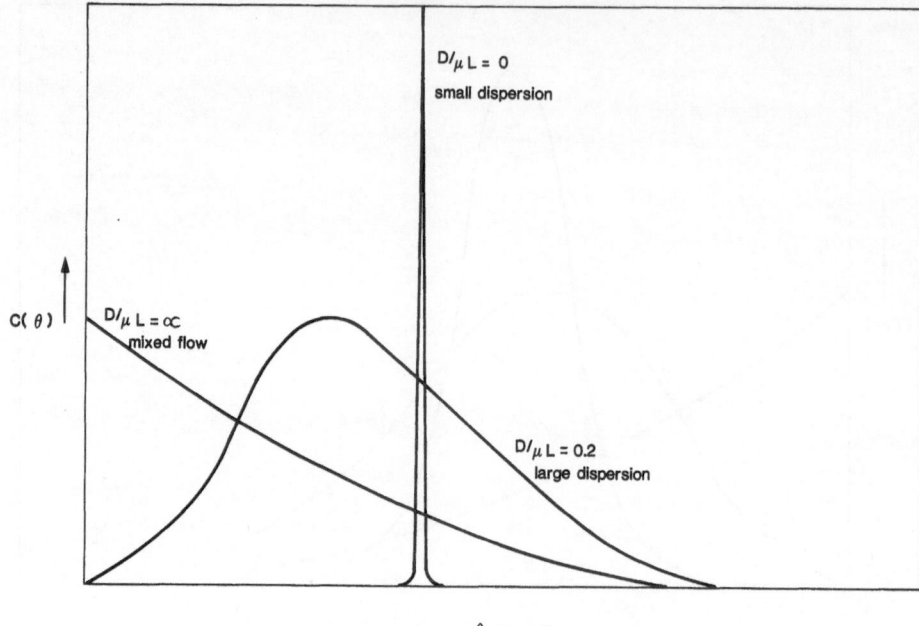

Figure 9.8. Examples of *C* curves, dispersed plug-flow model.

theory most process situations should be covered adequately by either of these two models, but in practice flows may deviate substantially from the ideal cases and these deviations are often caused by other factors (bypassing, channelling, stagnant regions, etc). When this is the case, the above two models often prove inadequate, and mixed models may need to be used.

9.4.3 *Mixed models*

Mixed models are combinations of several different types of flow element; three examples are given in Figure 9.9. The use of mixed models is fraught with difficulty and just because the data fit a model it does not mean that the model is correct. For example, the reactor systems shown in Figure 9.9 (*a*) and 9.9 (*b*) will give exactly the same physical tracer concentration *v.* time response Figure 9.9(*d*), but may give very different product concentrations at the exit. More complex models have correspondingly more variables and very often misleading information may result. Mixed models should, therefore, always be used with caution.

The model shown in Figure 9.9(*a*) is a simple example of a mixed model. It consists of an element of plug flow which causes the tracer pulse to the delayed by time t_1, before going into *n* stirred tank in series with a mean residence time of $\bar{t}$. This model may be described in terms of the three parameters, $\bar{t}, t_1$, and *n*.

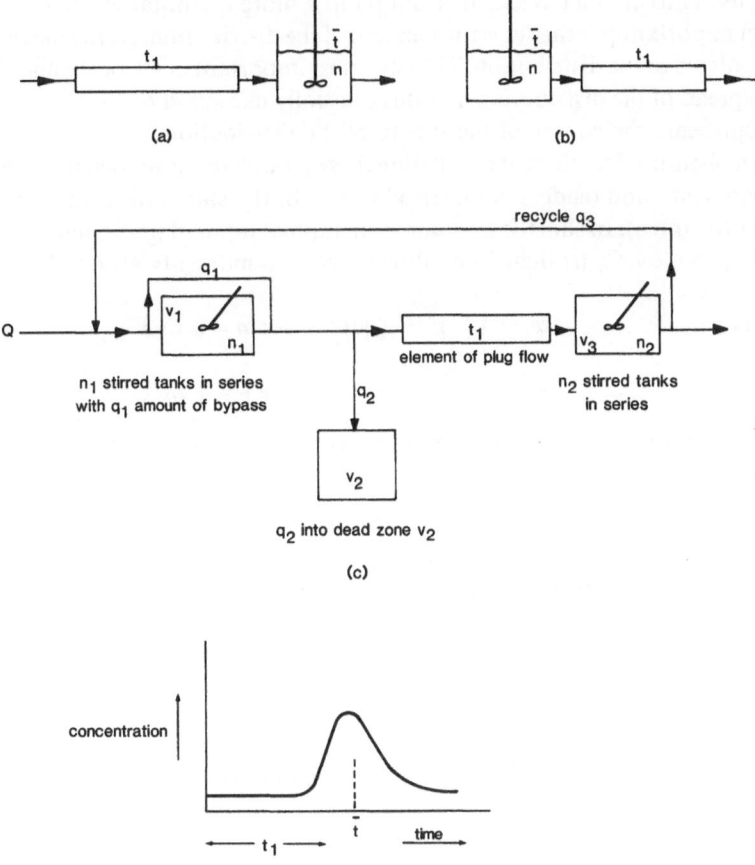

Figure 9.9. Examples of mixed models.

In dimensionless units the response of such a system to an impulse input is given by

$$C(\theta) = \frac{n^n}{(n-1)!}\left(\frac{t-t_1}{\bar{t}-t_1}\right)^{n-1} \exp - n\left(\frac{t-t_1}{\bar{t}-t_1}\right) \qquad (9.8)$$

(Compare with equation 9.6).
Similar and much more complex expressions can be derived for other mixed-model systems.

9.5 Calculation of parameters

As mentioned previously, simple comparison of tracer concentration curves with those expected for a system may provide sufficient information, but often it is useful to characterize a distribution in a more quantitative manner. The most important parameter is the location of the distribution, i.e. the mean time or centroid of the distribution. The next most important descriptive quantity is the spread of the distribution and this is usually measured by the variance σ^2. It represents the square of the spread of the distribution.

An often-used method of calculating these parameters is termed the Method of Moments, and readers are referred to one of the standard textbooks for a fuller treatment. In short, if we assume an impulse input of tracer and an outlet pulse given by $C_1(t)$ then computing successive moments about the origin

we have
$$\alpha_n = \int_0^\infty t^n \cdot C_1(t)\mathrm{d}t \qquad \text{for } n = 0 \text{ to } 3$$

and
$$A_n = \alpha_n/\alpha_0 \qquad \text{for } n = 1 \text{ to } 3.$$

Using Laplace transforms and the convolution integral[4-8] it can be shown that

$$\text{first moment about the origin} = A_1 = \alpha_1/\alpha_0$$
$$\text{second moment about the origin} = A_2 = \alpha_2/\alpha_0$$
$$\text{third moment about the origin} = A_3 = \alpha_3/\alpha_0$$

where

$$\alpha_0 = \int_0^\infty C\,\mathrm{d}t = \sum_{i=1}^n C_i \qquad \text{(for equally spaced data points)}$$

$$\alpha_1 = \int_0^\infty Ct\,\mathrm{d}t = \sum_{i=1}^n C_i t \qquad \text{(for equally spaced data points)}$$

$$\alpha_2 = \int_0^\infty Ct^2\,\mathrm{d}t = \sum_{i=1}^n C_i t^2 \qquad \text{(for equally spaced data points)}$$

$$\alpha_3 = \int_0^\infty Ct^3\,\mathrm{d}t = \sum_{i=1}^n C_i t^3 \qquad \text{(for equally spaced data points)}$$

Variance = second moment about the mean = $A_2' = A_2 - A_1^2$
The third moment about the mean, $A_3 = A_3 - 3A_2A_1 + 2A_1^3$

$$\text{The mean time } \bar{t} = A_1 \qquad (9.9)$$

$$\text{Variance } \sigma^2 = A_2 \qquad (9.10)$$

We can extend this treatment to our three non-ideal models as follows.

9.5.1 Stirred-tank equivalent model

In equation (9.6) we stated that

$$C(\theta) = \frac{n^n}{(n-1)!}\left(\frac{t}{\bar{t}}\right)^{n-1} e^{-n(t/\bar{t})}$$

This is a special form of the more general gamma distribution of probability, the moments of which are well known. The first three moments about $t = 0$ are:

$$M_1 = \bar{t}$$

$$M_2 = \bar{t}^2 \cdot \frac{n+1}{n}$$

$$M_3 = \bar{t}^2 \cdot \frac{(n+2)(n+1)}{n^2}$$

Using the same notation as used in (9.5) we have

$$\bar{t} = A_1$$

and

$$n = \frac{A_1^2}{(A_2 - A_1^2)} = \frac{A_1^2}{A_2'} \tag{9.11}$$

Thus by calculating the first two moments of the residence-time distribution, values of $\bar{t}$ and n can be calculated. Therefore, the experimental RTD can be described in terms of n stirred tanks in series with a mean residence time of $\bar{t}$.

9.5.2 Dispersed plug-flow model

We stated that

$$C(\theta) = \frac{1}{2\sqrt{\pi(D/\mu L)}} \exp\left\{ \frac{-(1 - t/\bar{t})^2}{4(D/\mu L)} \right\}$$

by comparing moments as in the stirred-tank model

$$\bar{t} = A_1$$

and

$$A_2' = 2\bar{t}^2(D/\mu L)\{1 - (D/\mu L)(1 - e^{-1/(D/\mu L)})\} \tag{9.12}$$

For a one-shot tracer input it is usually good enough[1] to approximate these equations to

$$\bar{t} = A_1$$

and

$$A_2' = 2\bar{t}^2(D/\mu L)$$

thus

$$\bar{t} = A_1$$

and

$$D/\mu L = \frac{A_2'}{2A_1^2}.$$

Therefore, by calculating moments, the mean residence time and inverse Peclet number may be calculated.

9.5.3 Simple mixed model

In our model above we stated that

$$C\theta = \frac{n^n}{(n-1)!}\left(\frac{t-t_1}{\bar{t}-t_1}\right)^{n-1} e^{-n(t-t_1)/(\bar{t}-t_1)}$$

The first three moments of this distribution about t_1 are

$$M_1 = \bar{t} - t_1$$

$$M_2 = (\bar{t}-t_1)^2\left(\frac{n+1}{n}\right)$$

$$M_3 = (\bar{t}-t_1)^3\frac{(n+2)(n+1)}{n^2}.$$

Rearranging and using the same notation as used in (9.5),

$$\bar{t} = A_1$$

$$t_1 = A_1 - \frac{2(A_2')^2}{A_3'}$$

$$n = \frac{4(A_2')^3}{(A_3')^2}.$$

Thus the method of moments technique can be used to fit the three simple non-ideal models we have discussed to experimental RTD data. It can also be used for the analysis of non-impulse inputs and more complex models than the ones proposed here. Other methods are available to determine t and n, and for a fuller treatment the reader is referred to one of the standard textbooks.

9.6 Diagnosing malfunctions of process equipment

We have indicated in the preceding sections how reactors may be modelled in terms of deviations from ideality and how to quantify these deviations in real terms. Reactor design is a vast subject involving the order of reactions, reaction rates, effects of temperature and pressure, homogeneity or otherwise of reactants, etc., and is far beyond the scope of this book. Let us instead assume that the reactor has been designed to carry out a specific function and it is not operating as it should. It could be a design fault, and by carrying out the appropriate calculations for the model of $\bar{t}$ and n, information can be gained to allow subsequent modification to the vessel. More often in the real situation some fault has developed in the reactor and determination of the C curve can give information on the type of fault which has developed and enable repairs to be planned and carried out effectively during shutdown. Often a drop in

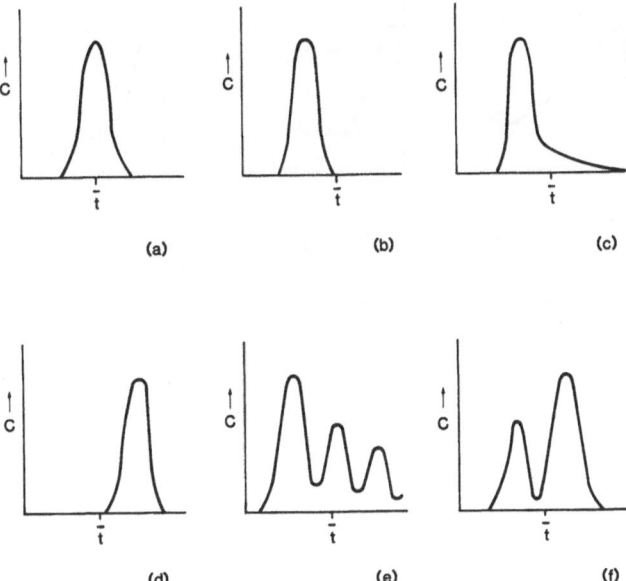

Figure 9.10. RTD illustrating malfunctions in plug-flow system.

performance leading to low conversion of reactants could be due to either a physical fault in the reactor, such as channelling through a catalyst bed, or a chemical fault such as poisoning of the catalyst in the bed. To identify which of the two is causing the loss of efficiency prior to shutdown is obviously of great importance. In the one case discharge of the catalyst and repacking of the bed is a sufficient and simple remedy, although extra maintenance effort may be required, for example, to maintain the discharged catalyst under an inert atmosphere to prevent oxidation. In the other case replacement of the catalyst involves a considerable financial investment in new catalyst and no precautions need be taken to prevent oxidation of the discharged catalyst. It is in this sort of 'trouble-shooting' situation, where determination of the RTD can give immediate, useful information on plant malfunction, that tracer techniques are most useful, and the case histories below concentrate on this category.

To illustrate the sort of effect we are looking for let us firstly consider the case of a reactor designed to operate as a simple plug-flow reactor, i.e. with low dispersion in the longitudinal direction. Figure 9.10 illustrates some of the faults which may occur in such a system, where $\bar{t}$ is the expected MRT.

Figure 9.10a is the expected trace with small amount of dispersion occurring and the expected MRT. Figure 9.10b shows the tracer arriving much earlier than expected, and indicates that the effective volume of the reactor is smaller than design, that is channelling is occurring and there must be dead zones within the reactor. This may not be too serious if the reactor is over-designed,

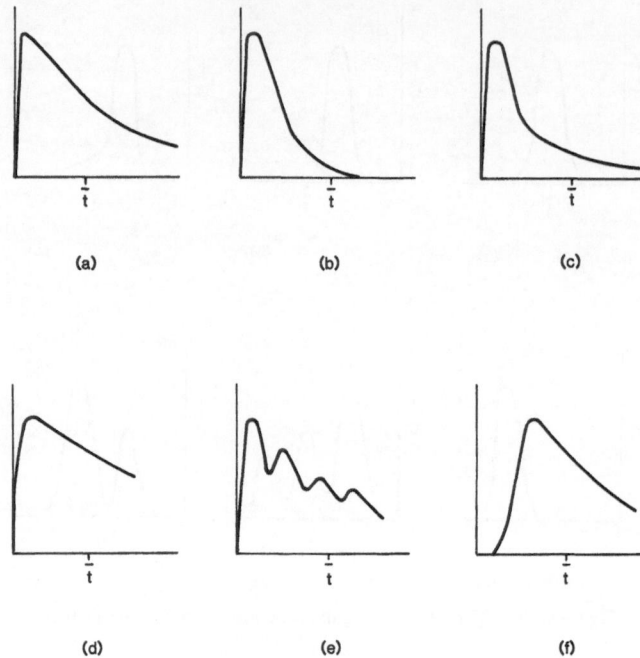

Figure 9.11.   RTD illustrating malfunctions in well-mixed system.

in that sufficient conversion may still be occurring, but this condition often worsens rapidly to an unacceptable degree. Figure 9.10c again shows the tracer arriving much earlier than expected, indicating channelling, but the long tail shows that some mixing is occurring into the relatively dead zones and that a considerable fraction of the reactants are staying in the reactor for a much longer time than expected, leading possibly to unwanted side reactions.

Figure 9.10d shows the tracer arriving at a much *later* time than expected, an unusual effect indicating that either the calculated volume or flowrate through the system is wrong or that the tracer is not following the reactants faithfully, due to absorption on the walls or internals of the vessel. Figure 9.10e shows that recirculation of the tracer is occurring within the vessel. If the mean time of the first tracer peak coincides with t then it indicates recirculation of the tracer in the process system external to the vessel, and the time difference between successive peaks gives the MRT of the external system. Figure 9.10f shows fluid channelling down two separate paths with different residence times in each channel.

If we now consider the other extreme of reactor type, that designed to achieve well-mixed flow, the problems listed above under the plug-flow reactor will show the concentration v. time patterns illustrated in Figure 9.11. They represent expected, early arrival, early arrival with long tail, late arrival and internal circulation respectively for the first five, and conclusions drawn from such traces are the same as for the plug-flow reactor. The sixth shows a time delay prior to arrival of the tracer in the expected pattern. The time delay could be due either to an instrument time lag or because the distance between the injection point and vessel is sufficiently long to introduce such a delay in the tracer entering the vessel.

9.7 Equipment arrangement for measurement of RTD using radiotracers

The usual arrangement of equipment is shown in Figure 9.12. Detectors, well shielded from the vessel and injection point, are positioned on the inlet and exit lines and their signal processed and fed to a high-speed recorder or tape recorder. If greater sensitivity is required, because of limitations on the amount of tracer which may be used or because exit lines are small, then multiple detectors can be mounted.

In some situations adequate shielding of the detectors may be difficult to arrange and the detectors may have to be positioned some distance from the vessel. The RTD must then be adjusted to take account of the transit time of the tracer in the lines if this is significant. However, with well-mixed reactors it

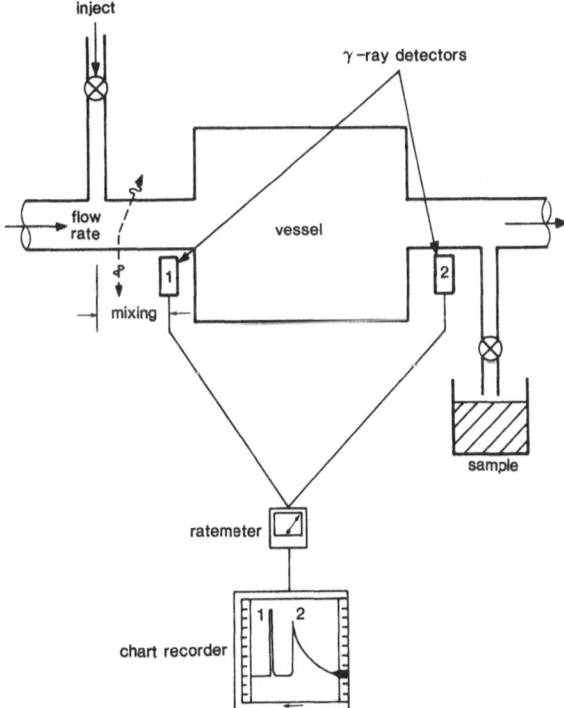

Figure 9.12. Residence-time measuring arrangement.

may be sufficient to place the unshielded detector on the wall of the vessel, since a characteristic of these reactors is that material leaving the vessel is representative of material in the reactor and vice versa.

Occasionally it may be necessary to take samples from the system because a chemically compatible beta-emitting tracer is being used, because information from 'mixing cup' samples is required, or simply because the vessel under investigation is too close to the next vessel in the system to allow shielding or separation of the detector from the vessel. Such samples should always be taken with care and the proper precautions employed.

On some occasions it may not be possible to use radioactive tracers to carry out RTD measurements, either because level control on the vessel is by nucleonic instrumentation or because there may not be time to obtain a licence to use radiotracers on site. It is also possible that the plant management may be reluctant to 'contaminate' their product even with short-lived tracers. To inject conventional tracers and assay samples by conventional techniques may require large quantities of tracer at levels unacceptable to the plant. In these situations the use of 'activable' tracers has often proved satisfactory. An element, not normally present in the process stream, is injected and samples taken for irradiation in a nuclear reactor. This produces a radioactive isotope of the element. The concentration of the element is then estimated by the usual counting techniques used for assay of radioactive materials. Examples of elements used as activable tracers in the form of organic or inorganic salts have included the following:

Manganese—detection limit 0.00005 micrograms
Indium—detection limit 0.001 micrograms
Vanadium—detection limit 0.0001 micrograms
Cobalt—detection limit 0.005 micrograms.

With such low detection limits, concentrations several factors above these levels are unlikely to cause any problem with product specifications and low, easily injected, quantities may be used.

9.8 Case histories

Case history 1:
Measurement of residence time in a heat-soak vessel, illustrating true plug flow

The problem being investigated was on a new process plant producing plastic. After the initial polymerization stages the material was formed into chips and then subjected to heat soaking for several hours to produce an increase in cross-linking resulting in an increase in molecular weight.

The heat soaking took place in two parallel vertical vessels V_3 and V_4, heat jacketed and fed from a batch tank V_1 via a screw feeder V_2. The system is shown in Figure 9.13. The heating was augmented by a stream of hot nitrogen

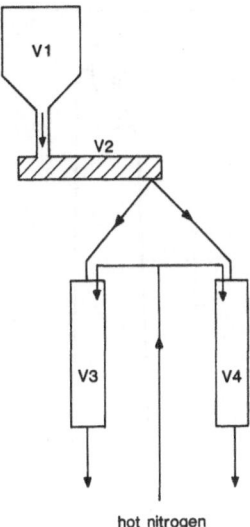

hot nitrogen

Figure 9.13. Polymer heat-soak system.

fed into the top of the reactors. The required increase in molecular weight was not occurring, and one theory was that bridging and caking of the polymer was occurring in the vessels, effectively reducing the volume and, therefore, the mean residence times.

In this case it was decided to use an 'activable' tracer, firstly because of nucleonic instrumentation downstream of the vessels and, secondly, because of the difficulty of labelling hard plastic chips with short-lived tracer. The only practical way of carrying out labelling of a sufficient number of chips would be by spraying tracer on to the surface and it was felt that abrasion would quickly remove the tracer and invalidate the result. A further consideration was that although the vessels were designed for plug flow, if significant bridging and caking was occurring then mixing would occur since they would act as baffles. Therefore it was decided to label one full batch of feed to a level of approximately 45 ppm with manganese acetate added at the liquid stage prior to extrusion and chipping. Such a level was perfectly acceptable to the plant, and ensured that the tracer would be detectable using the neutron activation techniques even if a large amount of dispersion occurred.

Injection took place as an extended step impulse of 6.2 hours from the batch feed vessel and when this was emptied normal feed was restored. Samples were taken at fifteen-minute intervals from the exits of V_2, V_3 and V_4 vessels by plant staff, irradiated (producing manganese-56), and counted. The RTD curves for vessels V_1, V_2 and V_3 are shown in Figure 9.14 (V_3 and V_4 responses were identical).

The results showed clearly that there was virtually no dispersion of the

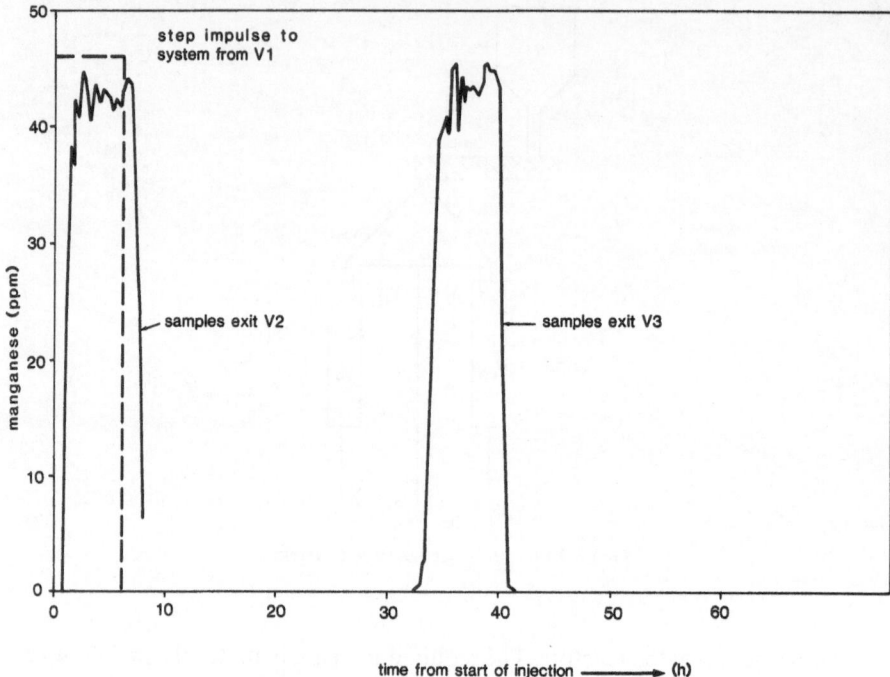

Figure 9.14. Tracer concentration curves—polymer heat-soak system.

tracer material, in fact a slight reduction in the width of the tracer response occurred between V_2 and V_3 which was attributable to a small increase in feed rate during the run. More importantly, the MRT of 33.2 h between the exit of V_1 and exit of V_3 coincided exactly with the calculated theoretical value, conclusively proving that the vessel was clean inside and avoiding unnecessary shutdown. The problem was eventually identified as being associated with the nitrogen stream and was cured by modifying this system.

One important spin-off of the measurement was that the plant management, knowing that the flow regime was perfect plug-flow, were able to calculate exactly the residence-time distribution in the vessel at any flowrate for material of a particular specification. This enabled them to divert the material to the appropriate receiver vessels at the correct time with minimum wastage between the cuts.

Case history 2 :
Measurement of residence time in a plug-flow reactor to determine the extent of solids deposition

Early arrival of a tracer peak indicates that dead zones and channelling are occurring, i.e. the effective volume of the reactor is reduced. An organic

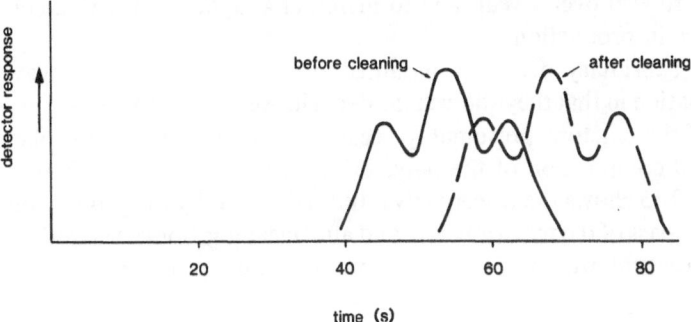

Figure 9.15. Detector response 'looking' at three passes of serpentine reactor, before and after cleaning.

chemical production unit was subject to substantial deposition of polymeric by-product material on the reactor walls, which made it necessary to shut down the system every six months so that polymer removal could be effected. The duration of each shutdown was two weeks, and while on-line, the feed rate had to be progressively reduced to allow sufficient reaction time as deposition built up. The total loss in production over the year amounted to some 20%. The unit consisted of two 15-cm diameter serpentine tubular reactors, bent into a cylindrical form, both immersed in heating baths. In an effort to avoid these shutdowns it was decided to control the amount of deposition by periodically purging the reactor with a solvent and to monitor the effectiveness of this treatment using a radiotracer residence-time measurement. The first measurement was carried out with the reactors in a semi-fouled condition.

A pulse of radiotracer was injected into the organic feed and two principal radiation detectors mounted at the inlet and exit monitored the passage of radiotracer through the reactors. Subsidiary detectors were mounted at the top of every third pass of the serpentine, unshielded so that one detector 'looked at' the three adjacent passes. From the response curves of the two principal detectors the mean residence time of the system was computed. The observed MRT, together with the known liquid feed to the reactor, facilitated the calculation of the free volume of the reactor. The reactor was then purged with solvent and the residence-time measurement repeated. The free volume of the reactor was again recalculated and found to be unchanged, implying that the purge had not been effective. Subsequently, several alternative types of purging operations were tried until the MRT indicated that the free volume of the reactor had increased, implying that the deposits had been substantially removed.

On the basis of these results, the most effective purging technique was selected and thereafter carried out on a regular basis. Adopting this procedure it was found possible to prolong the operating life of the reactor between shutdowns to well over a year, and to maintain a higher feed with a consequent increase in production.

The subsidiary detectors mounted on intermediate passes gave useful information in that they showed the deposits were easier to remove in the front part of the reactors, presumably because some difference in the chemical or physical composition of the polymer made it more susceptible to removal. Figure 9.15 shows the tracer arrival times detected by the probe mounted on the fifth pass of the reactor before and after cleaning. Some 25% increase in the residence time was seen to result from the cleaning process.

Case history 3:
Measurement of residence time in a well-mixed reactor, illustrating early arrival of tracer

As one aspect of research into design of a new process plant, residence-time distribution measurements were carried out on an existing reactor. The reaction was heterogeneous oxidation of a liquid, using air, giving a solid product, with very efficient stirring of the materials.

Based upon information from the original designers of the plant concerning the expected bed expansion due to entrainment of gas, an MRT of 60 minutes was calculated for the liquid/solid components. Using bromine-82 as a liquid tracer, the RTD shown in Figure 9.16 was obtained, with a MRT of 41 minutes and very good mixing occurring. The immediate conclusion was that large amounts of solid were being deposited in the reactor, effectively reducing the volume, although this was contrary to plant experience. Shortly afterwards the reactor was shut down, drained, opened and inspected and the absence of significant deposits confirmed. After start-up the RTD measurement was repeated using sodium-24 as a liquid tracer, with a very similar result.

It was then decided to carry out a level measurement on the reactor, using a sealed radioactive source, and the top level registered by the installed nucleonic gauge was confirmed as correct, but on extending the examination down the reactor a completely unexpected density profile was determined. Instead of a gradual increase in density down the reactor, the main bulk of the material had a density of approximately $0.25 \, \mathrm{g\,cm^{-3}}$ with extensive and sharply-defined areas of density $0.45 \, \mathrm{g\,cm^{-3}}$ around both sets of impellers. The mean density, in fact, was found to be only some two-thirds of the expected density, i.e. the mass of material in the reactor corresponded to the measured 40-minute MRT.

Most importantly, the work showed that 'ideal' systems studied in the laboratory and semi-technical plant stage were not amenable to scale-up to the full-size plant. Extensive modifications were made to the proposed operating conditions of the new plant and sizable savings in construction costs achieved because of reduction in reactor volumes.

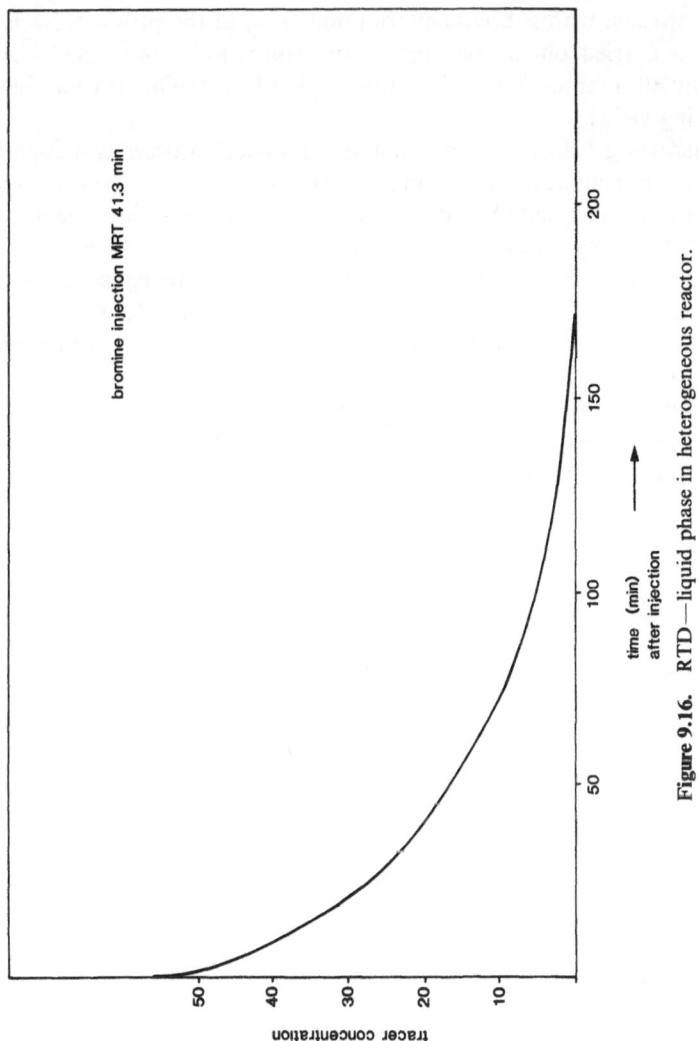

Figure 9.16. RTD—liquid phase in heterogeneous reactor.

F

Case history 4:
Measurement of residence time in a sterilizing vessel, illustrating early arrival of tracer with long tail

The success of many biological processes depends on complete sterilization of the feed materials fed to the reactor to prevent the growth of foreign microbes in the process, with subsequent contamination of the product. Such sterilization is carried out at elevated temperatures for a pre-determined time, typically 30 minutes. To achieve this, a plug-flow regime is desirable in the sterilizing vessel.

A sterilizing vessel for a continuous biological process was found to be inefficient in achieving the required degree of sterilization, and occasionally the whole product had to be discarded. It was decided to investigate the flow characteristics by injecting a radiotracer into the inlet and monitoring the tracer concentration at the exit. Figure 9.17 shows the concentration curve obtained. The mean residence time was calculated to be 25.9 minutes, not too far from the required thirty minutes. However, some 75% of the fluid was moving through the vessel in less than 30 minutes. The long tail, stretching beyond 60 minutes, extended the mean time.

It was obvious that the vessel was too wide for the throughput and that significant dispersion in the longitudinal direction was occurring. The

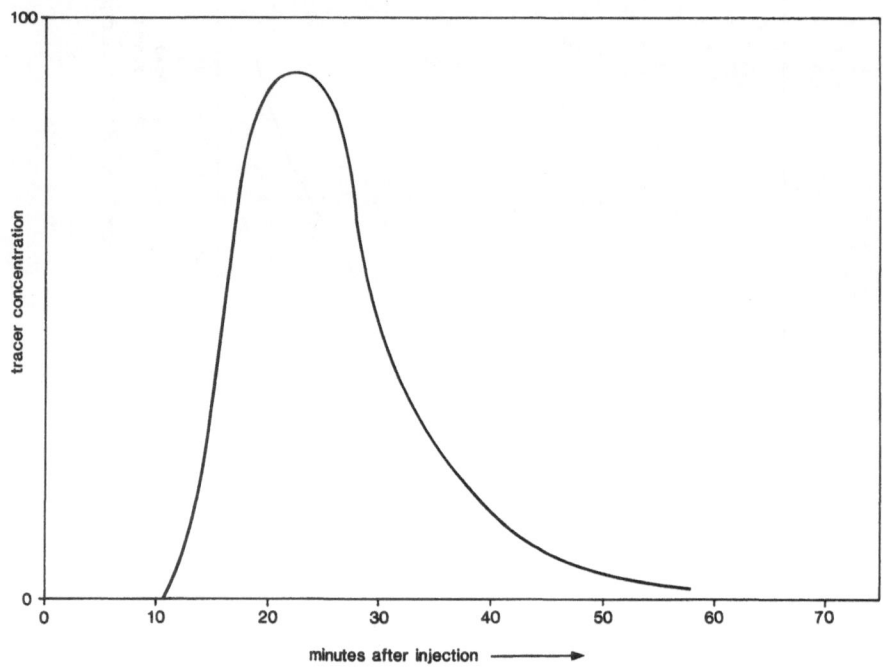

Figure 9.17.   Residence-time distribution—sterilizing vessel.

problem was cured by pre-fabricating a baffle system which was inserted into the vessel at the next shutdown, effectively increasing the length of the reactor and reducing the diameter.

Case history 5:
Measurement of residence time in a river system illustrating late arrival of tracer

As stated in Section 9.6, arrival of tracer later than expected is unusual if the system volume and flowrate is known, and can most often be attributed to instrumental delay or absorption of the tracer. However, accurate determination of volume is not always an easy matter, as the following example illustrates.

A Water Authority had constructed a water processing plant to supply the local townships. Water was abstracted from the river, which for a major portion of its length was bordered by a busy road carrying industrial as well as private traffic. Much of the industrial traffic was composed of tankers, full of petroleum products and a significant number carrying highly toxic chemicals. The Authority needed some idea of how long it could continue to safely abstract water if spillage into the river occurred from one of these tankers and how long contamination of the water would persist.

It was decided to measure the residence time distribution using tracer, injected into the river some thirty-five miles upstream of the abstraction plant. In practice, because of the length involved and the very large dilutions, the river was divided into two convenient stretches, the first of twenty miles covering the upper, faster-moving reaches and the second length of fifteen miles over the lower, slower-moving stretch. Separate injections were carried out into each length as pulses of about three minutes' duration, extending over the width of the river and at one-third depth.

Detection was at convenient intermediate positions roughly five miles apart, between the injection point and the end of each of the two lengths. Waterproofed scintillation detectors were suspended in the river at each detection position and the information from these was augmented by samples taken from the river at hourly intervals and radioassayed in the laboratory so that very low concentrations of tracer in the river could be measured and as complete a tracer concentration curve as possible defined.

As a bonus of the work, as well as the RTD curve being determined it was possible to measure the mean water flowrates at each detection position, using a 'total sample' method, and calculate mean depths of the river over the various sections. These measured flowrates showed good agreement with estimations using other techniques and confirmed that the tracer, bromine-82, was subject to virtually no absorption on the rocks and clay of the river bed.

However, times of travel were very different from those estimated; for example, an MRT of 160 hours between the first injection point and the water abstraction station was measured, compared with an estimated time of 30 to

36 hours. This was chiefly attributed to excessive ponding in two sections of the river, one an extensive area of marshland which was thought to be virtually stagnant but in fact took a considerable proportion of the total flow.

The large spread of the measured RTD was of particular interest. It showed that on some occasions following an accident it might not be necessary to close the abstraction station because much greater dilution of the contaminant would be achieved than was previously thought. Subsequent surveys based on the initial information have enabled a comprehensive modelling of the river hydrology under various river loadings.

Case history 6:
Measurement of residence time in a gas-phase reactor with recirculation

A large gas-phase reactor with external recirculation had operated for many months producing the same product. An urgent requirement arose to produce

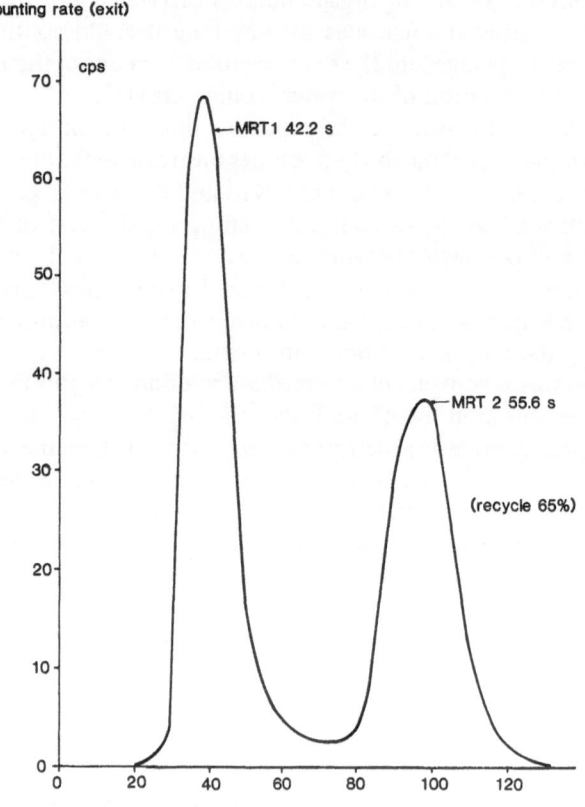

counting rate (exit)

time after injection of argon–41 tracer (s)

Figure 9.18. Primary and recycle response: continuous-process gas reactor.

a small batch of a specialized product which involved changing both reactants and the rate of recycled gas. Unfortunately, after such a long period of disuse both the automatic flow controller and the flow indicator on the recycle line were found to be inoperative and it was decided to adjust the flowrate using the manual stop valves on the exit and recycle lines. Residence-time measurements would be carried out at each stage of valve adjustment to measure the amount of recycle gas.

Operations were commenced with both valves fully open, a gas tracer, argon-41, was injected into the inlet line and the RTD at the reactor exit recorded. The valve on the exit line was then closed by one turn and the measurement repeated. A slight recycle pulse was noted on this occasion. Further turns of the valve were made and the quantity of recycle, calculated by comparison of the peak areas under the primary and recycle peaks, measured at each stage, until the required 65% recycle was achieved. Figure 9.18 shows the RTD at this rate.

The operation saved a plant shutdown at a critical time, saved many thousands of pounds, and most importantly, a valued customer was retained.

Case history 7:
Measurement of residence time in an effluent precipitator illustrating channelling

A large circular precipitator, illustrated diagrammatically in Figure 9.19, was not performing properly. The principle of operation was that water, carrying fine precipitates, was fed in at the top, percolated through to the base of the circular conical section, mixed with electrolyte fed in at the base from a ring main and formed a flocculant precipitate which stabilized at about half-way up the conical section. The clarified water spilled over into a collection trough and the precipitate was bled off to maintain its depth and density. It was found that the

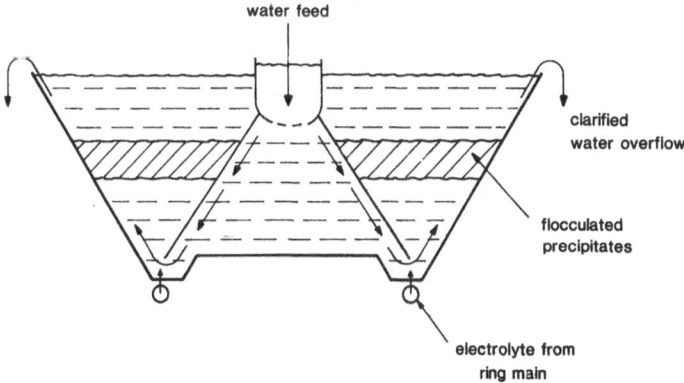

Figure 9.19. Sketch plan of water clarifier.

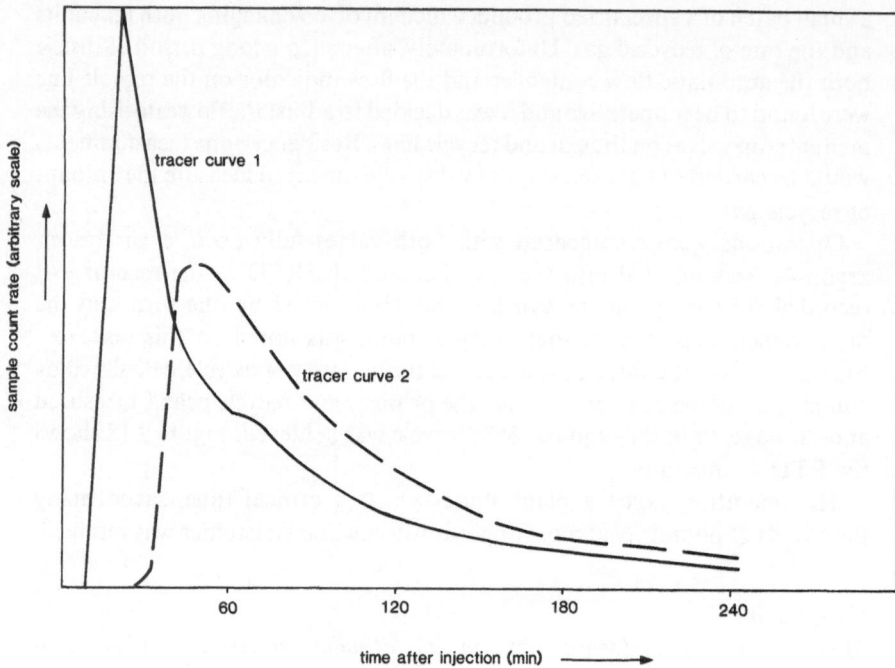

Figure 9.20.   Precipitator tracer concentration curves.

bed of floc was difficult to stabilize, was collecting in one particular area and tended to spill over with the clarified water.

Two possible explanations were suggested for the mal-operation. Firstly, that the electrolyte was not being injected at the same rate through the twelve inlet lines into the base from the ring main, causing high local build-up of flocculant and restricting flow of water. Secondly, that the flowrate of water was not equal to all sections, allowing flocculant to build up in areas of low flowrate. Initially, an attempt was made to measure the flowrate to each electrolyte injection point but it proved impossible to achieve sufficient accuracy in this. The small separation between each take-off, the fact that the flow was not turbulent and the long distance between the only available injection point and the ring main introduced large errors in the measurement.

It was therefore decided to investigate the flow pattern in two stages; involving injection of tracer into the feed water with the precipitator in a poor condition followed by similar injection after the precipitator had been cleaned and prior to formation of floc. A single pulse of tracer in aqueous solution was injected into the water feed line and water samples siphoned off from twelve equispaced positions just below the surface at the water overflow area. The two extremes of RTD found are shown in Figure 9.20.

Tracer Curve 1 was obtained in the area where there appeared to be minimum floc density, curve 2 from the area of maximum floc density. It was, therefore, confirmed that the floc pattern was associated with changes in flow: the higher the floc density the lower the mass of water flowing. It also appeared that the fault could worsen, i.e., as the flowrate dropped more floc would collect, causing a further drop in flow. The second series of measurements after cleaning of the precipitator produced identical RTD at each of the twelve measuring positions. It was, therefore, concluded that the fault lay with the electrolyte injection system. This was modified from a ring main system to twelve individual lines from a series of split headers, leading to a more even feed rate and a stable flocculant bed. A final measurement of residence times gave virtually identical responses at each of the twelve positions, confirming the effectiveness of the modifications.

References

1. Levenspiel, O. (1972) *Chemical Reaction Engineering.* 2nd edn., Wiley, New York.
2. Mecklenburgh, J. C. and Hartlands S. (1975) *The Theory of Backmixing.* Wiley, New York.
3. Ljunggren, K. (1967) Review of the use of radioactive tracers for evaluating parameters pertaining to the flow of material in plant and natural systems. In *Radioisotope Tracers in Industry and Geophysics,* IAEA, Vienna.
4. Levenspiel, O., and Smith, W. K. (1957) Notes on the diffusion type model for the longitudinal mixing of fluids inflow. *Chem. Eng. Sci.* **6**, 227.
5. Van Der Laan, E. Th. (1958) Notes on the diffusion-type model for the longitudinal mixing in flow (O. Levenspiel and W. K. Smith). *Chem Eng. Sci.* **7**, 187.
6. Aris, R. (1959) Notes on the diffusion-type model for the longitudinal mixing in flow (Levenspiel, Smith and Van Der Laan). *Chem. Eng. Sci.* **9**, 226.
7. Bischoff, K. B. (1960) Notes on the diffusion-type model for longitudinal mixing in flow. *Chem Eng. Sci.* **12**, 69.
8. Bischoff, K. B. and Levenspiel, O. (1962) Fluid dispersion—generalisation and comparison of mathematical models I, II. *Chem Eng. Sci.* **17**, 245, 247.

10 Leakage detection

R. ROPER

10.1 Introduction

Radiotracer techniques are frequently used for the detection of leaks in process plants (Chapter 1, Table 1.1). In modern, highly complex, large-capacity chemical plants the necessity to minimize expensive down-time has led to increased use of these methods. This chapter will look at the relative merits of radiotracer techniques as compared with other leak detection methods, such as search gas methods, thermal imaging and ultrasonics. Although the subject is examined with the chemical process industry in mind, it should be emphasized that the techniques are equally applicable to other process industries.

10.2 Leak detection techniques

Methods of leak detection can be conveniently divided into tests carried out on-line during production and those performed off-line during a maintenance shutdown or on new equipment prior to commissioning. Several of the techniques are appropriate to both situations, although the ease of performance and the sensitivity achieved are not necessarily the same in both circumstances. On-line tests can also be further subdivided into the investigation of leakage to atmosphere from a pressurized system, atmospheric ingress into evacuated, sealed systems and internal leakage in closed systems, e.g. leakage from the tube side to shell side of a heat exchanger. This latter type of problem is the most troublesome since in general the only information available about any suspected leak is that accumulated during the normal running of the system, i.e. temperatures, pressures, levels and possibly chemical analysis of the process streams. Whilst these data are very useful, they very rarely provide conclusive evidence of a leak. It is in this field of on-line internal leakage that radiotracer techniques are most important because of their versatility, sensitivity, ease of application and the unequivocal nature of the results.

10.2.1 On-line internal leakage

Generally on-line leakage testing is carried out by injecting a radioactive tracer into the process stream from which the leakage originates, and seeking the presence of that tracer in adjacent streams or in the environment, either by sample analysis or by external monitoring. Where external detection is employed leakages of 0.1% of the total flowrate can generally be measured, whilst if greater sensitivity is required then leakages of as little as 0.001% of the total flow rate can be detected using simple analysis.

Systems which may be tested are numerous and include the following examples:

(a) Leakage across the inlet baffle plate of a baffle plate exchanger
(b) Tube-side to shell-side leakage in a heat exchanger
(c) Reboiler steam or condenser cooling water leakage to plant product
(d) Leakage past isolation or relief valves
(e) Liquid entrainment, i.e. carry-over of liquid into gas streams
(f) Leakage into the subsoil from underground pipes.

10.2.2 On-line testing for leakage to atmosphere

Several techniques, some simple and others more sophisticated, are used for the detection of leakage into the atmosphere of process materials. These techniques can be classified as follows.

Visual inspection. This is an on-going process carried out by all industrial personnel, process, safety, maintenance and management, either looking for or accidentally discovering leaks of process fluids or gas.

Chemical reagent tests. Some process gases give chemical reactions with a simple reagent, e.g. leaking ammonia gas may be detected by its reaction with hydrogen chloride which produces dense white fumes of ammonium chloride.

Thermographic tests. Either simple temperature measurements or sophisticated thermal imaging equipment may be used to detect leaks in lagged sections of plant if the escaping material raises or lowers the temperature of the lagging appreciably from the ambient level.

Ultrasonic detection. All leaks of liquid or gas generate noise in either the sonic or ultrasonic region, the strength and frequency of the signal being a function of the differential pressure and the size and geometry of the hole. The sensitivity of this method will depend on whether the background noise can be reduced to an acceptable level. Special contact probes can be used to examine valves for passing of liquid, whilst directional probes can be used to examine local or remote areas of the plant.

Search gas methods. A search gas, such as a gaseous radioisotope, helium or a chlorinated hydrocarbon, is injected into the system under test and any leakage from the flanged joints or welds can be detected by external monitoring. The detection equipment depends upon the choice of tracer (radiation detection equipment, helium mass spectrometer or electron capture gauge, as appropriate).

10.2.3 *On-line testing for atmospheric ingress*

On operating low-pressure plant where atmospheric ingress is a problem, the technique employed is a variation of the search gas method. The suspect joints are shrouded with tracer and then samples of the gas at the plant ejectors are assayed for tracer content. The presence of the tracer in the samples confirms the presence of a leak, and its size may be estimated from the tracer concentration.

10.2.4 *Off-line leakage testing*

Several techniques are commonly used.

Bubble test. The most simple methods are the bubble test and use of a vacuum box. The traditional soapy water method of detecting leaks under pressure can be made more sensitive and effective using various proprietary preparations.

Ultrasonic detection. As described in section 10.2.2, commercial ultrasonic leak detectors are available, using directional microphones to pinpoint with a high degree of accuracy the source of sound which they pick up and hence the location of the leak.

Search gas methods. The tracer gases described in section 10.2.2 may be used. Checking the tube bundle of a heat exchanger is a good example. The search gas is introduced into the shell side of an isolated heat exchanger and then, with an appropriate detector, the tracer is sought for in the tubes, after a suitable period of accumulation, or at the tube plate for any leakage through faulty welds. Experience shows that the non-radioactive tracer helium is generally to be preferred for this type of off-line leakage testing. There is no need for radiological precautions and it is not subject to the contamination problems that can be encountered when using chlorinated hydrocarbons, where many industrial solvents can cause confusion. Both helium and radiotracer gas techniques are highly sensitive, unequivocal and can be used at elevated temperatures when circumstances demand that off-line leakage tests are carried out under comparable conditions to those at which the plant normally operates.

Table 10.1 Comparative sensitivities of leak detection techniques*

Technique	Leakage rate (At $cm^3 s^{-1}$) On-line	Off-line
Visual	10^{-1} to 10^{-2}	
Bubble test		10^{-3}
Ultrasonic detection	10^{-2}	10^{-3}
Reactive gas	10^{-2} to 10^{-3}	
Hot wire thermal conductivity	10^{-4}*	10^{-4}
Positive emission	10^{-5}*	10^{-5}
Electron capture gauge	10^{-10}*	10^{-10}
Radiotracer	10^{-10}	10^{-10}
Helium mass spectrometer (over pressure mode)	10^{-8}	10^{-8}
Helium mass spectrometer (vacuum mode)	10^{-10}	10^{-10}

*These sensitivities are the ideal sensitivities and it should be emphasized that when used in on-line situations they may be subject to substantial reductions as a result of interference by various process materials.

10.2.5 Sensitivity of leak detection techniques

Comparison of the sensitivities of the various techniques is difficult since each test will depend on local factors such as the geometry of the leak and the degree of dilution in the surrounding air or other medium. Table 10.1 gives a comparison of the sensitivity of which the various techniques are capable, given suitable conditions.

The leakage rates are expressed in units of atmosphere cubic centimetres per second (at $cm^3 s^{-1}$). This is defined as the amount of gas which will cause a pressure rise of one atmosphere in a volume of one cubic centimetre in one second.

The on-line radiotracer leakage rates quoted apply only to leaks to the atmosphere or to atmospheric ingress. As mentioned earlier, the sensitivity of radiotracer tests for internal leakages between process streams is 0.1% of the total flow rate using external detection and 0.001% of the total flow rate using sample analysis techniques.

10.3 Description of radiotracer techniques

There are three basic methods of leak detection using radioisotope tracers:

(1) by measurement of flow rate
(2) by residence-time measurement
(3) by 'direct' tracer methods.

A description of these techniques is given, together with the circumstances in which they are used. Also described are examples where modifications to, and combinations of, the basic techniques are made for specific applications.

10.3.1 Leak detection by flowrate measurement

The simplest and most straightforward method of leak detection uses either the pulse velocity technique or the dilution technique, as described in Chapter 8, to measure the leakage flow rate directly. The most common uses of this method are (a) measurement of leakage flow rate through relief valves to flare systems, and (b) measurement of leakage flow rates past isolation valves.

The experimental method is as shown in Figure 10.1. A sharp pulse of radioactive material is injected into the process stream, upstream of the relief valve, and two detectors, appropriately placed on the line to the flare stack, downstream of the valve, measure the velocity of the leakage material. This linear velocity can be converted to volume flow rate with a knowledge of the mean diameter of the section of pipe between the two detectors. The volume flow rate, ($Q\,\mathrm{m^3\,h^{-1}}$), is given by

$$Q = 0.556 \times d^2 \times L/t \qquad (10.1)$$

where d, the diameter, is in inches, L is in feet and t is in seconds.

This method gives an unequivocal, quantified measurement of the leakage flow rate.

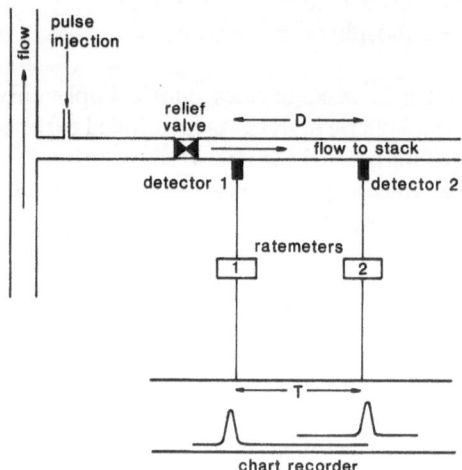

mean velocity = D/T

Figure 10.1. Leak detection unit using the pulse-velocity technique.

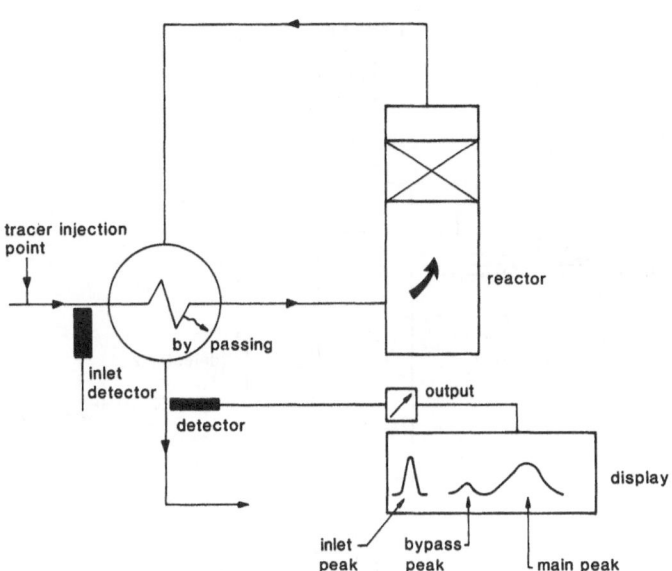

Figure 10.2. Experimental method for leakage detection using residence-time measurement technique.

10.3.2 *Leak detection by residence-time measurement*

A second approach to leak detection is to use the residence-time measurement technique as described in Chapter 9. This method involves the examination of the C curve for a subsidiary peak in the residence-time distribution.

Figure 10.2 shows the experimental method. Again a sharp pulse of a suitable radioactive tracer is injected into the process stream and its passage monitored through the vessel by the externally-mounted radiation detectors. The first probe shows the activity entering the vessel and the second detector records the activity leaving the vessel. This response from the second detector is the C curve from which the residence time would be calculated. Any leakage or bypassing past the central baffle would be indicated by a subsidiary peak preceding the main peak (see Figure 10.2). Since the outlet detector monitors the total activity injected, then the leakage rate, expressed as a percentage of the total flow rate, is the area of the subsidiary peak expressed as a percentage of the sum of the areas of the subsidiary peak and the main peak.

10.3.3 *Leak detection by 'direct' tracer techniques*

A further category of leak detection may be referred to as 'direct' tracer techniques. These techniques are probably the most common and involve the injection of a suitable radiotracer into the process stream which is suspected of

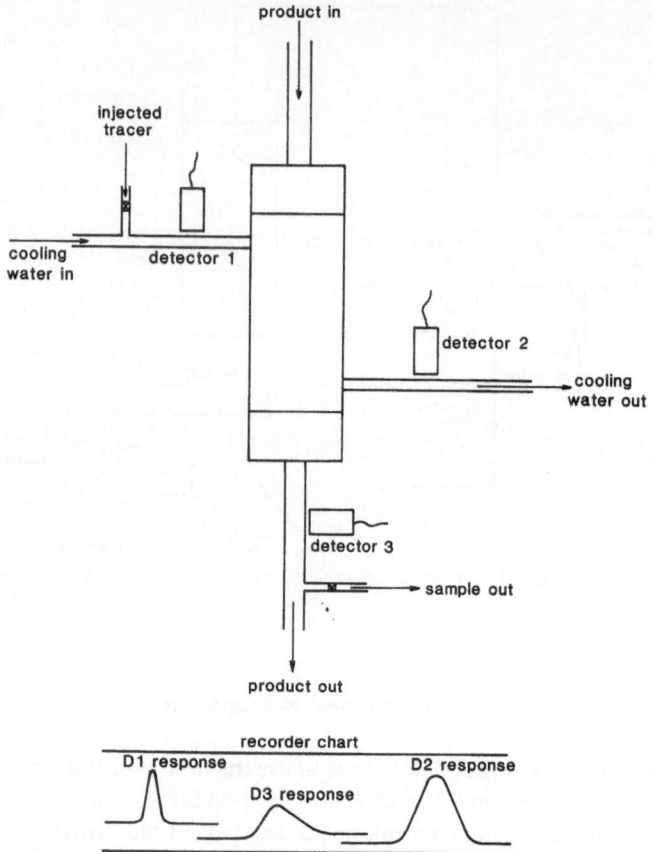

Figure 10.3. Leakage detection using 'direct' tracer techniques.

leaking, and seeking the presence of any leaked tracer in the countercurrent stream. This can be done either by sampling the stream and assaying for radiotracer content, or by taking advantage of the penetrating radiation emitted by the radioisotope tracer and using sensitive radiation detectors mounted externally on the associated pipework. The system shown in Figure 10.3 can be used to demonstrate both principles.

(a) *Leak detection using external radiation detectors.* The sharp pulse of activity is injected into the inlet cooling water and detectors 1, 2 and 3 are positioned as shown to monitor its passage through the exchanger. Detectors 1 and 2 show the inlet and outlet responses, whilst detector 3 will only respond if there is any leakage from the shell side to the tube side of the exchanger. Typical detector responses are also shown in Figure 10.3. Calculation of the amount of leakage is made by comparison of the respective areas under the main inlet peak and the leak peak.

Several factors can affect the calculation of the leak size and corrections should be made for the following, if necessary. (1) Different detector efficiencies—it is not always possible, or even desirable, to have all the detectors with the same efficiency and each detector must be calibrated prior to the experiment so that the areas under the peak can be corrected appropriately. (2) Detector geometry—if the lines carrying the medium under investigation are of different size, wall thickness, etc., then the volume of material producing the response at the detector may be different or reduced by the extra metal of the wall. An appropriate correction must be made. (3) Difference in material flow rates—the detector response is dependent on the time the radioactive material is in front of it and is consequently dependent on the material velocity. Detector response is measured in counts per second, or, if the detector efficiency is 100%, it is measured in disintegrations per second. If in this latter case we consider M becquerels of activity moving past a detector in one second we would expect to register M disintegrations in that second. However, if it moves past in two seconds we should register $2M$ disintegrations, i.e. the count rate is inversely proportional to the velocity. Hence if the velocities of the relevant streams are different then a correction needs to be made.

As stated in section 10.2.1, leakages of approximately 0.1% of the total flow rate can be measured using this technique. However, care must be exercised when using this method, as confusion can be caused by extraneous responses at the leak detector from adjacent pipework or vessels carrying the injected material after it has left the vessel under inspection. In closely-confined congested areas on modern plants, it is generally desirable to surround the highly sensitive leak detector with lead shielding so that it is unresponsive to possible extraneous signals.

(b) *Leak detection by sample analysis.* Here a known quantity of radioactive material (A MBq) is injected into the inlet water feed, and samples of the product are taken over a period of time. Ideally the length of the sampling time should be such as to span the entire 'leak-peak'. These samples are then assayed for tracer content using a highly efficient, large-volume, sodium iodide well-counting system as described in Chapter 8. This method gives much greater sensitivity with leak detection limits of approximately 0.001% of the total flow rate, dependent on the flow rates involved. The formula for the calculation of the leak size is derived as follows.

Assuming A megabecquerels (MBq) of radioactivity is injected into the inlet water feed of W te h^{-1}, then if the leakage rate is L te h^{-1} the quantity of activity injected into the product via the leak would be $A \times L/W$ MBq.

Let the concentration of tracer in the product stream be $C_n(t)$, then

$$A \times L/W = P \int_0^\infty C_n(t) \mathrm{d}t \qquad (10.2)$$

where $C_n(t)$ is the time-dependent concentration of the tracer in the product stream in MBq per tonne and P te h^{-1} is the product rate. The integral in equation (10.2) is the area under the concentration/time curve, and to evaluate this integral we take samples of the product and measure the counts per second in each sample. Hence we need to know the relationship between concentration and countrate.

Assume that the sample size is S^1 te and the activity in the sample is a MBq, then the number of disintegrations per second (dps) in the sample is

$$n = 10^6 \times a \text{ (dps)}.$$

If the efficiency of the detector is $E\%$, then the countrate (cps) will be

$$C = E/100 \times n = 10^4 \times E \times a \text{ (cps)}.$$

Since the concentration of activity in the sample is a/S^1 MBq/te, then

$$\text{concentration} = a/S^1 = C/E \times 10^{-4} \times 1/S^1$$

i.e.
$$\text{concentration} = C/E \times 10^{-1} \times 1/S$$

where S is the sample weight in kilograms.

Substituting for the concentration in equation (10.2) gives

$$A \times L/W = P/E \times 10^{-1} \times 1/S \int_0^\infty C\,dt \qquad (10.3)$$

We now need to evaluate the area under the countrate/time curve. In the normal experiment we take evenly-spaced samples each of duration T minutes, i.e. we approximate to the integral by summation.

Thus $\quad A \times L/W = P/E \times 10^{-1} \times 1/E \times 1/S \times 1/60 \times \sum CT$

Hence
$$L = \frac{\sum CT \times P \times W}{S \times E \times A} \times 1.67 \times 10^{-3} \text{ te h}^{-1} \qquad (10.4)$$

10.3.4 *Liquid entrainment detection using 'direct' tracer techniques*

Liquid droplets entrained in the gas streams from process vessels can cause severe problems on chemical plants. For example, severe damage to a gas compressor can be caused by liquid transferred in the gas stream in an oil fractionation system. Also, in distillation processes the carry-over of non-volatile metallic compounds in the overhead stream from a column can cause catalyst poisoning in the subsequent catalytic cracking units. Consequently, to be able to detect the presence of liquid carry-over is extremely important. It is of even greater importance if the amount of carry-over can be quantified and correlated with process flow rates, such as feed rate or reflux rate to a distillation column, since then effective alterations to the relevant process streams could result in the operation of the system in such a manner that the carry-over is either eliminated or reduced to an acceptable level.

Gross carry-over can sometimes be investigated using the gamma-ray absorption techniques as described in Chapter 13. However, liquid entrainment is generally at a level which is not amenable to this technique and can only be investigated using radioactive tracers. The general principle of the method is basically the same as the direct tracer technique using sample analysis. A suitable radioactive tracer is injected into the process stream which is suspected of being carried over and samples extracted from the relevant gaseous stream are assayed for tracer content. The measured concentration of activity in the samples together with the relevant feed and overheads rates permits the level of entrainment to be calculated.

In some systems it is not possible to extract samples from the gaseous stream and detection using the normal external detector system is also unsuitable. Such a system is encountered in a waste-heat boiler system generating steam from a boiler feed-water supply. Sodium salts, carried over with liquid droplets in the steam, deposit on the associated pipework downstream and cause corrosion. A novel technique has been derived to detect carry-over in these circumstances. Sodium-24, as sodium carbonate, is injected into the boiler feed-water, and the process gas lines at exit from the steam generator are examined for deposited sodium. The theory behind this method of testing is that the leaked water will carry with it some of the dissolved sodium into the gas stream. As the water droplets impinge on the hot walls of the exit pipework, the water will evaporate, leaving the active sodium deposited on the walls, leading to a continuing increase in count rate through the walls.

10.3.5 Underground leakage detection

Radiotracer techniques are extremely useful in the detection of leaks in underground pipes because the leaked radioactive material can be detected without direct contact with the pipe. However, it must be appreciated that there is a limit to the depth of material through which the radiation can be detected (see Chapter 13 on gamma-ray absorption techniques). The maximum depth is generally considered to be between 30 and 36 inches depending on the material.

The method of detection in each case is dependent on the particular circumstances. These methods can be split into two distinct categories: (a) static and (b) dynamic.

(a) *Static method.* The section of underground pipe under investigation is filled with a radioactive tracer, such as sodium-24 as sodium carbonate. The pipe is then left under pressure for a period of time for the tracer to leak into the subsoil. Then the line is flushed out, and radiation detectors, either scintillation or Geiger, are used to detect the tracer in the ground. This may be done either by seeking along the surface, if the pipe is not buried at too low a level, or alternatively by inserting detectors in strategically-placed boreholes. If leakage is indicated then this can be confirmed by taking groundwater samples and assaying them for tracer content. It is not always necessary to use liquid tracers in pipes containing liquids, especially if the line is to be drained before any tests. A gas tracer may be injected and any leakage detected at the surface with the appropriate detectors. The use of gas will usually shorten the experimental time, as the gas will percolate to the surface more quickly than a liquid.

For longer lengths of pipe where boreholes are not appropriate, an alternative approach is needed. One method is to detect the leak from the inside of the pipe. In this instance a plug of radioactive solution which covers the full cross-sectional area of the pipe is passed through the pipe. At a discrete time or distance behind the activity plug a 'pig' (see Chapter 11) carrying sensitive radiation and recording equipment is carried along the pipe by the liquid stream. This both locates leaks and records their distances from the inlet point. The position of the 'pig' may be accurately determined using external gamma-ray sources as distance markers. An alternative system is to winch the 'pig' along the line and record the distances using a cable length meter.

(b) *Dynamic methods.* Basically these are applications of the pulse-velocity technique. A pulse of activity is injected into the line and the velocity along the underground pipe is measured, either with radiation detectors at the ground surface (if the pipe is less than three feet below the surface) or with detectors in discretely placed boreholes for greater depths. The velocity of the liquid changes sharply as it passes the leak. Comparison of the apparent velocity in the leak section with velocities in the sections either side of the leak section indicates the position of the leak within a few metres. Examples are given in the case histories below.

10.4 Detection equipment

The detection equipment used for both types of external leak detection, i.e. the pulse-velocity measurement and 'direct' techniques, generally comprises two-inch sodium iodide scintillation detectors coupled to appropriate transportable electronic equipment (see Chapter 3). For the sample analysis technique the counting equipment is generally a high-efficiency, large-volume sodium iodide well-type scintillation detector. Since, in this latter case, counting is performed under laboratory conditions, the electronics are usually more sophisticated, incorporating pulse-height analysis to identify the radioisotope of interest.

It is difficult to be specific about the injection equipment required as it depends on the physical nature of the stream to be injected into, the pressure of the stream under investigation, its temperature and perhaps its toxicity. In general, however, a hand-operated hydraulic pump can be used for liquid injections into liquid streams up to pressures of approximately 1000 psig, whilst for gas streams the radioactive gas is injected with an inert backing gas, such as nitrogen, from a cylinder of pressure exceeding that in the line. Special injection systems are required for high-pressure liquid and gas systems. Sampling can also be subject to the same hazards, e.g. temperature, pressure and toxicity. Care and thought need to be exercised at all times when sampling chemicals, irrespective of whether they contain radioactive tracers or not.

10.5 Case histories

The following case histories cover a broad spectrum of the application of radiotracer leak detection techniques on chemical plants. Some of the examples are simple, straightforward applications, whilst others are novel variations. The scope for the application of such techniques is enormous and is limited only by the ingenuity and imagination of the user.

10.5.1 *Detection of leakage into product using the pulse-velocity flow measurement technique : ammonia converter investigation.*

Figure 10.4 illustrates how a leakage problem can be investigated using the pulse-velocity flow measurement technique. Temperature measurements and chemical analysis of the gas at exit from an ammonia converter showed abnormalities which may have resulted from (a) poor catalyst conversion, (b) a leaking bellows within the converter, or (c) leakage through a supposedly isolated bypass valve. Quick action was required to identify the cause of the problem so that product of acceptable specification could be re-established.

A sharp pulse of radioactive gas (krypton-85 in this case) was injected into the interchanger bypass line. Four detectors, appropriately placed on the lines as shown in Figure 10.4, were used to check for a leak at the bottom bellows and to measure (a) the gas velocity at exit from the converter and (b) the gas velocity after the junction with the converter bypass valve line. The measured gas velocities were $37\,\mathrm{ft\,s^{-1}}$ and $57\,\mathrm{ft\,s^{-1}}$ for the sections of line between detectors $D2/D3$ and $D3/D4$ respectively. From these measurements and a knowledge of the internal pipe diameters the volume flow rates were calculated. The data showed that approximately 35% of the total gas flow rate to the product cooler was leaking through the converter bypass valve. The tests showed that the bottom bellows was also leaking, and this leak actually facilitated the velocity measurements in that the exit pulses were as sharp as the inlet pulses. If there had been no leak at the bottom bellows the exit pulses would have been less well defined, as the radiotracer would have had to pass through the catalyst bed. Measurement of the gas velocity would not then have been as accurate.

The measurements showed that there were two causes of the malfunction, which necessitated an immediate shutdown. Because of the precise diagnosis of the two faults, maintenance effort was concentrated on the problem areas and the shutdown time was minimized.

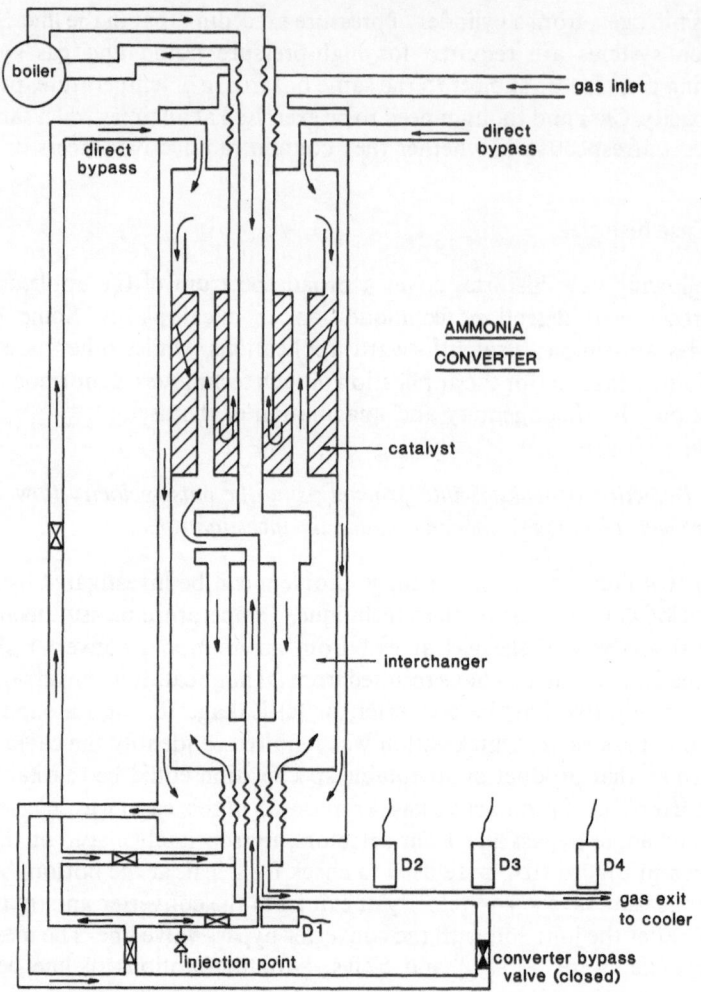

Figure 10.4.   Detection of bypass flow to product using the pulse-velocity technique.

10.5.2 *Detection of leaks in a factory drainage system using 'activable tracer' techniques*

The collapse of a substantial floor area in a large storage building revealed the accumulation of a large amount of water beneath the building. This water appeared to have leaked from the factory drainage system. A schematic diagram of the drainage system is shown in Figure 10.5. A 'background' sample of the leaked water was first analysed using neutron activation analysis techniques so that suitable trace elements could be identified. Five selected tracers, In, V, Co, Mn and I, none of which was found to be present in the water,

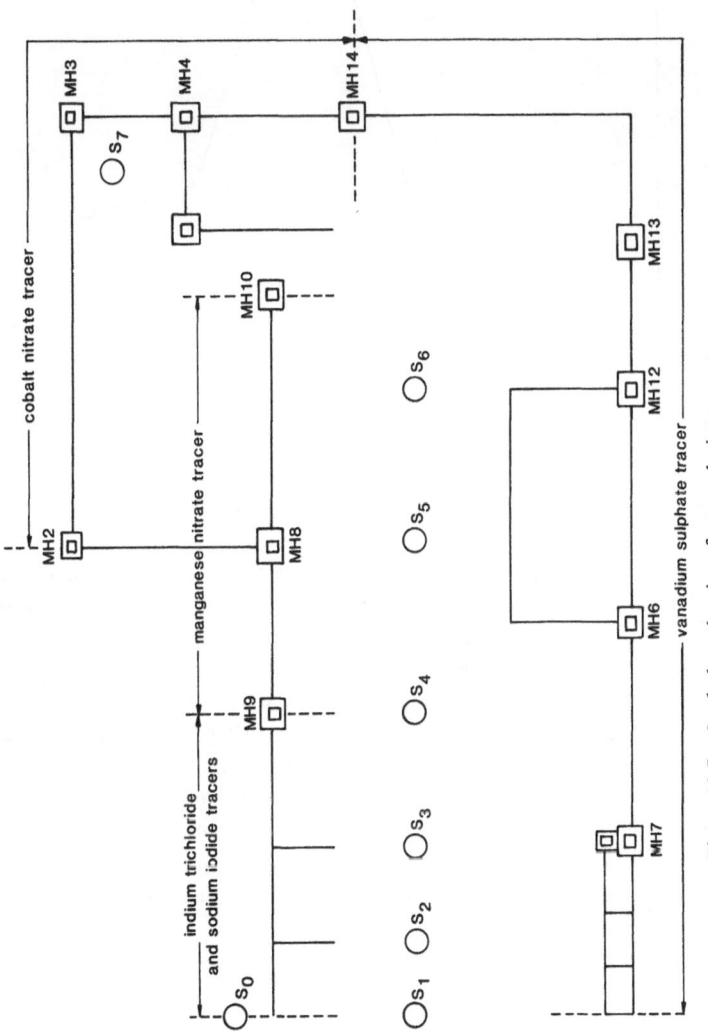

Figure 10.5. Leak detection in a factory drainage system.

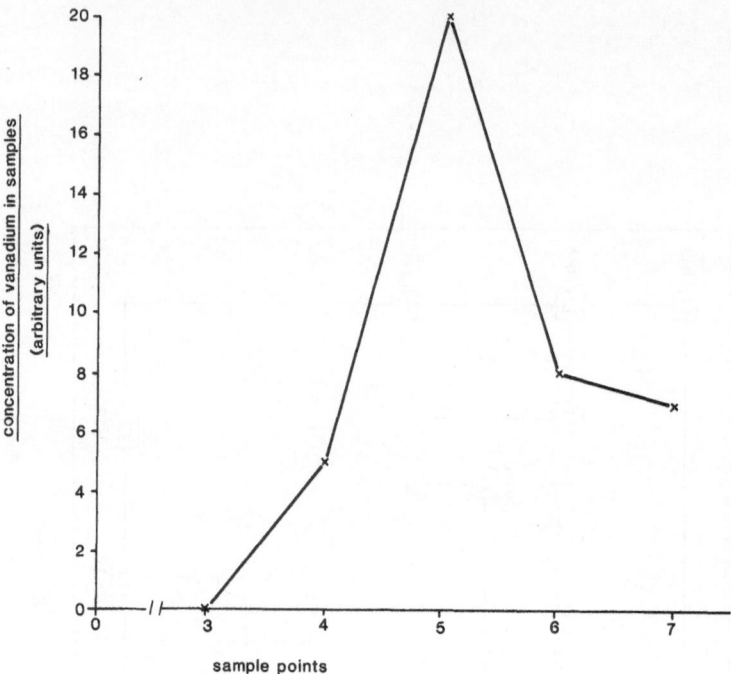

Figure 10.6. Maximum concentration of vanadium at each sample point.

were selected. The drainage system was divided into four sections which could be physically isolated from each other. The drains were emptied and water from beneath the building was pumped out as far as practicable. Each of the four drain sections was then filled with a solution of the appropriate water-soluble tracer. The tracer was introduced slowly whilst filling the drain with water from a fire hose in order to ensure that it was dispersed throughout the whole section of the drain. The drains were then left isolated for a period of 24 h. Samples were then taken from the eight strategically selected points (Figure 10.5) and all the samples were analysed by neutron activation analysis.

No cobalt or manganese was found in any of the samples. Vanadium was found in the samples taken from sample points S4, S5, S6 and S7. Figure 10.6 shows the maximum concentration found at each sample point. The distribution of concentrations suggested a leak in the section of drain in the vicinity of sample point S5. Indium and iodine were also found in samples from sample points S0, S1, S2 and S3 which indicated the presence of a second leak. Figure 10.7 shows the maximum concentration of these two tracers at each sample point. In a subsequent test of the suspect areas the area of leakage was better defined so that repair work could be carried out with a minimum of excavation and expense.

'Activable tracers' were used in this investigation so that several tests could be carried out simultaneously over a long period of time.

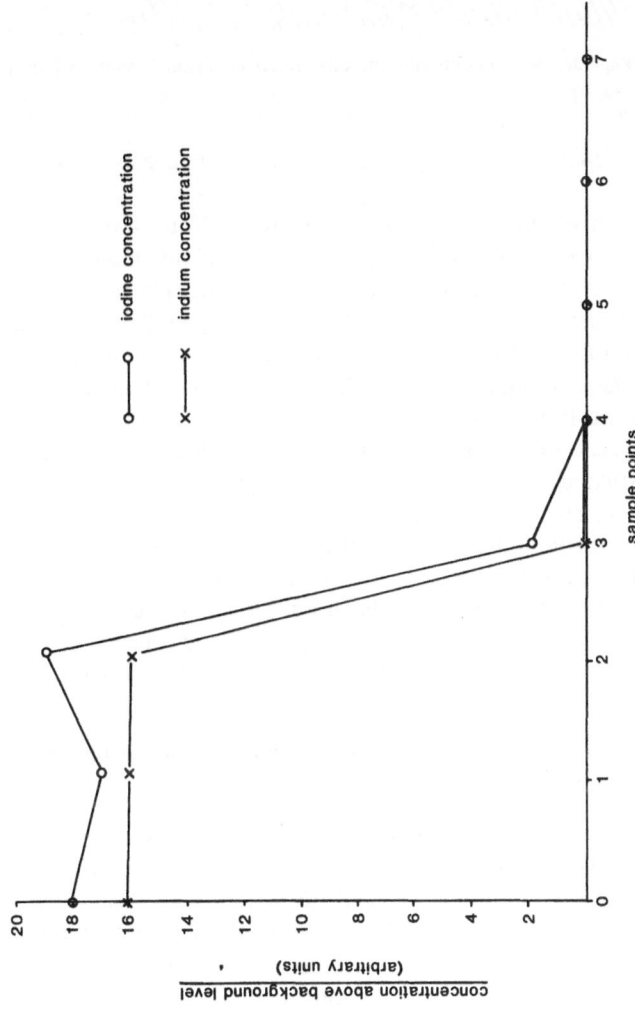

Figure 10.7. Maximum concentrations of iodine and indium at each sample point.

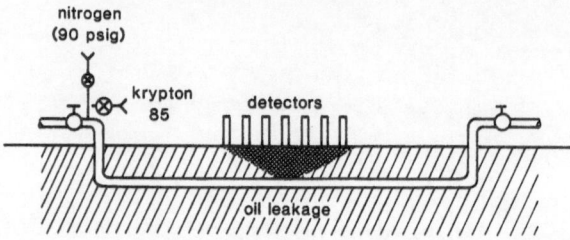

Figure 10.8.   Leakage detection in underground oil-transfer line using a gaseous radiotracer.

10.5.3 *Leakage detection in underground pipes using a gaseous radiotracer*

Oil contamination of the ground, extending over 300 to 400 yards at the surface, was discovered above an oil transfer pipeline, indicative of an underground leak. The rate at which the area of contamination had grown suggested that the leak responsible would be relatively small. Pinpointing the exact location of the leak by visual inspection was extremely difficult because of the extent of the oil contamination, so leak testing was carried out during a planned shutdown. Figure 10.8 illustrates the problem.

The one-mile section of pipe, which was three feet below the surface, was drained of oil, isolated and then pressurized with nitrogen to 90 psig. A pulse of krypton-85 was injected at one end. This travelled towards the leak in the pipe and then diffused through the ground to the surface after approximately three hours. It was detected using portable scintillation probes. The location of the leak was established and obviated the need for extensive and expensive excavations. The use of a gaseous tracer instead of a liquid tracer shortened the duration of the experimental work and hence reduced the down-time.

10.5.4 *Internal leakage detection on a bank of heat exchangers using the 'direct' tracer technique with external detectors*

Laboratory tests indicated a leak of reactor feed into the product stream in a feed/effluent exchanger system which consisted of five exchangers in series. The requirement was to determine which, if any, of the exchangers was leaking so that the plant downtime for exchanger repair could be kept to a minimum. A leak of 1–2% of the total flow of 780 IGPM was suspected but, in fact, any leak greater than 0.5% would have been detrimental to the product stream. Figure 10.9 shows a schematic diagram of the system. The bank of five exchangers pre-heats a feed stream to the reactor (on the shell side) with the return hot effluent (on the tube side).

A radioactive tracer, 185 MBq of bromine-82 as *p*-dibromobenzene, was injected into the liquid feed to the exchangers and its arrival and passage through the exchangers monitored by externally-mounted radiation detectors

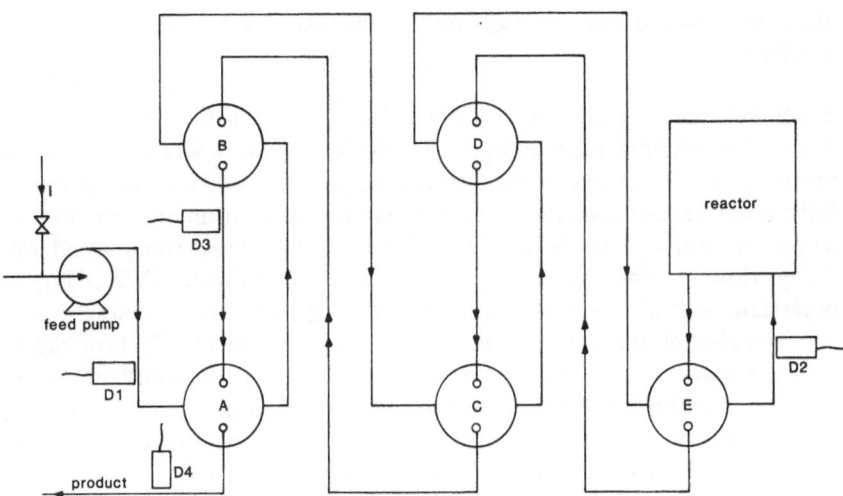

Figure 10.9. Schematic diagram of heat exchanger systems showing injection and detector positions.

on the feed inlet and exit lines. An additional detector (D4) was mounted on the final product line to the air cooler to detect any leakage of material between shell and tube sides of the exchangers. After the initial test had confirmed the presence of a large leak, further tests were carried out with detector D4 appropriately positioned between the exchangers to optimize sensitivity of detection for each exchanger. Figure 10.10 shows the initial responses which indicated a leak of approximately 2.5% of feed into the product. Subsequent tests showed that exchanger A was leaking and there was no leakage from exchangers B, C, D and E. The minimum detectable leakages for these exchangers was 0.4% of the feed rate.

As a bonus, additional information about the residence times through the exchangers was also obtained from these tests.

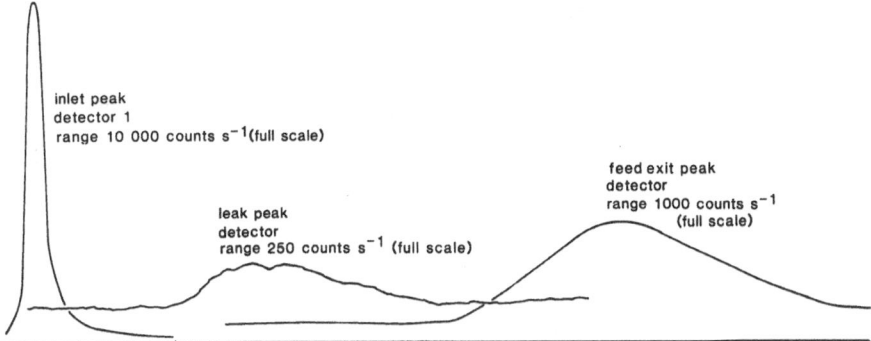

Figure 10.10. Feed inlet, feed exit and leak detector responses.

10.5.5 *Detection of water leakage into product using 'direct' tracer sampling technique*

Routine chemical analysis of the product from a batch distillation unit showed that it was contaminated with water. Figure 10.11 shows a schematic diagram of the system. As can be seen the water in the product could have arisen due to leakage of (a) steam from the reboiler, (b) cooling water from the condenser or (c) cooling water from the product cooler. To determine which vessel was faulty, three experiments were carried out (1) bromine-82 (370 MBq) as potassium bromide was injected into the cooling water to the product cooler and samples of the product assayed for tracer content; (2) bromine-82 (370 MBq) was injected into the condenser cooling-water feed and samples of the product exit pump A were removed and assayed for tracer content; and (3) 18.5 GBq of tritiated water was injected into the steam feed to the reboiler and samples from both the column bottoms and overheads were taken and assayed for tracer content. The results of the tests were as follows.

Test 1. The samples of product showed no trace of radioactivity, hence there was no leakage in the product cooler. Using the formula as derived in section 10.3.2 and the flow rates as shown in the diagram, we evaluated the minimum detectable leak limit:

$$\text{minimum detectable leak, } L = \frac{1 \times 50 \times 1 \times 2}{2 \times 20 \times 370} \times 1.67 \times 10^{-3} \text{te h}^{-1}$$

$$= 11 \text{ g h}^{-1}$$

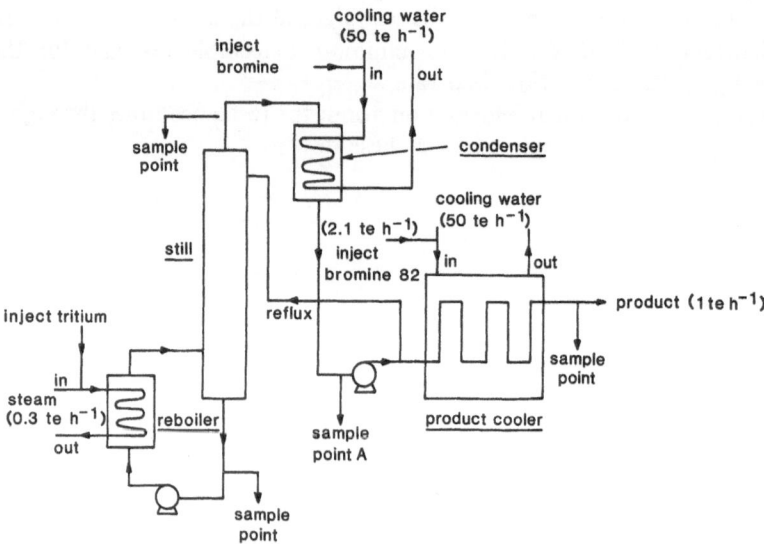

Figure 10.11. Detection of water into product leak using 'direct' tracer sampling technique.

where the sample weight, $S = 2\,\text{kg}$, the efficiency, $E = 20\%$ and the minimum radiation detectable in the samples expressed as counts per second was 1.

Test 2. Again there was no tracer in the samples and the condenser was therefore not leaking either. The minimum detectable leakage was again calculated in the same manner:

$$\text{minimum detectable leak, } L, = \frac{1 \times 50 \times 2.1 \times 2}{2 \times 20 \times 370} \times 1.67 \times 10^{-3}\,\text{te}\,\text{h}^{-1}$$

$$= 24\,\text{g}\,\text{h}^{-1}$$

Test 3. The column bottoms samples contained tritium, confirming a leak of steam from the reboiler. No tracer was found in the overheads samples which were taken over a period of thirty minutes after injection. This suggested that the water was held up in the still for longer than the sampling time. Since the still was operating without reflux at the time, this was a distinct possibility. (A subsequent test on another occasion showed tracer in the overheads when the still was operating normally, with reflux.) Since the activity was leaking into a 'hold-up' volume in the bottom of the still, the normal leak formula was not applicable. The leak size was evaluated as follows. The leaked activity into the still bottom was $L/Q \times A$ where L was the leak flow rate, Q the steam flow rate and A the injected activity. Assuming good mixing, which was most likely since there was a high circulation rate, the activity in the sample was the leaked activity $\times r/R$, where r was the sample size and R the 'hold-up' volume (both in m^3).

Since we used a beta-emitter and analysed the samples using liquid-scintillation counting techniques, the radioactivity in the samples was measured directly in disintegrations per second.

$$\text{Disintegrations per second, } D = \text{activity in samples} \times 10^6$$

$$= L/Q \times A \times r/R \times 10^6$$

$$\text{Hence the leakage rate, } L, = \frac{D \times Q \times R}{A \times r \times 10^6}\,\text{te}\,\text{h}^{-1}$$

Since the mean activity in each sample was 342 dpm, the sample size $0.5\,\text{cm}^3$, the 'hold-up' volume $7.5\,\text{m}^3$ and the steam flow rate $0.3\,\text{te}\,\text{h}^{-1}$, then

$$\text{leakage rate} = \frac{5.7 \times 0.3 \times 7.5}{18.5 \times 10^3 \times 0.5 \times 10^{-6} \times 10^6}\,\text{te}\,\text{h}^{-1}$$

$$= 1.386 \times 10^{-3}\,\text{te}\,\text{h}^{-1}$$
$$= 1.4\,\text{kg}\,\text{h}^{-1}.$$

Since the overheads rate was $2.66\,\text{te}\,\text{h}^{-1}$, then expressed as a percentage the leak was $1.4 \times 10^{-3}/2.66 \times 100 = 0.05\%$ of the overheads rate.

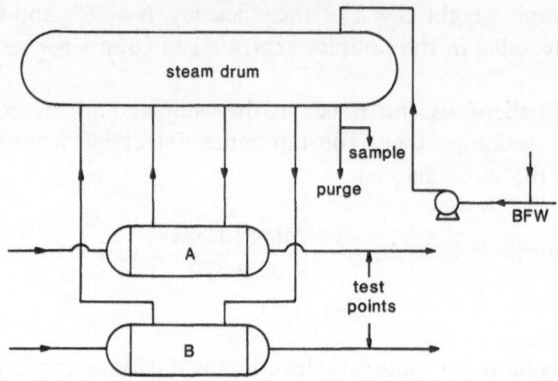

Figure 10.12. Internal leakage detection using radiotracer deposit technique.

10.5.6 *Internal leakage detection using tracer deposit technique*

Figure 10.12 shows a schematic diagram of the waste-heat boiler system where leakage was suspected. Two points required clarification. Firstly: was there a leak, and if so, in which vessel was it located? Secondly: if a leak was found to be present, what was its magnitude? To solve the two problems, two different experimental approaches were adopted.

Firstly, 3.7 GBq of sodium-24, as sodium carbonate, were injected into the boiler feed-water to the steam drum and the process-gas exit lines from boilers A and B were examined, using a hand-held collimated scintillation counter to detect build-up of tracer in the line due to internal leakage within the boilers. The theory behind the test was that any leakage occurring would carry with it some of the dissolved sodium into the gas stream. As the water droplets impinged on the hot walls of the exit pipes, the water would evaporate, leaving the sodium on the walls. This would give rise to a continuing increase in the measured radiation through the walls. Table 10.2 shows the measured

Table 10.2 Radiation in cps at process gas exit lines.

Time (GMT)	Boiler A (cps)	Boiler B (cps)
11.10	140	152
11.20	2057	440
11.24	3421	
11.26	4367	
11.29	5612	420
11.44	11 226	
13.28	21 000	358
15.38	18 670	340

Note that the initial count at the injection time, 11.10 hours, was a background radiation count and the small number of counts above background at B boiler exit was due to radiation from the material in the boiler.

radiation in counts per second on the process gas exit lines of each waste-heat boiler. As can be seen, boiler A indicated leakage and boiler B, no leakage.

The next task was to roughly quantify the leakage rate. To accomplish this, the fall in the tracer concentration of the circulating water was measured. The concentration of tracer in the feed water is affected by the addition of make-up water which compensates for purge losses, gland seal losses and the unknown boiler leakage.

Assuming the boiler feed-water volume to be constant and that the tracer concentration is C, then the change in concentration, dC, varies as $C\,dt$, i.e. $dC/C = -Y\,dt$, where Y is a constant equal to the loss per hour, u, divided by the volume, v.

Hence

$$dC/C = -u/v\,dt$$

therefore

$$\int_{c_0}^{c_t} \frac{dC}{C} = \frac{u}{v} \int_0^t dt$$

i.e.

$$\ln C_0 - \ln C_t = (u/v)t.$$

It must be remembered that the radiotracer has its own natural decay which will also affect the concentration. Full account of this must therefore be taken in the calculation of the leak rate. Using the data of Table 10.2, two samples of the circulating water taken at 11.40 and 13.35 h had count rates of 9478 and 1470 respectively when corrected for decay time. Hence $\ln 9478 - \ln 1470 = u \times 1.92/70$ where the volume was $70\,m^3$. The total water loss rate was therefore $68\,m^3$. Since purge losses at the time were known to be about $5\,m^3\,h^{-1}$ the leakage rate from the boiler was approximately $60\,m^3\,h^{-1}$.

10.5.7 Detection of bypassing/channelling on a sulphuric acid plant converter

The conversion efficiency of a sulphuric acid plant converter was below specification. During a plant shutdown, inspection of the converter revealed that the separation plates between the beds were damaged. Several of the triangular sections between beds 2A/2B and 3A/3B, as designated in Figure 10.13, were found to be displaced. These were replaced and clamped in position. After the subsequent start-up, although some improvement in the SO_2 to SO_3 conversion had been effected, gas analysis of the process streams indicated that conversion losses were still unacceptably high. Either of two malfunctions might have been responsible: (a) further displacement of the separation plates or (b) channelling through the beds. A series of experiments was designed to investigate the possibilities. Figure 10.13 shows the converter, with injection and detection points indicated for the tests carried out.

For the initial test krypton-85 (74 GBq) was injected into the gas inlet line to bed 3B at position I1 and detectors D1, D2 and D3 were positioned at 3B bed inlet, 3A bed exit and 3B bed exit respectively. Figure 10.14 shows the detector

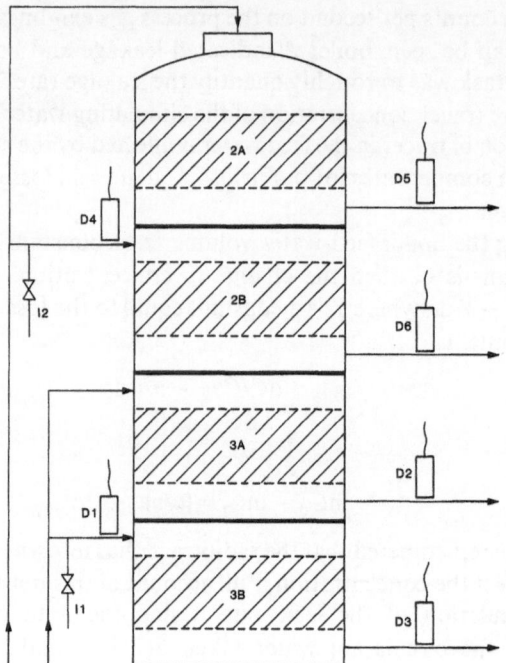

Figure 10.13. Sulphuric acid converter showing injection and detector positions.

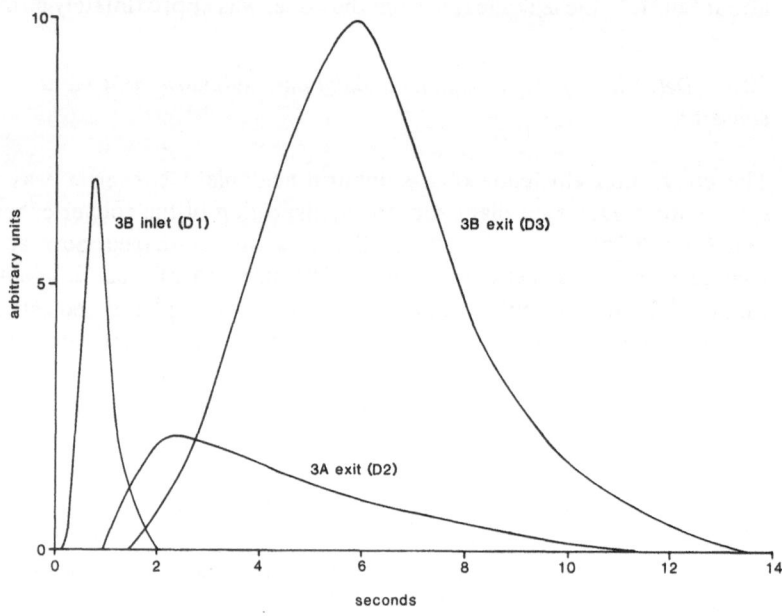

Figure 10.14. Detector responses for 3A/3B bed bypassing.

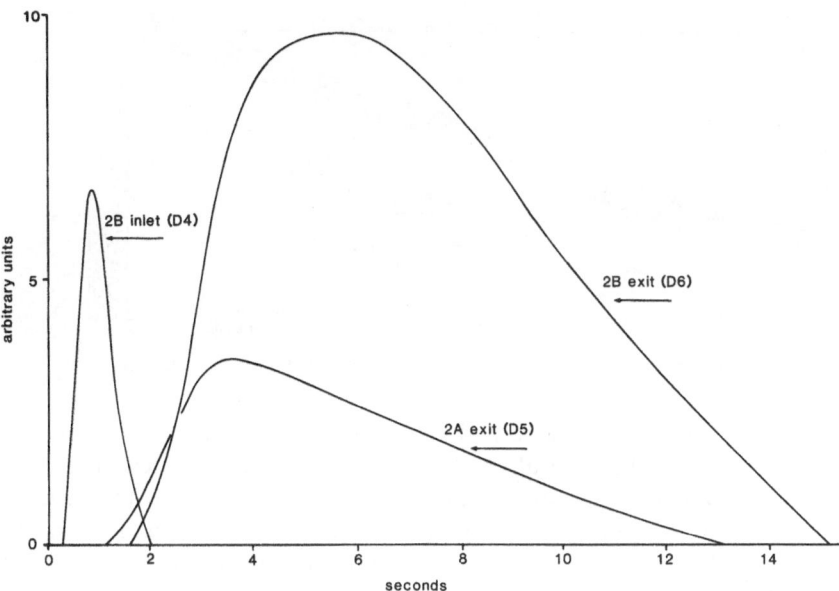

Figure 10.15.   Detector responses for 2A/2B bed bypassing.

responses, uncorrected for detector efficiency and for flowrate differences as described in section 10.3.3. Examination of the detector responses showed a large peak from the 3A bed exit line, the leading edge of which preceded the 3B bed exit peak by approximately 0.5 seconds. This peak, due to leaked tracer, can only arise from tracer bypassing from 3B inlet to 3A exit due to lifting or displacement of the triangular separation plates. After correcting for the relevant gas velocities and detector efficiencies, the leakage rate was calculated to be 20% of the feed to 3B bed.

A similar test carried out between beds 2A and 2B by injecting into 2B inlet at I2, with detectors D4, D5 and D6 at 2B inlet, 2A exit and 2B exit respectively, again showed leakage across the separator plates (see Figure 10.15 for detector responses). Calculations showed that 40% of the feed to 2B bed was leaking into 2A exit line. The tests showed that either the repairs to the separator plates had been inadequate or further plates had been damaged or dislodged during the start-up. Consequently a further shutdown was necessary to remedy the problem.

10.5.8 *Detection of leakage from a brine reservoir into the drainage system*

A reservoir was used to store brine from subterranean salt-caverns used for ethylene storage. A leak into the reservoir drainage sump was observed during the transfer of brine from the caverns to the reservoir. The reservoir was

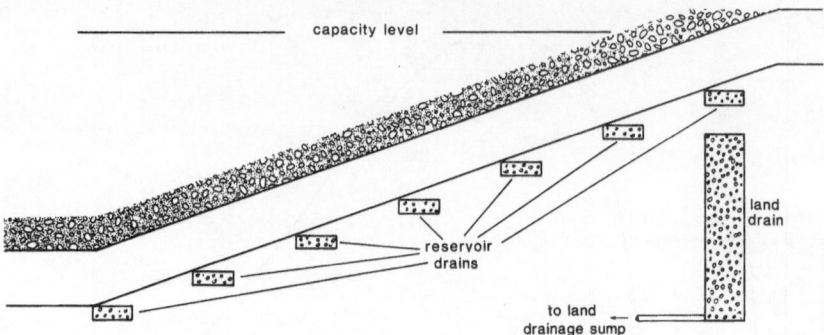

Figure 10.16. Construction and drainage system of brine reservoir.

approximately 100 m square and was constructed of non-porous clay of minimum depth 4 ft. The walls had a slope of 1 in 3, and to prevent the clay drying out, with the subsequent danger of cracking, they were covered with an 18-inch layer of porous slag. The reservoir's depth was 16 ft and its capacity 80 000 m^3. Figure 10.16 shows the construction and drainage systems. There were two drainage systems:

(a) The reservoir drains which consisted of seven 3 ft by 1 ft gravel beds which completely surrounded the reservoir beneath the sloping walls and which were interconnected every 50 ft. At the lowest point the lowest drain was connected to the reservoir drainage sump. Any brine which leaked through the wall into the sump was pumped from here back into the reservoir.
(b) The land drainage system which was to protect the reservoir wall against groundwater flowing down the hill. The water from this drain flowed into the local river by the sewerage system.

During commissioning of the reservoir the flow into the reservoir sump was minimal until the depth of brine reached approximately 10 ft, when the flow suddenly increased to about 20 gallon per minute. The material flowing into the sump was brine of density 1.1 g cm^{-3}. This was consistent with a leak in the reservoir wall at about the 10 ft level, since it was known that the brine layered with a surface density of 1.1 g cm^{-3} and a bottom density of 1.2 g cm^{-3}. The level was lowered to 9 ft and the flow to the sump decreased over several days. It was at this stage that it was decided to locate the leak using a radioactive tracer technique.

To locate the leak, radioactive bromine-82 as ammonium bromide was injected at the brine surface over two sides of the reservoir, and the leaked activity was detected at the sump by (a) taking samples and assaying them for tracer content, and (b) with a scintillation probe in the outlet drain, connected to a ratemeter and a strip chart recorder. Whilst the latter technique indicated the appearance of the activity it was not quantitative due to the build-up of

Table 10.3 Injections and results of reservoir leakage tests.

Test	Injection site	Tracer used	Activity in samples at sump		Peak activity (cps per litre)
			Time after injection		
			1st appearance	Peak	
1	ABCDE	2.2 GBq in water	23 min	88 min	2100
2	CD	370 MBq in water	—	—	—
3	CB	370 MBq in water	—	—	—
4	AB	296 MBq in brine*	60 min	150 min	15
5	BC	296 MBq in brine*	—	—	—
6	CD	296 MBq in brine*	—	—	—
7	DE	296 MBq in brine*	—	—	—
8	EF	296 MBq in brine*	—	—	—
9	EFGHA	925 MBq in brine*	35 min	115 min	300
10	AH	222 MBq in brine*	—	—	—
11	GH	222 MBq in brine*	—	—	—
12	EFG	444 MBq in brine*	—	—	—
13	HAB	555 MBq in brine+	25 min	112 min	157
14	IAJ	740 MBq in brine+	19 min	32 min	160
				75 min	420
15	AK	740 MBq in brine+	16 min	28 min	380
				71	760

Note:-* Brine density was 1.1 g cm^{-3}
$^{+}$ Brine density was 1.2 g cm^{-3}

activity in the sump and the variation of the sump level. Several injections were made over a period of eight days. A total of 8.325 GBq of activity was put into the reservoir, but because of the natural decay there was at no time more than 4.625 GBq actually present. Thus, in the 40 000 m^3 of brine there was approximately 1% of the permitted drinking water concentration (ICRP recommendations) of bromine-82.

Table 10.3 lists the injections carried out (positions as indicated in Figure 10.17) and the results obtained. The initial injections along the north and west sides of the reservoir showed a positive leak with activity appearing as shown in Figure 10.18. The second injections at CD and CB, as shown, gave no indication of leakage. Further injections at AB, BC, CD, DE and EF gave activity at the sump but only a very small amount compared with that which would have been expected to come directly from the leak. It was concluded that the leaking section had probably not been directly tested at this stage and that movement of the activity from the injection site by wind dispersion was effecting the results. Subsequently, solutions were made up in brine of density 1.1 g cm^{-3} to ensure that the movement of the injected material was downwards. The other two sides of the reservoir were then tested and, as can be seen, only the first injection gave a positive result. This, once again, was much smaller than expected. The evidence was now pointing to the leak being adjacent to corner A and so the next injection was into HAB. This injected material was made up in dense brine from the bottom of the reservoir (density 1.2 g cm^{-3}). Activity appeared after approximately 25 minutes, as in

G

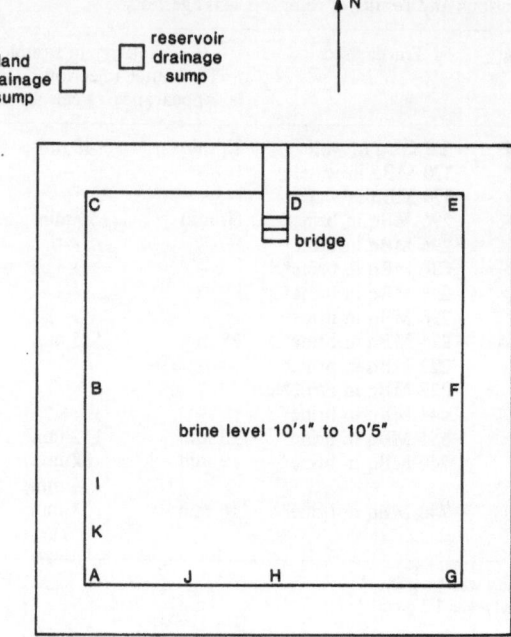

Figure 10.17. Brine reservoir showing injection areas.

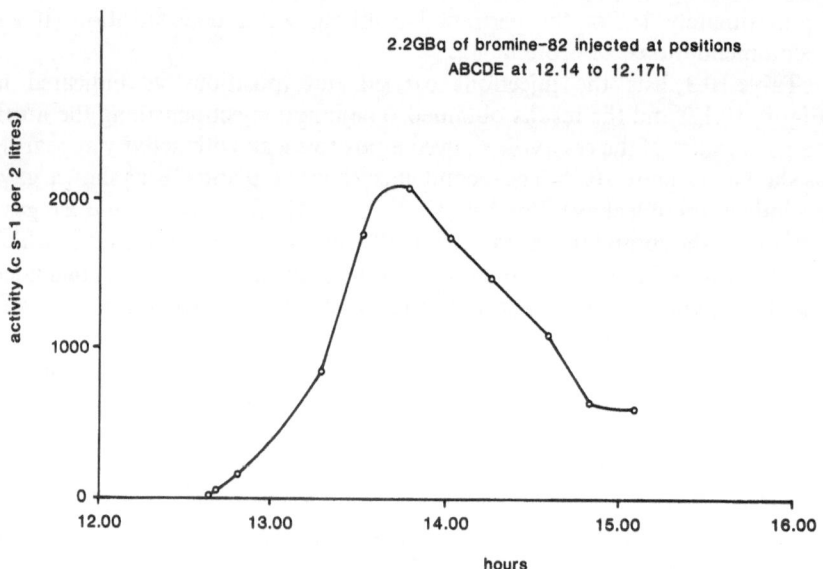

Figure 10.18. Activity in reservoir drainage sump from initial injection.

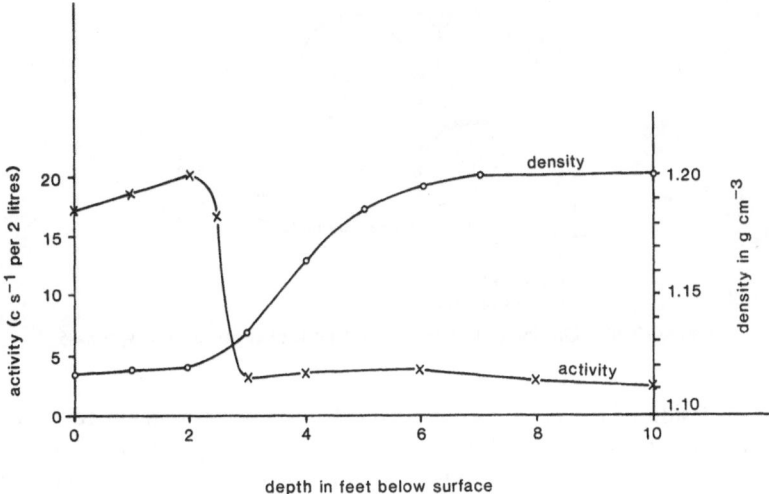

Figure 10.19.   Activity and density as a function of depth below the surface.

the first test. Two final injections were made at *IAJ*, where *IA* and *AJ* were quarter sides of the reservoir, and at *KA* which was 15 m long. Again these solutions were in brine of density 1.2 g cm^{-3}. Two activity peaks were observed at the reservoir drain from each injection. These were similar in relative shape and size, indicating the possible presence of two leaks in section *KA*, or more likely, two leakage flow paths from a single leak to the drain sump.

Samples of brine were now taken from the reservoir at various depths, analysed for density and activity and the results compared with measurements on samples from the sump. Figure 10.19 shows the densities and activities observed as a function of depth below the surface. Samples from the sump had activity and density of 11.0 cps in 2 litres and 1.129 g cm^{-3} respectively. As can be seen from Figure 10.19, these correspond to depths of 2 ft 7 ins and 2 ft 9 ins, which suggested that the leak was not at 10 ft from the bottom of the reservoir but rather at 7 ft 6 ins.

It was concluded, and subsequently verified, that the leak was close to the south-west corner (*A*) along the west side (section *KA*) at a height of 7 ft 6 ins $\pm$ 1 ft above the bottom of the reservoir.

10.5.9 *Off-line investigation of heat exchanger for leakage*

A leak from the tube side to the shell side of a heat exchanger, giving product contamination, had been diagnosed by conventional radiotracer techniques. During a planned shutdown a hydraulic test had not pinpointed the leak, so a tracer test was carried out. Figure 10.20 outlines the experimental procedure.

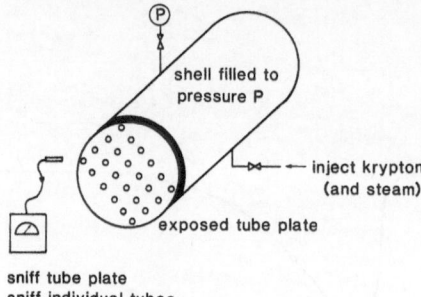

Figure 10.20.   Off-line leak detection of heat exchanger using krypton-85.

The shell side of the exchanger was filled initially with steam to raise the temperature to the vicinity of the normal operating temperature. At this point rubber bungs were inserted into both ends of the tubes. Krypton-85 was then bled into the shell side until it was uniformly distributed throughout and the pressure was held at a test pressure of 100 psig. Using a 'sniffer' probe, containing a Geiger tube, the tube plate was examined for leakage of krypton. After allowing a suitable period of time for accumulation of any leakage into the tubes, (approximately 15 h in this case) each tube was individually tested.

A leak was located at a tube plate weld round one of the tubes which was repaired and retested before the exchanger was put back into commission.

11 Miscellaneous radiotracer applications

G. REED

11.1 Mixing and blending studies

Mixing and blending studies may be carried out both on batch systems and on continuous-flow systems.

11.1.1 *Blending*

The term 'blending' is usually applied to systems in which no reaction takes place, for example, the addition of a small amount of sub-specification material to a quantity of material which exceeds specification requirements giving product still within specification.

With batch blending systems the main considerations are usually to carry out the operation as quickly as possible or with as little expenditure of energy as possible. They are often carried out in tanks with an internal stirrer or by circulating the mixture through an external pump and returning it to the tank. With either system the investigation of the blending operation using radiotracers is the same. The radiotracer, preferably short-lived, is added to the tank containing the materials to be blended in a miscible form. Manganese-56, as acetate, bromine-82 as potassium bromide or sodium-24 as sodium carbonate are frequently used for aqueous systems; bromine-82 as paradibromobenzene or manganese-56 as naphthanate for organic systems. If the quantities of material are large or they are intended for distribution to the public within a short time, then a low concentration of tracer is both desirable and necessary. Samples are taken at discrete time intervals and the activity of the samples estimated to give a tracer concentration curve. To avoid the time-consuming counting process, and particularly if the blending time is expected to be short, it is more convenient to determine the tracer concentration curve using externally-mounted detectors. Figure 11.1 shows suitable arrangements of such detectors, and a typical tracer concentration v. time curve.

If an external detection system of $2'' \times 2''$ scintillation detectors is being used, a tracer concentration of 5–20 MBq per m^3 will ensure an adequate count rate and a doserate level on the outside of the tank of less than $7.5\ \mu$Gy per hour.

Consequently there need be no restriction on process personnel working near the tank. If the sampling technique is used then the quantities of tracer used may be reduced by a factor of more than a hundred, depending on the sensitivity of the counting system.

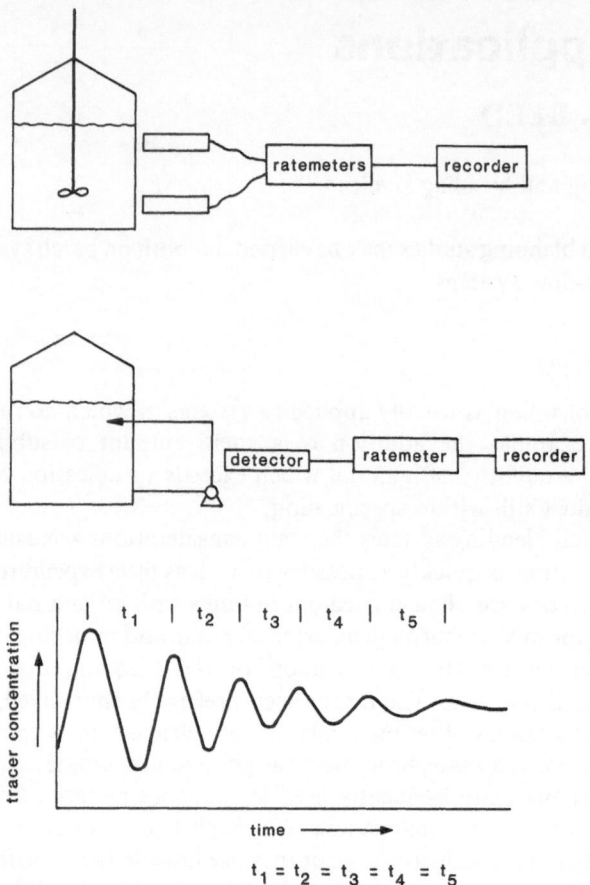

Figure 11.1. Equipment arrangement and typical tracer concentration curve for blending study.

The treatment of the results of the blending measurements using external detectors may be simplified. It is often sufficient to determine the time required for the amplitude of the successive concentration peaks to reduce to 10% of the amplitude of the first peak, indicating 90% dispersion of the tracer. Under the same conditions, double this period of time will be indicative of 99% dispersion and triple the time, 99.9% blending achieved. With the sampling technique when successive samples give the same count rate, due allowance being made for statistical errors, then mixing is complete.

Many blending-time measurements have been carried out, often with surprising results. As expected, it is far more difficult to blend highly viscous liquids than those of low viscosity, but on several occasions it has been found that the calculated blending time was far in excess of that actually required, particularly when internal stirrers are used in conjunction with side-supported baffles in tanks. Blending in such tanks was obviously far more efficient than

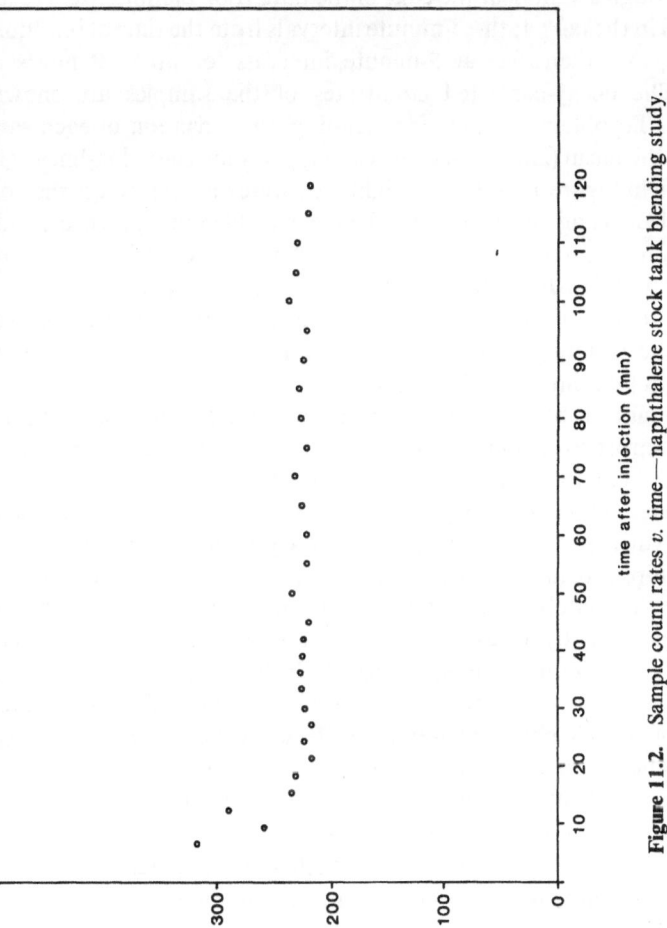

Figure 11.2. Sample count rates *v.* time—naphthalene stock tank blending study.

expected, and considerable savings have been achieved by reducing the blending period originally specified.

An example of this, which also illustrates a slightly different approach to the treatment of results, concerned blending of batches of naphthalene into a large heated stock tank using an internal paddle. The blending time was thought to be about 90 minutes. Bromine-82 as paradibromobenzene was injected into the tank through a vent near the edge of the tank roof. Samples were taken of the material in the tank at three-minute intervals from the time of injection for 45 minutes, and thereafter at 5-minute intervals for up to 2 hours after injection. The decay-corrected countrates of the samples are shown in Figure 11.2. To obtain the time for blending, the deviation of each sample count from the mean (226 counts per second) was plotted on log-linear graph paper shown in Figure 11.3. Two straight lines were drawn through the points, the first of decreasing deviation from the mean as blending proceeds, and the second at constant deviation. The standard deviation of the sample count is $4 \, \mathrm{c \, s^{-1}}$, that is 1.8%. It can be seen from the graph that this value is easily reached in about 30 minutes, when the tracer is 98% mixed in the vessel contents. The blending time following addition of batches was, therefore, reduced from 90 minutes to 30 minutes.

However, not all blending problems are concerned with simple tanks. A requirement arose to examine the blending efficiency of a newly-designed and commissioned unit for blending polymer chips. The blending process was via a complex system of valves and gravity hoppers, capable of producing a wide range of polymer specifications from a few basic polymers. With such a process where the chips are finally melted and extruded it is, of course, essential that the chips are intimately blended if variations in the polymer thread are to be avoided. Consequently it was necessary to examine small samples, taken from a total charge of 80 tonnes, in some detail. Each chip had a mean weight of 0.00987 grams and it was calculated that a sample weight of 100 g, containing in excess of 10 000 chips, would be sufficiently accurate statistically to determine whether blending was adequate in the unit.

The physical difficulty of examining each chip for the presence of a tracer in almost a hundred samples was considerable and it was decided in this case to use an activable tracer rather than a radioactive tracer. A ten-tonne batch of polymer was prepared in the normal manner and indium trichloride added at the liquid stage to produce an even concentration of 10 ppm of indium in the polymer. This was then chipped and transferred into one of the feed hoppers of the blending system. Eighty tonnes of non-labelled material was loaded into the remainder of the system and the blending process started. Samples were taken at ten-minute intervals throughout the operation and exactly 100 g of each sample separated — the total number of chips in each 100-g sample being known accurately from the mean weight of each chip. These 100-g samples were then irradiated in the core of a nuclear reactor producing, in those containing indium, a small amount of indium-116 m. As it happened, there was

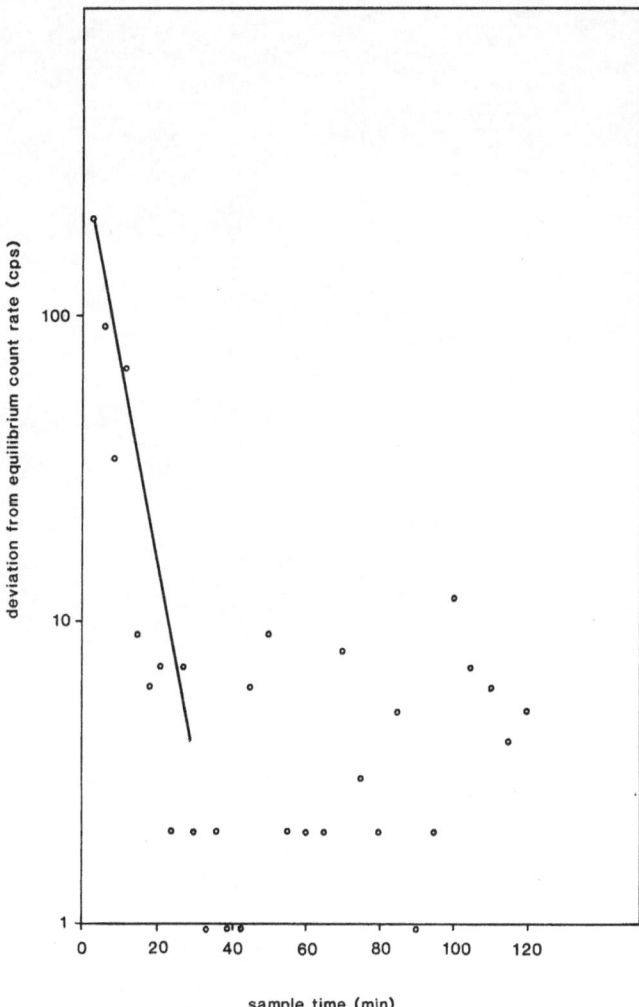

Figure 11.3. Plot of sample count rate deviation from mean *v.* time—naphthalene stock tank blending study

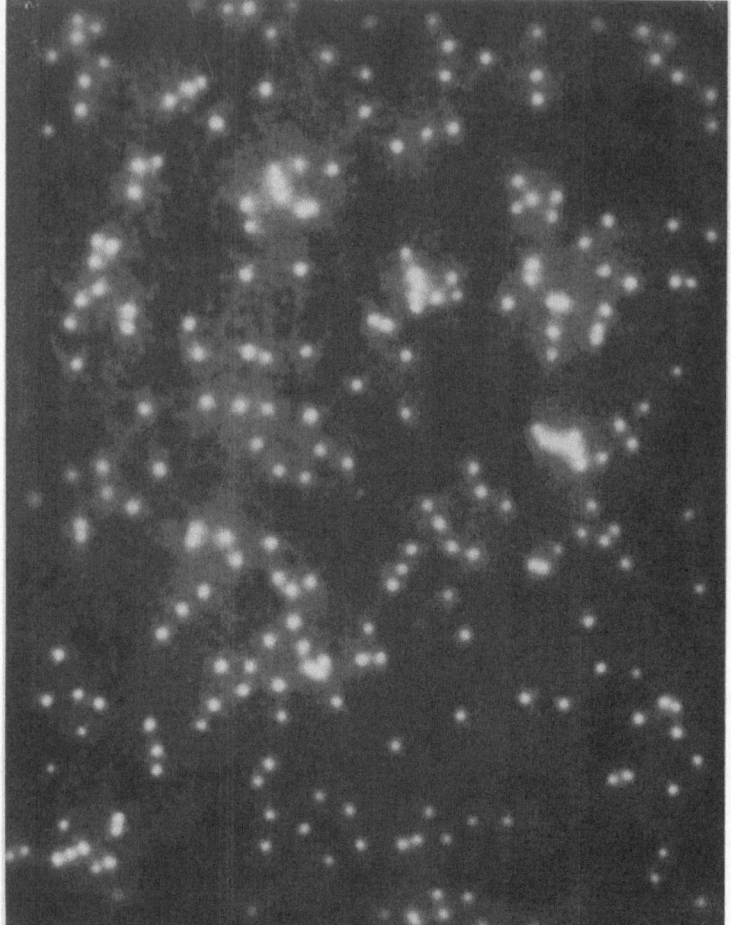

Figure 11.4.   Indium-labelled polymer chips in a matrix of chips containing only manganese.

a small quantity of manganese in all the chips which produced manganese-56. The irradiated samples were then spread over a photographic film in a single layer and left in contact for 30 minutes. A typical autoradiograph produced from the film exposure is shown in Figure 11.4. The bright spots correspond to each of the indium- and manganese-containing chips and are quite easy to identify and quantify. The duller spots correspond to the chips which contain only manganese but it was not necessary to count these since the total number of chips was known from the sample weight.

It was found that the spread of the number of labelled chips in the hundred or so samples lay within a tight band and that the blender was operating within the design specification. As a bonus, the residence time through the system was also determined, using the labelled chips as a tracer and sampling over the whole period from start of injection to final passage of tracer.

11.1.2 *Mixing studies*

As mentioned above, the term 'mixing' is usually applied to systems where fluids are mixed and reaction takes place between them; blending is where no reaction takes place. The reader is referred to one of the standard textbooks on chemical engineering for a treatment of the theory of mixing, and only a few of the more salient points will be considered here.

The Residence-Time Distribution of the material in the system is obviously of importance, particularly where extremely fast reactions are taking place. Short mixing times are essential to achieve uniformity of composition and an acceptable product quality. Similarly, where large volumes are involved and mixing is inefficient with consequent stagnant regions, then it is important that we should be able to compare the behaviour of the system with a model to identify deviations from design criteria.

Such models are of two types, either short-term or long-term. Short-term models are based on the stirred-tank model described in Chapter 9 with complete recirculation imposed. Since they are short-term models, it is really immaterial whether the reaction is taking place in a batch reactor or a continuous throughflow reactor—the time period of interest is too small to be significantly affected by the throughput of tracer. Such models are adequate to characterize the types of mixing profiles shown in Figure 11.5 when a tracer is mixed into the tank.

The longer-term models are based on a double stirred tank consisting of a main stirred tank which takes the majority of flow, representing the 'active part' of the reactor, and a subsidiary tank into which some of the flow is diverted, representing the stagnant regions. These models are rather complex and should be used with caution.

A second important factor, particularly with fast reactions, is the size of the reacting clumps of molecules. Single molecules of one reactant next to a single

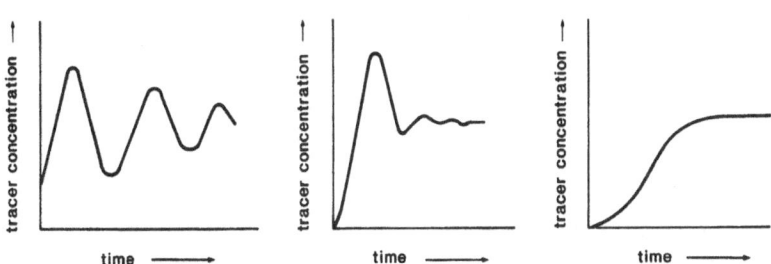

Figure 11.5. Tracer concentration profiles which may be characterized by the short-term model.

molecule of the second reactant have a high chance of reacting together. Large groups of molecules of the two reactants adjacent to each other can only react at the group boundaries and the reaction proceeds less efficiently. This is obviously less important if the reaction rate is low since the groups have time to break and then react. For fast reaction rates, however, a great deal of variation in product quality can result from such poor contact. This is particularly so if the product of the reaction is a solid. Large aggregates of the solid molecules grow at the boundaries and the reaction then stops until breakdown occurs and the reactants again come into contact.

For such fast reactions, therefore, efficient mixing is important for two reasons, firstly to ensure that the two reactants are dispersed and in close contact with each other, effectively making the reaction zone as large as possible, and secondly, to ensure that the product is dispersed and maintained in a fine form.

This point was illustrated in work carried out on a plant producing a plastic polymer by catalytic action on the monomer in a batch reactor. Smaller plants producing the product had been in operation for several years, each producing material to a very tight product specification, dependent entirely on the initial catalyst distribution in the monomer. For economic reasons it was decided that the five smaller plants should be replaced by one about five times the size. It was expected that achieving the same product specification would prove extremely difficult, because of the difficulty in reproducing the same mixing pattern in the larger reactors. In fact, it was realized that with the larger reactors it would not be possible to achieve the desired pattern simply by using a larger model of the same type of stirrer and that a novel design must be used in conjunction with a number of side-mounted baffles.

The first stage, therefore, in the design of the new plant was to characterize the mixing pattern in the old-type small reactors. A problem arose because the new reactors had to be engineered to something approaching their final mixing configuration before the plant came on-line, thus avoiding costly delays. It was therefore decided to carry out all of the work with water in the reactors, and make the theoretical allowances for differences in viscosity, etc.

One of the old reactors was charged with water, the stirrer started at the usual speed and a quantity of sodium-24 in aqueous solution injected quite simply through an open manhole at the top. Sodium iodide scintillation detectors were positioned external to the vessel, near the top level, two-thirds level, one-third level, and the base. All four detectors gave similar cyclical concentration v. time traces and all four steadied out to $\pm 2\%$ of the final mean value within about 35 seconds. The fact that all four showed the same type of trace indicated that the tracer was distributed in the vertical direction very quickly.

The next stage in the investigation was to use the pilot plant reactor to carry out a similar radiotracer run, and at this point the differences in the proposed stirrer system and the old stirrer system became obvious. Many runs were

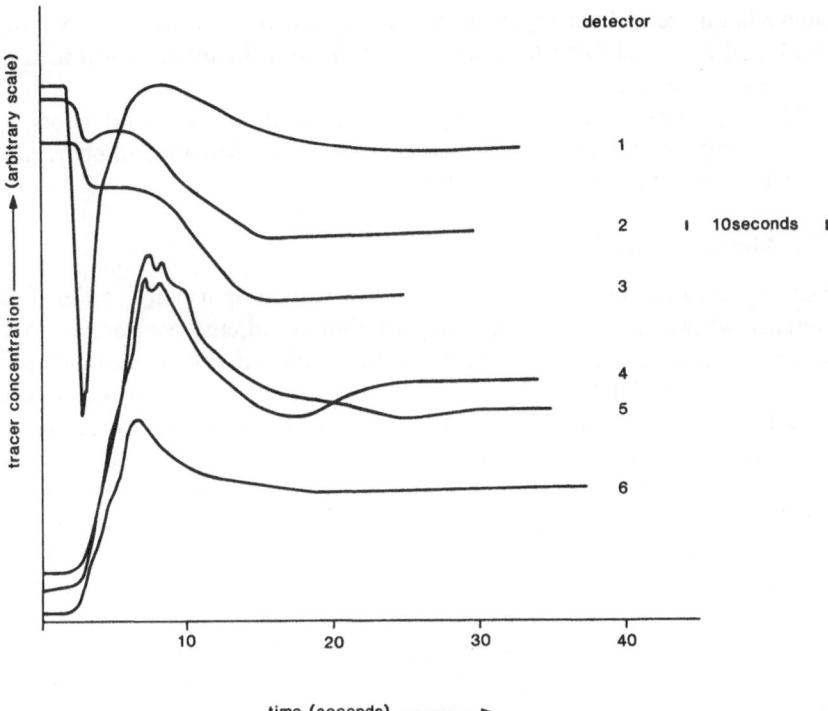

Figure 11.6. Tracer concentration curve variations observed in polymer production vessel.

carried out with different charged volumes, varying stirrer speeds and a variety of baffle configurations. Long and short baffles were used in combinations of up to six of each. Several important conclusions were reached concerning optimum baffle length and position. For example, it was found that if the baffles were too long the swirl velocity at the top of the vessel was reduced to such an extent that there was very little agitation of the vessel surface. Conversely, of course, with no baffles there was virtually no conversion of swirl velocity into vertical velocity and distribution down the vessel took an inordinate length of time. In addition it was critical that the top of the baffles reached the liquid surface. If this was not so then the material above the baffle did not mix well with the material below and the average mixing deteriorated.

These conclusions were put into practice on the full-scale vessel on the new plant, sufficient supports being located to provide the option of extending the baffle lengths if necessary and the radiotracer experiments repeated. In fact, it was found that the direct scale-up from the pilot plant, that is the charge height to baffle length ratio, was correct and that minor tuning in terms of stirrer speed was all that was required to achieve the desired mixing time and product quality. However, the mixing pattern was quite different. Rather than a consistent cyclical pattern being observed, as in the old-type reactor, those shown in Figure 11.6 were typical, Detector 1 being the detector located at the top liquid level and Detector 6 at the bottom with the others positioned at intermediate points.

The expenditure of a relatively small amount of time and effort in determining the required mixing characteristics was, therefore, amply repaid in achieving a smooth plant start-up.

11.2 Measurement of volume

The use of radioisotope tracers to measure volume is usually confined to systems where conventional instrumentation or alternative radioisotope sealed source techniques cannot be effectively employed. With clean fluids and regular-shaped volumes the need rarely arises, since even if no instruments are installed on the vessel gamma-transmission or neutron- or gamma-backscatter techniques can usually establish interfaces, making the calculation of volume a simple task. It is in the situation where significant depths of irregularly shaped solid deposits have occurred, or the volume to be measured consists of several interconnected smaller volumes, or access is too difficult for other instruments to be used, that the technique is most often employed.

The method is based on an isotope dilution technique. A known quantity of radiotracer is dispersed into and thoroughly mixed with the volume to be measured. Samples are then taken of the fluid and the tracer concentration estimated.

If a volume of tracer V_1 and concentration C_1 is injected, giving a concentration of C_2 in the fluid under investigation then

$$C_1 V_1 = C_2 (V_1 + V_2)$$

where V_2 is the volume of fluid. In general, V_2 is much greater than V_1, therefore

$$V_2 = \frac{C_1 V_1}{C_2}$$

It is important, of course, to ensure that the tracer being used is physically compatible with the fluid under investigation. If the system is of two components, then the tracer chosen should be overwhelmingly soluble in the component being investigated, to reduce errors.

The most common use of the method in industry is in the estimation of the quantity of mercury in electrolytic cells producing chlorine. In these cells a very thin layer of mercury is used as the cathode, carbon as the anode and molten sodium chloride forms the electrolyte. Sodium liberated at the cathode forms an amalgam with the mercury, and estimation of the quantity of mercury present in the cell is, therefore, difficult. A known quantity of mercury-203, a gamma-emitting isotope of mercury, half-life 47 days, is added to the cell, allowed to mix with the mercury and amalgam and then the mixture sampled. Assay of the mercury sample after breaking down the amalgam then allows estimation of the total mercury both in the free state and as the amalgam.

A variation of this technique has also been successfully employed in estimating the quantity of sludge in crude oil stock tanks. Many crude oils form heavy waxy deposits, and after some years the depth of these build up, often to several feet, reducing the useful volume of the tanks. The level of sludge is usually not constant across the tank, huge mounds and deep depressions being created. Two problems arise, firstly the estimation of useful volume within the tank, and secondly estimation of deposit volume when the tanks are to be cleaned. Such cleaning is often carried out using expensive emulsifying agents; to order too little of the chemical means ineffective cleaning or delay, too much is wasteful. A measured quantity of a short-lived tracer in an organic form, such as bromine-82 as paradibromobenzene, is added to the crude oil and mixed thoroughly in the oil. The mixing is usually carried out by circulating the oil through an external pump and may take a considerable length of time. Samples are taken during the mixing period until a series of successive samples show the same count rate, after correction for decay, indicating that mixing is complete. The difference between the volume of oil calculated from the dilution of the tracer, and physical measurement of depth in the tank and diameter of the tank is, of course, the volume of wax deposit.

A similar technique was used when a problem arose concerning disposal of a highly toxic chemical. The compound was found to be deposited on the sides and bottom of a tank after supernatant liquid had been drained off. Arrangements had to be made for disposal which, of course, necessitated determining the quantity involved. In this case, a gas tracer, argon-41, was injected into the closed tank, the paddle started to assist dispersion and the volume of gas in the tank calculated from the dilution of the argon. The chemical was subsequently dissolved in a solvent, drummed off and disposed of by controlled incineration.

Often useful information on volumes of systems may be gained incidentally, for example, during measurements of flow rates in closed circulating cooling-water systems. The tracer used is eventually mixed evenly throughout the system; the exchangers, pipework, cooler and buffer ponds. Estimation of the tracer concentration allows the volume to be calculated if the total amount of tracer injected is known. Suitable corrections may have to be made for purges from the system and decay of the tracer.

A more complex example, not of reactor or system volume but volume of one component of a system was encountered in a process where it was necessary to study the fate of expensive palladium catalyst. Initially the palladium is in a palladous form and is reduced to palladium during the reaction. It may then be reoxidized to palladous by cupric ions to remain as a useful catalyst in solution. As palladium metal it will deposit on reactor solids or plate out in metal surfaces, for example, heat exchangers, where it is not only useless as a catalyst, but cannot be recovered until plant shutdown.

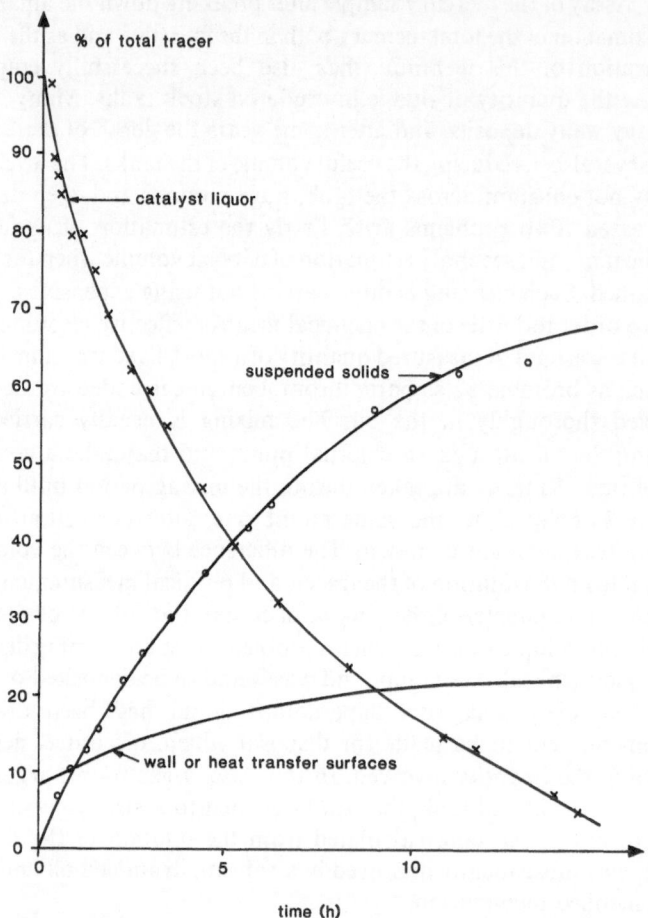

Figure 11.7. Distribution of radiotracer in reactor.

To study the fate of palladium, palladium-109, half-life 13 h, was injected at a constant rate into the ingoing liquid reactant. Mixing in this reactor was rapid. Samples of reactor liquid comprising largely acid and suspended solids were taken at discrete intervals of time. The samples were filtered, the suspended solids dissolved to give a homogeneous solution and then both the filtrate and dissolved solids counted to determine their tracer concentrations.

The plot of tracer content versus time for liquid and suspended solids is shown in Figure 11.7. At any instant the sum of the amount of tracer in the liquid and that in the solids, subtracted from the total injected, must equal that on the walls, thus giving the lower curve. The conclusion was that under the particular conditions of operation about 20–25% of palladium was deposited on the walls and lost to the reaction. 70–75% eventually goes to suspended solids and may be regenerated. The half-life of palladium in the form when it was effective as a catalyst was deduced to be four hours. This information could not have been obtained in any other way than by a radioactive technique.

11.3 Ventilation studies using radioisotopes

The design of effective ventilation systems is extremely difficult. Many factors have to be taken into account, not simply the amount of air being exchanged or ventilated. The position of inlet and exit ducts, the shape of the ducts, the streamlining of surfaces, etc., all play an important part. Even when all known factors have been considered and allowed for in the design it is necessary to prove that the design and construction have been carried out correctly, particularly when toxic or explosive materials are present. It is, therefore, becoming more common to supplement the traditional visual smoke-bomb tests, hot-wire anemometer measurements, etc., with more detailed and sensitive studies. Further impetus has been given by recent legislative requirements to carry out hazard studies on dangerous operations such as rupture of process vessels containing noxious materials in enclosed spaces. The rate at which the noxious material clears from the process area is one important aspect of such studies.

In the nuclear industry such hazard studies and the necessity for effective containment and efficient ventilation have been a requirement for many years, due to the high radiotoxicity of many isotopes. Because of the high sensitivity of the various types of detectors, alpha, beta and gamma, such studies in general presented few problems and considerable expertise in the technique arose. This expertise translates to the process plant environment with few modifications except that, of course, a radiotracer must be introduced, rather than being present 'naturally'.

Other tracers and detector systems have been and still are being used. Chlorinated hydrocarbons, as gaseous tracers, and electron capture or thermal conductivity detectors have enjoyed considerable success, but suffer from the major drawback that they are non-specific. They respond to a wide variety of materials which may be present in the work area, such as solvents, oil vapours and even cigarette smoke. Radiation detectors are, of course, quite specific and radioactive tracer techniques are now tending to be used in preference to other non-radioactive techniques.

The method involves injecting a radioactive gas into the area under study and monitoring the gas concentrations over a period of time using detectors positioned at points of interest. The injection may take place as a sharp pulse to simulate say fracture of a pipe, or as an extended injection. On occasions it may be of interest to flood the whole area with radiotracer with the extraction system isolated and then to switch the ventilation on and study the decay of the tracer. Whichever method is used, it is important that a sufficiently high concentration of the tracer is injected initially to allow the desired sensitivity to be achieved. Factors which affect this are the detector background countrate, the detector efficiency, detector shielding and the radioisotope used. It is often easier to expose the detector system to a known concentration of tracer to quantify these factors rather than to try to calculate them. Often relatively high concentrations of tracer may be needed, and here the ability to inject and detect remote from the area under investigation is an advantage.

The use of the technique to investigate small systems can be illustrated by reference to Figure 11.8, a motor car. With both floor and window vents operating, the traces shown in the figure are obtained. The air change rate at front floor level is much greater than that at front face level, principally because of the larger vent area at this position. The combined flows ensure an even faster change rate at the rear seat face level as air flows to the exit vents.

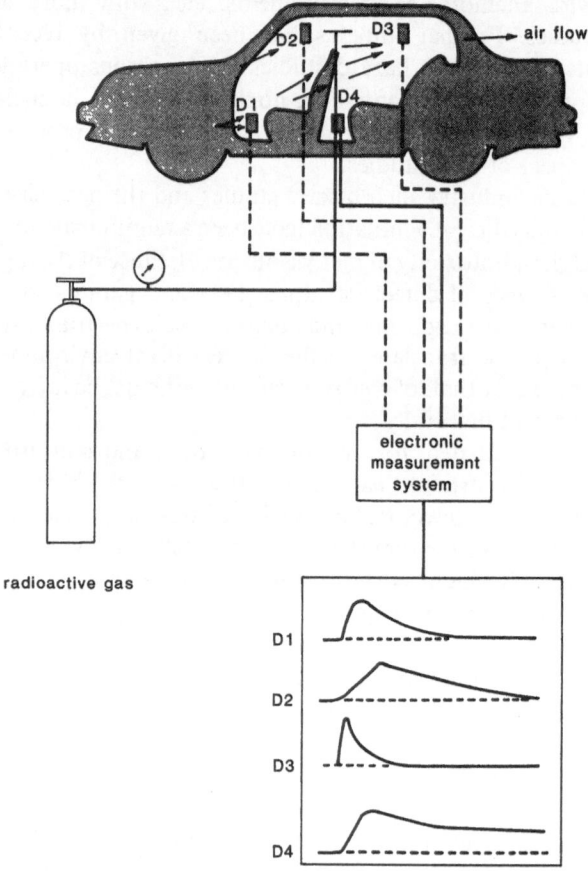

Figure 11.8. Ventilation study—motor car industry.

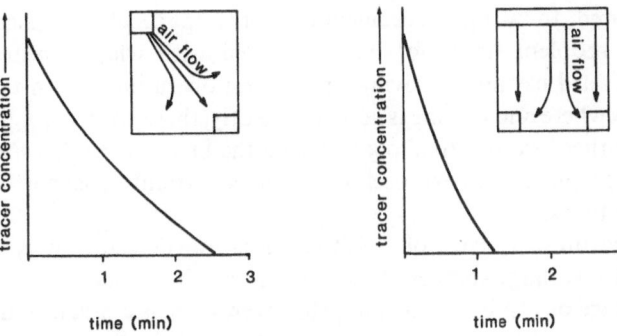

Figure 11.9. A radiotracer study of turbulent and laminar flow ventilation systems.

At rear seat floor level, however, the air is much more stagnant and an extremely low changeover rate is apparent. Similar studies are carried out by most major manufacturers.

The radiotracer technique was used recently to confirm the ventilation system design for a new production unit producing a highly toxic chemical. Before the ventilation system orders were placed a scale model was constructed with the two alternative types of ventilation being considered incorporated. The cheaper system consisted of large inlet ducts along one wall at ceiling height, and exit ducts along the opposite floor. The alternative system involved constructing a false ceiling with large-area diffusers inset, and exit ducts along all four walls at floor level. Both involved the same air flow rate into the room. For the same initial concentration of radiotracer in the room prior to switching on the ventilation, the two tracer concentration curves shown in Figure 11.9 were obtained. The first system created turbulent conditions in the room and consequently took a much longer time to clear than the second system which operated under laminar flow conditions. It was found that although the initial costs of the diffuser system were greater, these would be more than offset by the lower operating costs (heating of the air) because the same ventilation efficiency could be achieved at a lower flowrate.

11.4 Line pigging

The labelling of line pigs and spheres with radioisotopes to assist in locating them is probably one of the oldest industrial applications of radiotracers. With the introduction of new legislation in the 1950s the practice fell into disuse. Far greater control was exercised in the purchase, use and accounting for radioactive material, and pigging operators became aware of the hazards associated with the large sources of iridium-192 and cobalt-60 they were using.

In addition, specialized pigs became common, operating on magnetic or other principles, encouraged by the proliferation of pipelines carrying chemicals, gas and oil. However, these pigs are expensive and often have to be manufactured to a tight specification for a particular application. On occasions problems may not be anticipated with what appears to be a straightforward cleaning, proving or gauging operation. When the first pig sticks somewhere and the line is needed urgently, there is often not time to hire or buy another 'transmitting' pig to locate the first one. In this situation the radioisotope pig can prove useful, and it is certainly cheaper than other equivalent types.

With the present types of sensitive radiation detectors it is no longer necessary to use large sources of radiation if the pipes under investigation are on the surface or just below ground. However, there is a practical limit to the depth of pipes which may be investigated. Below about 1-m depth of concrete or soil, the gamma-radiation from the source is attenuated to such an extent that detection of the passage of the pig carrying the source is very difficult. In this situation it is necessary to use detectors at convenient exposed positions on the line or to excavate holes near the line and monitor the pig passing or failing to appear. By reducing the distance between the excavations and sending further pigs down the line, the position of, for example, a blockage may be pinpointed with greater accuracy.

With the greater availability of short-lived radiotracers these are now preferred, rather than the longer-lived sealed sources used previously. For short sections of line up to about 8 km in length, manganese-56 (half-life 2.6 h) is used. For longer sections sodium-24 (half-life 14.7 h) is suitable. Quantities of radioactive material are determined by the size of the line, the fluid in the line and the expected velocity of the pig. 1 GBq of sodium-24 would be adequate for the 140-cm cleaning pig shown in Figure 11.10 (courtesy of General Descaling Co. Ltd. of Worksop) in an oil-filled line, 0.01 GBq would be sufficient in a 10-cm pig in a gas line.

The isotope, in a small sealed capsule, is inserted into a threaded hollow tube welded onto the pig, and the tube plugged. Handling during insertion and recovery is minimal and managed by one radiation worker. Detectors may vary in sensitivity depending on the type of information required. If the line is above ground and the pig therefore easy to locate, then simple Geiger tubes will suffice. If the passage of the pig needs to be monitored at several locations, and is possibly travelling at a considerable velocity, then more sensitive scintillation detectors may be required. A record of the passage may be made by feeding a signal to a recorder or causing a signal light to be illuminated if the associated ratemeter exceeds a preset level.

Such a system has been used with great success on a liquid ammonia transfer line. A similar line in Canada collapsed due to the adjacent soil freezing and expanding, and the line is now routinely checked using a gauging pig. As it happens the line passes under three roads, a railway line and a golf course. The pig velocity reaches some 90 km h^{-1} in the passage down the line and its exit from each underground section is monitored using a small scintillation detector/ratemeter combination strapped to the line. The ratemeters are set to

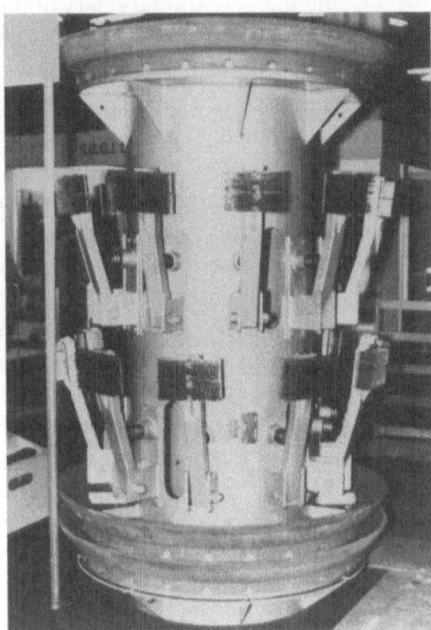

Figure 11.10   140-cm line-cleaning pig (courtesy of General Descaling Co. Ltd., Worksop).

alarm at a signal strength of twice the natural radiation background. Using this system one man can comfortably carry out the radiation aspects of the work rather than the five who would be needed if the alarms were not used.

Isotope-labelled pigs have been used in a wide variety of situations but one feature common to all has been the ease with which the arrival of the pig at its final destination has been identified. In a number of lines it has been preceded by a large quantity of debris, often sufficient to fill the pig catcher and prevent entry into the catching system. A detector positioned some distance upstream of the catcher will alarm and may considerably reduce time wasted in waiting for a pig which, in fact, has already reached its final position.

11.5 Corrosion and wear studies

One of the main considerations taken into account by the engineer when designing plant is the correct choice of the materials of construction. Even so one of the major maintenance costs is the replacement of items which have been eroded or corroded, and this does not necessarily reflect on the quality of the original material specification. Quite simply in many cases the ideal material may not be available or a conscious decision may be taken on an economic basis to use a cheaper material and replace at regular intervals. Such decisions inherently assume a particular rate of erosion or corrosion which may be wildly in error, placing the operational safety of the plant in jeopardy, disrupting maintenance schedules or necessitating costly shutdown time while replacements are ordered.

Many factors can affect the assumed corrosion rates, minor changes in process material composition, a change of particle size in slurries or variations in operating conditions at steady state, shutdown or start-up. All can lead to increased general corrosion levels or rapid local corrosion at sensitive spots.

The use of radioisotopes to measure severe corrosion and erosion both on-line and off-line is described elsewhere in Chapters 13 and 14. However, in many cases less severe corrosion is of interest or early warning of the onset of corrosion is required and the incorporation of radioisotopes in the plant matrix may provide the required sensitivity, supplementing or replacing more established techniques.

The principle itself is quite a simple one. Basically a small area of the plant item of interest or a coupon of the metal under investigation is made radioactive and placed in contact with the plant stream. As erosion or corrosion occurs either the decrease in activity of the radioactive item is measured, using an adjacent detector, or the increase in radioactivity of the process stream due to uptake of corrosion products is estimated.

11.5.1 *Neutron activation of samples*

This latter method has been used for many years to study wear of piston rings in combustion engines, and the effect of oils and oil additives on such wear. The general method was to irradiate the whole of the top piston ring (the one subject to most wear) in a nuclear reactor, producing by neutron bombardment a range of isotopes from the main and minor constituents of the steel. Neutron irradiation produces the radioactive isotopes throughout the whole of the sample because of the high range of neutrons in steel and other non-hydrogenous materials. Iron-55, iron-59, chromium-51, cobalt-60, phosphorus-32, silicon-31 and manganese-56 were the principal isotopes produced, and before transport the short-lived ones were allowed to decay, leaving P-32 and Fe-59 as beta-emitters and Fe-59 and Co-60 as gamma-emitters. In practice the gamma-emitters were detected in the oil rather than the beta-emitters, since phosphorus tended to concentrate at the grain boundaries of the metal and its activity in the oil was not truly representative of the amount of corrosion occurring. Initially, samples were abstracted from the circulating oil system and counted but the detecting system soon progressed to the use of scintillation counters located inside a cell with the oil circulating through. The increased sensitivity reduced errors, allowed automation of the system and greatly reduced the time needed to detect very low levels of corrosion; fractions of a milligram loss from the ring being readily detectable.

The same principle may be used to study corrosion problems in areas where access is very difficult. For example, with modern downhole tools, neutron irradiated steel coupons may be located at the bottom of oil risers, in oil production wells where corrosion is a particular problem. The iron-59 activity produced by 4 weeks irradiation in a nuclear reactor at a neutron flux of 10^{12} neutrons cm^{-2} s^{-1} is about 2.6 MBq per g. After allowing a month for the majority of the short-lived isotopes to decay, some 1.7 MBq per g of iron-59 will be left. Assuming a total surface area of the coupon of 10^3 cm^2 and a production rate from the well of 10^4 barrels per day, a corrosion loss of 0.1 microns per day from the steel surface may be easily detected by measuring the Fe-59 concentration in the product. The efficiency of corrosion inhibitors can, therefore, be studied by measuring the wear prior to and after coating of the well risers with the inhibitors while the well is in production.

11.5.2 Thin-layer activation of samples

One of the difficulties associated with neutron irradiation is that the whole of the sample in the neutron beam is made radioactive, producing large amounts of appropriate isotopes, most of which are inaccessible to corrosion until the top layers have been removed. The high activities make handling and transport correspondingly more difficult. Conversely, of course, the fact that the sample is radioactive throughout means that it has a longer life with respect to any corrosion work being carried out. However, the handling and transport difficulties can to a large extent be overcome if only a small portion of the sample is made radioactive, and this is achievable by irradiation with an ion beam.

Of particular use in wear studies of steel is the production of cobalt-56 (half-life 77 days, principal gamma-energy 0.85 MeV) by irradiation of the steel with a beam of high-energy protons giving the reaction ^{56}Fe (p, n) ^{56}Co. The depth of activation is dependent on the initial proton beam energy and may be calculated with a fair degree of certainty, errors in the calculation being associated with uncertainties in the proton energy and the flatness of the surface. When it is necessary to know the actual depth accurately, a common technique is to irradiate a stack of thin iron foils under the same beam conditions, separate the stack and measure the activity on each foil, until the activated depth and activity profile is established.

The usual depths chosen for investigations are between 25 and 300 microns with a total cobalt-56 content of a few MBq or less. With external detectors positioned adjacent to the irradiated section, changes of count rate (after correction for decay) equivalent to erosion or wear of about 1% of the total depth may be detected. Where it is possible to circulate debris-carrying fluid through a detection cell, or sample the fluid and count on a more sensitive detection system, the sensitivity of the technique may be improved by several orders of magnitude.

However, such sampling techniques are often more suited to the laboratory

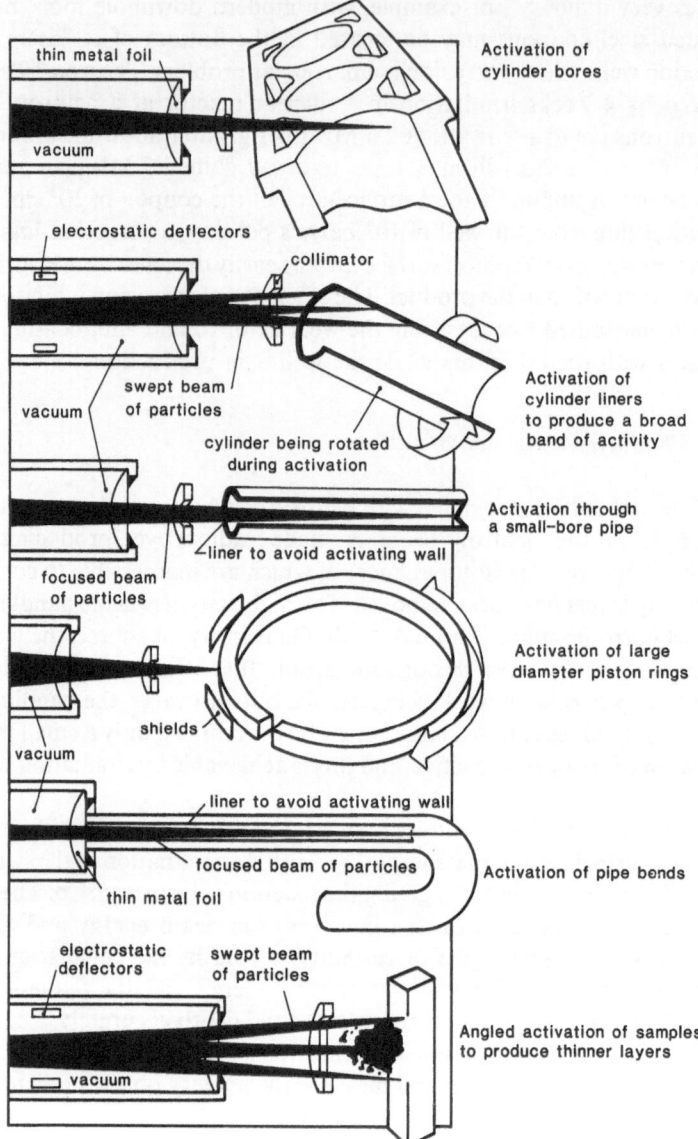

Figure 11.11.   Variation of active area by beam manipulation (schematic). After Asher *et al.* (1983).

or research area in that they require a considerable input from trained staff and expertise in specialized instrumentation. Consequently, most plant applications of thin-layer activation and neutron activation techniques rely on detectors located adjacent to the activated object, using the following method.

Duplicate samples are irradiated under identical conditions, which are controlled in the case of thin-layer activation to determine the depth of the activated layer according to relationships established by the foil stack calibration method. After irradiation the samples are allowed to cool for a period to allow short-lived isotopes to decay. One sample is then installed in the plant and the other in a mock-up which simulates the plant environment in terms of nature and thickness of surrounding constructional materials (vessel or pipe walls, lagging, inert simulated process materials, distance, etc.) and their attenuating effects on the radioactivity. Small jigs are required to locate precisely the external detection equipment on plant and on the control sample. Normally a conventional waterproofed scintillation counter with the appropriate scaler/ratemeter/timer/recorder system is used. Radioactivity loss can then be monitored continuously with periodic counter calibration using the duplicate control sample which has not been exposed to the process fluid. Measurements are possible over several half-lives of the principal isotope being monitored. The use of the duplicate control sample avoids the necessity for preliminary determination of the effective half-life of the mixture of isotopes produced in the sample and gives an automatic correction for changes in the efficiency of the detector arrangement, although much effort is now being expended in improving the temperature stability of scintillation counters.

Some methods of producing specific areas of producing specific areas of activation are shown schematically in Figure 11.11.

Currently several corrosion investigations are being carried out within Imperial Chemical Industries PLC by the Physics and Radioisotope Services Group. Some involve insertion of thin-layer activated and neutron-activated coupons into process items though specially designed gland systems. Similar studies using TLA bends and straight sections of pipework are in progress, and others involve long-term investigation of corrosion in heat exchanger systems using neutron-activated ferrules inserted into the heat exchanger tubes. It is expected that this type of operation will be a major growth area over the next few years.

References

Sailer, S. (1982) Radionuclide technology applications in development and production in the automobile industry. In *Industrial Application of Radioisotopes and Radiation Technology*, IAEA, Vienna, 433.

Deterding, J. H. and Dyson, A. (1954) Radioactive isotopes for measuring piston ring wear. *The Engineer*, 442.

Askouri, N. A., Chen, N. S., Ettinger, K. V., Fremlin, J. H. and Nowotny, R. (1975) On-line wear monitoring by surface activation. *Int. J. Appl. Rad. Isotopes* 26, 61–70.

Niiler, A and Caldwell, S. E. (1976) The 56 Fe (p, n) 56 Co reaction in steel wear measurement. *Nucl. Inst. Metds.* 138, 179–183.

Asher, J. and Conlon, T. W. (1982) Radioactive techniques in corrosion monitoring. In *Corrosion Monitoring in the Oil Petrochemical and Process Industries*, Oyez Longman, London, 91.

Asher, J., Conlon, T. W., Tofield, B. C. and Wilkins, N. J. M (1983) A new plant corrosion monitoring technique. In *On-Line Monitoring of Continuous Process Plants*, S. C. I/Ellis Horwood, Chichester.

12 Sealed-source applications

J. S. CHARLTON

12.1 Introduction

Previous chapters have served to demonstrate the versatility of radioactive tracer techniques in investigating a wide range of plant problems. Case histories have amply demonstrated that in many situations the use of a tracer is not only the best solution but the *only* solution. However, useful as they are, radiotracer techniques suffer from one significant disadvantage: by definition, radioactive materials are injected into the plant and therefore become dispersed in the process material. In other words the process material becomes mildly contaminated with radioactive material. Of course, by careful choice of a tracer with short half-life, by exercising judgement over the amount of tracer injected and by ensuring that the radioactive material is diluted by the vastly greater volume of process material, the level of contamination, and therefore, the hazard, can be made negligibly small (Chapter 7). Nevertheless, each radiotracer investigation requires detailed prior planning. The logistics of carrying out the measurement need careful consideration and, additionally, one must invariably satisfy national or local environmental authorities that the proposed work is safe.

Precise legal requirements vary considerably around the world but in certain countries it can be very time-consuming to obtain the necessary permission to carry out tracer work. The large number of tracer jobs completed by just one group in a typical year (Chapter 1) is clear indication that this type of work can be, and is, carried out on a regular basis—but there are occasions when the urgency of the production problem requires immediate action.

Fortunately there is another category of radioisotope technique—the sealed-source technique, which can be carried out at short notice and which can often provide insights into plant problems which are just as valuable as those provided by radiotracer methods. Additionally, sealed-source techniques can often be used in conjunction with radiotracers to provide complementary views of the causes of a plant problem.

The comparative ease with which a sealed-source technique can be employed stems from the fact that with sealed sources there is no contamination of the process material. The radioactive isotope remains permanently encapsulated throughout the course of the measurement. Penetrating radiations from the source capsule are directed at the vessel or material of interest, and, by examining the way in which the transmitted or scattered radiation is modified, it is possible to draw conclusions about the vessel and its contents.

At this point it is perhaps useful to dispel a popular myth concerning the use of sealed sources: it should be noted that the interaction of the radiation with the contents of the vessel does *not* make them radioactive. One might perhaps reinforce this point by considering the parallel with the diagnostic use of X-rays. X-rays are passed directly through the human body to irradiate a photographic plate, but neither the body or the radiograph becomes radioactive in the process.

The single exception to this rule occurs when neutron sources are being used (Chapter 15). Neutrons are capable of inducing radioactivity in the material which they irradiate, but the intensity of neutron sources commonly used for process investigation means that this induced radioactivity is at a very low level indeed.

In the next few chapters, we will be examining some important sealed-source techniques involving both the absorption and the scattering of different types of radiation. Before doing this, it will be useful to discuss sealed sources in more detail and to say something about the criteria used to select a particular source for a particular application.

12.2 Types of sealed source

12.2.1 Introduction

In using radioisotopes as tracers the chemical form of the tracer and its compatibility with the process stream were of great importance in selecting a particular tracer for a particular job. Sealed-source applications are different. Here the chemical form is not of concern, since the radioactive material is encapsulated. In general, one is chiefly interested in five properties of the source:

(a) The type of radiation emitted
(b) Its energy
(c) The intensity of the emitted radiation
(d) The half-life of the isotope
(e) The cost of the source.

Implicitly, one will also be concerned to ensure that the construction of the source is such that it is capable of withstanding the (sometimes harsh) environmental conditions under which the source will be used without suffering leakage of the radioactive material. To produce sealed sources which combine rugged construction with the other desirable properties required by the user is a specialized task. To comply with international safety standards

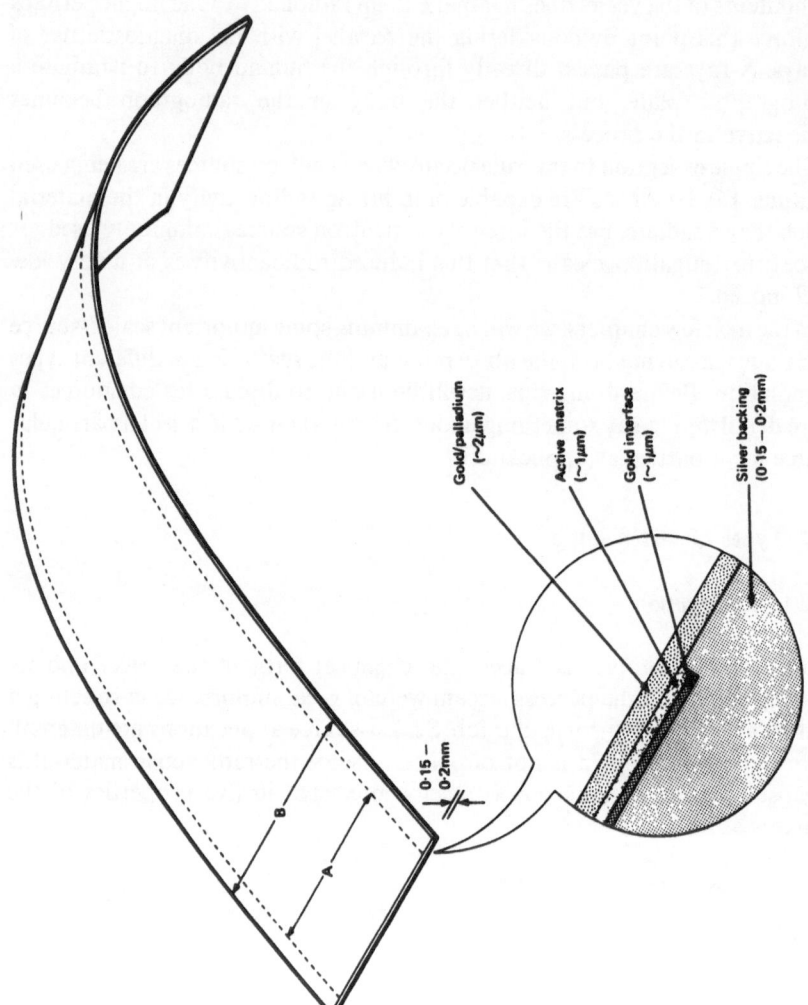

Figure 12.1. Americium-241 alpha foil (by kind permission, Amersham International PLC).

the source and its ability to stand up to taxing environmental conditions must be fully proven.

Sealed-source manufacture is thus a task for the expert and there is a comparatively small number of companies worldwide which are able to supply sealed sources on a commercial basis. One such is Amersham International PLC of the UK and it is with their kind permission that data on several types of source are reproduced in this chapter[1]. Designs vary considerably, depending upon the nature of the radiation of interest.

12.2.2 Alpha-particle sources

As described in Chapter 2, the alpha-particle is a very weakly penetrating radiation. Even very energetic alphas of several MeV are absorbed by a centimetre or so of air as a result of the intense interaction of the doubly-charged particles with the molecular electrons.

The design of an alpha-emitting source is, therefore, especially difficult. The source must emit an appropriate intensity of alpha-particles and at the same time the design must ensure that the radioisotope remains fully encapsulated. This means that the encapsulation of the radioactive material must be extremely thin, or total absorption of the alphas would occur. Such encapsulation can be achieved by several means. One of the more effective source designs is shown in Figure 12.1.

The alpha-emitter is rendered as immobile as possible by applying it as a stable compound dispersed in a gold or silver matrix sintered at high temperature. This is supported on a backing of silver with a front cover of gold alloy, 0.003 mm thick. This type of foil source is particularly useful: the foils, which typically are manufactured in lengths up to 1 m, can be readily subdivided for individual applications since the exposed edge of active material after cutting is of such small area that there is minimal risk of contamination.

Several alpha-emitters are available in this form. Two of the more useful are listed in Table 12.1.

12.2.3 Beta-particle sources

As we saw in Chapter 2, beta-emission is the most common of all forms of radioactive decay. A wide variety of beta-particle emitters is available. In

Table 12.1 Useful alpha-particle sources

Isotope	Half-life	Alpha-energy (MeV)
Americium-241	458 y	5.5
Radium-226	1620 y	4.5–7.7

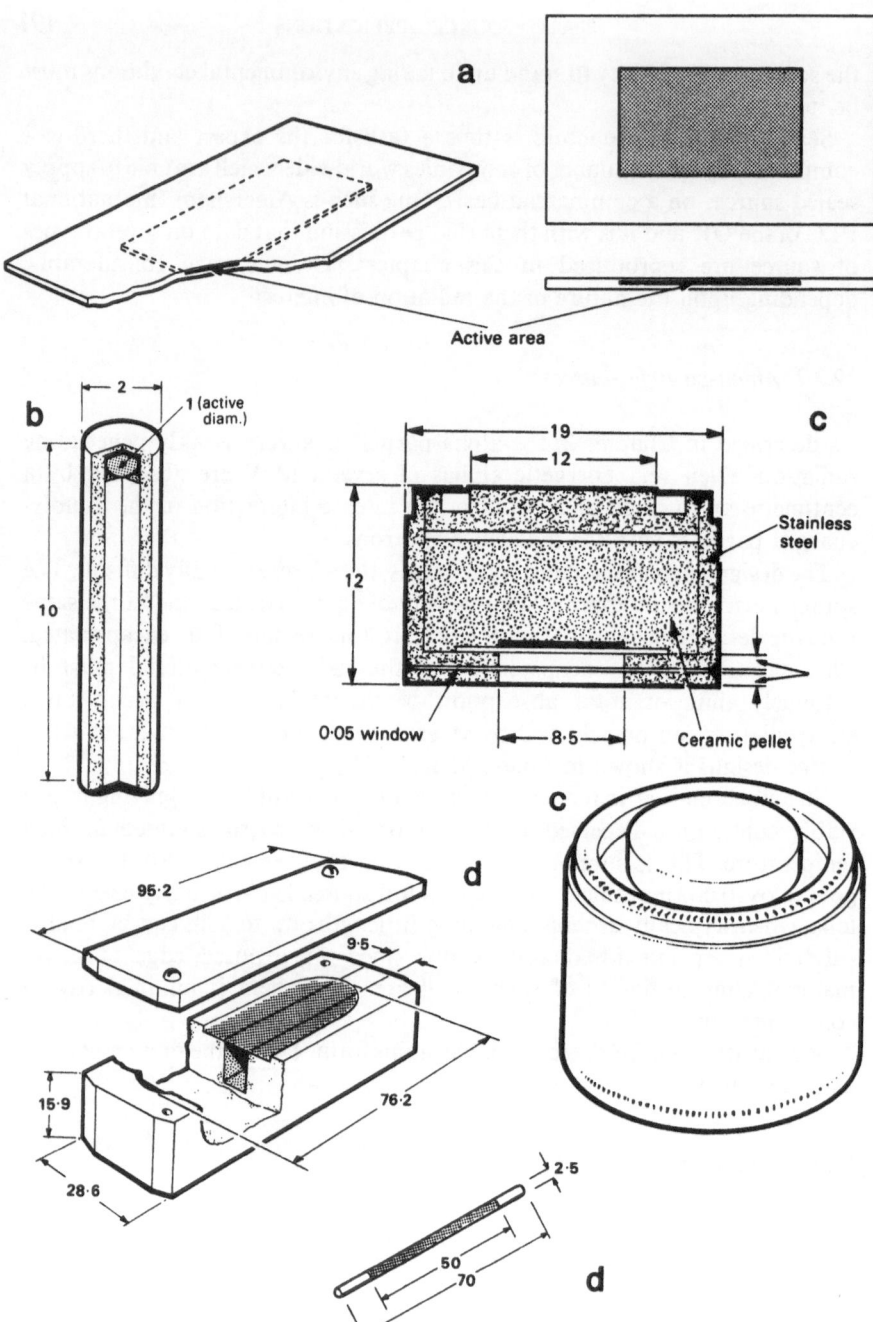

Figure 12.2. Various types of beta-particle source. (*a*) Extended area sources (promethium-147 and thallium-204); (*b*) point sources (strontium-90); (*c*) construction of a typical strontium-90 disc source; (*d*) krypton-85 sources for industrial use. By kind permission of Amersham International PLC.

Table 12.2 Some useful beta-particle sources

Radioisotope	β-energy (max. MeV)	Half-life
Carbon-14	0.159	5730 y
Krypton-85	0.67	10.6 y
Nickel-63	0.067	92 y
Promethium-147	0.23	2.6 y
Ruthenium/rhodium-106	3.6	1.0 y
Strontium/yttrium-90	2.27	28 y
Thallium-204	0.77	3.76 y

addition, the betas from different sources are emitted with widely different energies ranging from a few keV (cf. tritium with a maximum beta-energy of 0.018 MeV) up to several MeV (ruthenium-106—rhodium-106 has a maximum beta-energy of 3.6 MeV).

There is a correspondingly wide range of construction of sources available. As in the case of alpha-particles, foil sources are extensively used. The radioisotope as a stable compound is mixed with metallic silver, compressed, sintered and welded into a silver containment which can then be extended by rolling to form strips or discs. The foil may be used as an extended area source (Figure 12.2a), in which case it is usually desirable to give further corrosion resistance by adding a rolled-on gold protective layer, or small discs of foil may be incorporated into stainless steel containers to give, effectively, point-sources of radiation (Figure 12.2b).

For extreme resistance to corrosion and temperature ceramic or glass-bead sources offer the best choice. Strontium-90 sources, for example, are made by firing strontium-90 titanate on to the surface of a ceramic base of non-radioactive strontium titanate and protecting the active layer with an additional coating of glaze. This method is suitable for a source of extended area (Figure 12.2c). Strontium-90 glass in stainless steel encapsulation is more appropriate for a point source.

Sources in which the beta-emitter is incorporated as a gas are also useful for certain applications. This is particularly true of the radioisotope krypton-85 which has a beta-energy suitable for several industrial users. The gas may be hermetically sealed inside a thin-walled tube or inside a container fitted with a metal foil window (Figure 12.2d). For sources of high activity, the gas is absorbed on activated charcoal to avoid the necessity of filling at a pressure greater than atmospheric.

Useful beta-particle sources are listed in Table 12.2.

12.2.4 *Gamma-ray sources*

Gamma-ray and other photon sources are the sources most commonly used for industrial applications. There are very many different types and designs of source available but it is possible to draw a broad dividing line between 'low-energy' photon sources (by which we mean that the photon emissions are lower than about 200 keV) and 'high-energy' sources of energy 0.5 MeV and greater. By an unfortunate quirk of nature there are few commercially viable sources with emissions in the energy region between these two groups.

Some commonly used low-energy photon sources are listed in Table 12.3. The construction of these sources varies considerably and depends inevitably upon the application for which the source is intended.

The disc design (Figure 12.3a) is extremely versatile. The isotope can be present in several forms: as an inert compound incorporated into a rolled aluminium foil; as an electrodeposit on some appropriate backing, or incorporated in an ion-exchange resin sealed within a metal capsule. In all cases the outer containment of the source (usually stainless steel) is provided with a thin exit window for the photons. For very low-energy photons, the window is constructed from low-density, low-atomic number elements such as beryllium or titanium.

Table 12.3 Low-energy photon sources in common use

Isotope	Principal photon energy (keV)	Half-life	Remarks
Americium-241	60	458 y	The long half-life makes this type of source very useful for continuous gauging applications.
Cadmium-109	22	453 days	A versatile source useful for counter calibration and for X-ray fluorescence analysis
Cobalt-57	122, 136	270 days	Useful for Mössbauer spectroscopy studies
Iron-55	5.9	2.7 y	Very useful for low-energy X-ray fluorescence analysis
Lead-210	47	22 y	X-ray fluorescence applications. Also a useful energy standard
Plutonium-238	12–17	86 y	X-ray fluorescence applications
Thallium-170	84	127 days	Specially useful for low-energy radiographic purposes
Tritium/ zirconium	5–9	12.26 y	Bremsstrahlung sources. The energy spectra can be modified by admixture of the radioisotope with different elements
Strontium-90/ aluminium	60–150	28 y	
Promethium-147/ aluminium	12–45	2.6 y	

Point sources are desirable for certain applications where spatial definition is important. Here the specific activity of the radioactive isotope is clearly important since it is this which limits the photon output which can be obtained from a source of given diameter. The isotope as a pellet or bead is contained within a stainless-steel capsule.

There is a further type of low-energy photon source in common usage—the so-called bremsstrahlung source. This makes use of the phenomenon, described in Chapter 2, whereby the slowing down of beta-particles in matter gives rise to the emission of electromagnetic radiation. Accordingly, bremsstrahlung sources may be constructed by mixing a beta-emitter with an appropriate secondary material in which the slowing-down process can occur. By varying the composition of the secondary material the energy of the emitted bremsstrahlung spectrum may likewise be altered[2].

A typical bremsstrahlung source is illustrated in Figure 12.3b. In this case, the radioisotope promethium-147, as the oxide, is sintered with aluminium, rolled into a foil and mounted in a metal holder. The mica window ensures high output intensity of the relatively low-energy photons.

There is yet a further type of low-energy photon source which is extremely useful in applications where the photon energy is critical (X-ray fluorescence analysis is probably the best example). It is possible, within limits, to select the photon energy by employing a so-called 'source-target assembly' (Figure 12.3c). Here, the primary photon emitter is a long-lived isotope—such as the 458-year americium-241. This is constructed in an annular form such that the 60 keV gamma-rays are directed at a target element which is selected for the particular application. The americium gammas excite fluorescent X-rays from the target material, and it is these photons which are emitted from the source-target assembly with negligible interference from the primary gammas. This device makes possible long half-life sources of high spectral purity—particularly useful for instrumental applications.

The construction of high-energy gamma-ray sources, though no less rigorously controlled is free from many of the difficulties associated with those sources previously described because of the highly penetrating nature of the radiation. Self-absorption in the source material itself is rarely a significant problem and the encapsulation of the source can also be made very robust. A brief look at a chart of the known radionuclides reveals a very large number of isotopes emitting high-energy gamma rays. However, a comparatively small number are suitable for use as sealed sources. The chief constraints are *half-life*, which must be adequately long; *availability*, which in turn influences cost; and *spectral purity*—it is not generally desirable to have a spectrum complicated by interfering gamma-ray peaks of significantly different energy. The most widely-used high-energy sources are listed in Table 12.4.

Sources for industrial use (with the exception of the very large sources used for radiation processing) are usually of the 'point' source variety and comprise beads or pellets of the radioisotope, in an appropriately inert form,

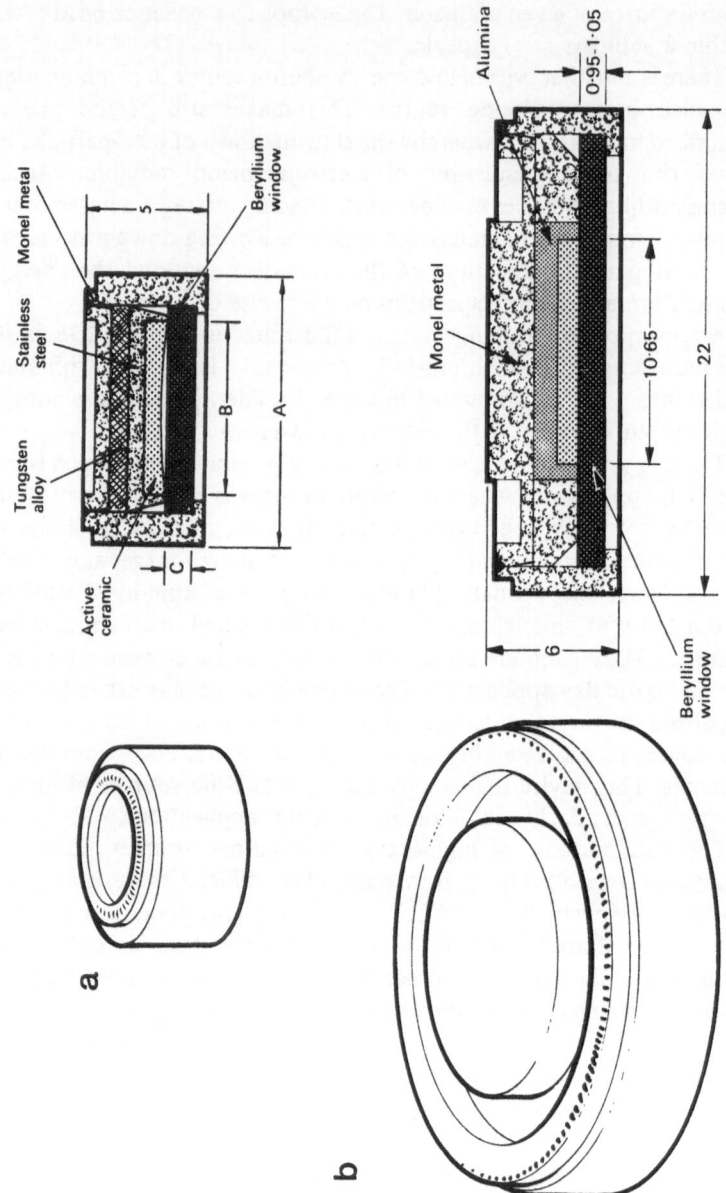

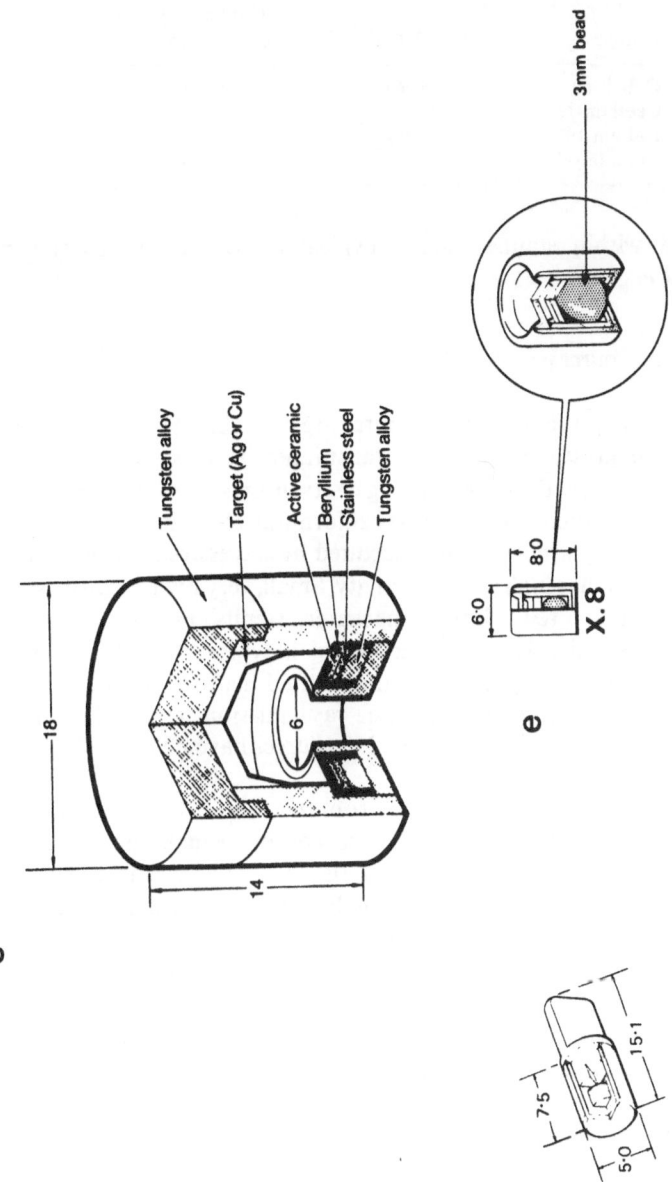

Figure 12.3. Various types of gamma-ray source. (a) Low-energy gamma disc source; (b) promethium-147 bremsstrahlung source; (c) annular source—target assembly; (d) cobalt-60; (e) caesium-137. By kind permission of Amersham International PLC.

Table 12.4 High-energy gamma-ray sources in common use

Isotope	Half-life	Gamma-ray energy (MeV)
Cobalt-60	5.26 y	1.7, 1.33
Caesium-137	30 y	0.66
Iridium-192	74 days	0.3–0.6
Antimony-124	60 days	0.6–2.1

encapsulated within stainless steel. Typical constructions are shown in Figure 12.3d, e.

12.2.5 Neutron sources

Basically, two categories of neutron source are available. The first comprises a relatively small group of nuclides: transuranic isotopes, which decay by spontaneous fission, with accompanying neutron emission. Of these, that most commonly encountered is the isotope californium-252. This has a half-life of 2.65 years and, being artificially manufactured by successive neutron capture in a high-flux reactor, has high specific activity. Small, very high-intensity sources can thus be manufactured — 1 mg of the material emits approximately 5×10^9 neutrons per second[3]. The construction of a typical source is shown in Figure 12.4. The energy spectrum of the emitted neutrons is relatively 'soft' and, in addition, the associated gamma-ray emission is low. Thus from a radiological point of view, californium-252 sources have much to recommend them.

There are, however, disadvantages. From the routine user's point of view, perhaps the most serious problem is the high cost of the material. A less serious objection for most practical purposes is the fact that there is a measure of uncertainty associated with the half-life of the sources. This arises because, as a result of the complex production process, several different spontaneous-fission isotopes are usually present. This can present problems if the sources are to be used in critical long-term gauging applications.

In the second category of source, neutrons are emitted via a two-stage process. The sources comprise a light element, such as beryllium, intimately mixed with a radioisotope which emits a gamma-ray or alpha-particle. The interaction of the primary radiation with the beryllium causes neutrons to be emitted. The so-called 'gamma-n' sources have much to recommend them for research purposes, producing neutrons with a very small energy spread. However, this is not usually an important consideration for the average user. Furthermore, there is a positive disadvantage inherent in this type of source: that of the high-level gamma-radiation.

For this reason alpha-n sources are usually preferred for industrial applications. Arguably the most useful of these is the so-called americium/

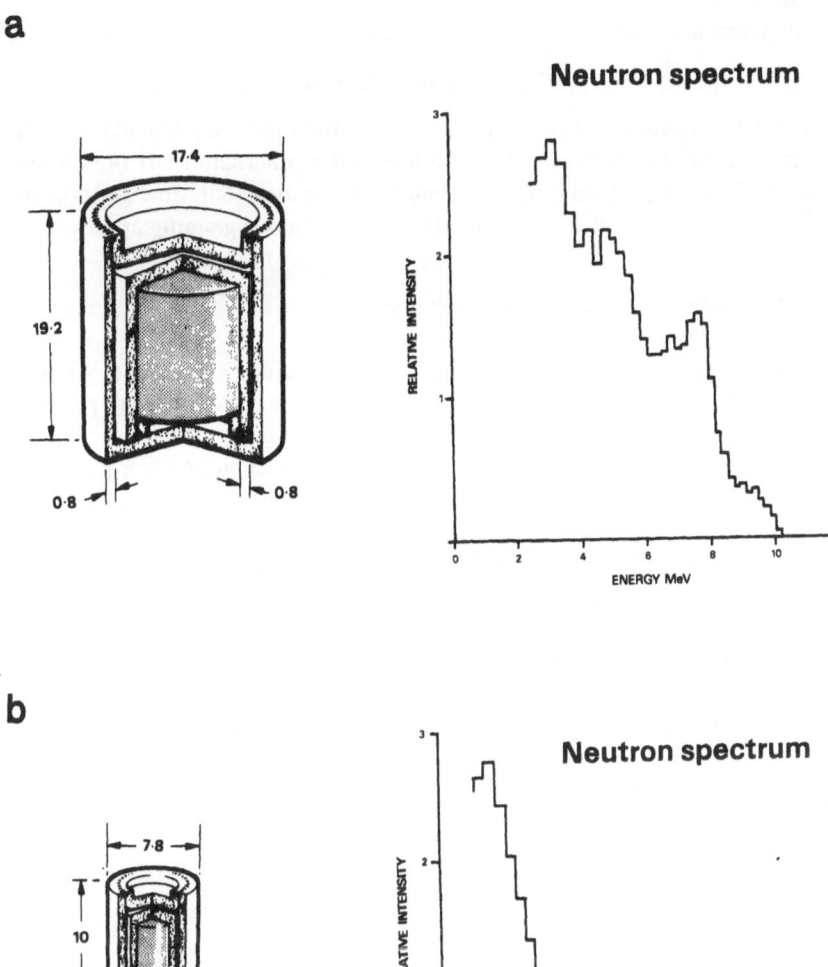

Figure 12.4. Commercially available radioisotope neutron sources. (a) Americium-241/beryllium neutron sources. [241]Am/Be source made and measured at The Radiochemical Centre using a stilbene crystal and pulse shape discrimination. Neutron spectrum after Lorch, E. A. (1973) *Int. J. Appl. Radiat. Isotopes* **24**, 588–589. (b) Californium-252: source made and measured at The Radiochemical Centre using a stilbene crystal and pulse shape discrimination. Neutron spectrum after Lorch, E. A. (1973) *Int. J. Appl. Radiat. Isotopes* **24**, 590.

beryllium neutron source. A typical source construction is shown in Figure 12.4.

Neutrons are emitted by the nuclear process

$$^9_4\text{Be} + ^4_2\alpha \rightarrow ^{12}_6\text{C} + \text{n}.$$

The level of gamma-ray emission is low and the source is particularly useful because of the long half-life of the alpha-emitter americium-241 (458 years). This renders the neutron output extremely stable over long time periods and such sources are, therefore, ideally suited to continuous gauging applications.

12.3 Selection of a measurement technique

As we have seen, a wide variety of radioisotope sources is available. We have also seen (Chapter 3) that several radiation detection techniques are possible. Given this variety it is worth taking some time to discuss how one sets about the task of selecting an appropriate measurement technique for a particular application. This choice is, of course, clearly dependent upon the nature of the problem which is to be tackled.

In some cases, the nature of the problem to be studied is such that the experimental method is uniquely defined: in others a number of different and, on first sight, equally valid, approaches may be possible. Careful evaluation of the competitive methods available is then needed.

As a starting-point, the radioisotope practitioner should give consideration to the physical phenomena which can be harnessed for use, such as:

(a) Alpha-particle absorption
(b) Beta-particle absorption
(c) Beta-particle scattering
(d) X-ray fluorescence
(e) Low-energy gamma/X-ray absorption
(f) Gamma-ray absorption
(g) Gamma-ray scattering
(h) Neutron absorption
(i) Neutron moderation
(j) Neutron activation.

(This list is not fully comprehensive. However, it is estimated that the phenomena listed form the basis of well over 90% of industrial applications.)

Phenomena (a)–(e) are most commonly used for analytical and gauging applications. Because the radiations involved are of low energy, techniques based upon these phenomena cannot be directly applied on operating plant without some control being exercised over the system to be studied. For example, if X-ray fluorescence is to be used to analyse the components of a process stream, it is usually necessary to divert a portion of that stream through a sample cell, specially constructed with a thin 'window' capable of transmitting the low-energy photons (Chapter 14). Similarly, if beta-scattering is used to measure the thickness of extruded plastic the geometry of the source-sample-detector configuration must be carefully controlled (Chapter 14).

In contrast, techniques based upon the last five phenomena listed do not in general require any modification to the plant. Neutrons and high-energy gamma-rays are easily capable of penetrating the (usually) substantial thicknesses of process vessels, and this facilitates direct, on-site measurements.

This latter category of techniques is therefore of greatest value in providing rapid and convenient answers to urgent plant problems and so we shall concentrate attention in this area.

For the sake of example, let us suppose that a production unit, because of the failure of an installed level gauge, requires urgent information about the level inside a process vessel. Let us further suppose that the vessel is of 8-m diameter, with walls of thickness 25 mm. The vessel operates at elevated temperature and is covered with thermal insulation 150 mm in thickness.

Three level-measuring techniques commonly used are gamma-ray absorption, neutron backscattering and gamma-ray backscattering. The experienced practitioner would immediately rule out gamma-ray backscattering. He is aware that this technique is relatively insensitive for thick-walled vessels. He is also aware that the technique is geometry-dependent and that irregularities in the thermal insulation could completely invalidate the measurement. Neutron backscatter, too, is an unlikely candidate, the problem in this case being the thickness of the thermal insulation which limits the closeness of approach of the measuring gauge to the liquid in the vessel. In contrast, the gamma-ray absorption technique is not subject to any of the above limitations and in this case represents the best practical means of carrying out the measurement. The basic equipment configuration for gamma-ray absorption measurements is discussed in detail in Chapter 13.

Having established the technique, thought must then be given to the source–detector combination best suited to perform the measurement. Here, the practitioner will be guided by the general principle (stemming from radiological considerations) that the source size should be kept as small as possible, consistent with obtaining the required information on a realistic time-scale. Because he requires only a spot-measurement of level, long-term stability of the detector is not an important consideration. It is likely, therefore, that a scintillation counter will be selected as the detector because of its high efficiency (Chapter 3).

In order for radiation to be transmitted through the thick walls of the vessel, a high-energy gamma-ray source must be selected. It is likely that the choice will be between caesium-137 and cobalt-60 (Table 12.4) because of the ready availability of these sources, and of the two types, cobalt-60 would probably be preferred in this instance because of the greater penetrating power of its 1.3-MeV gamma-rays.

How large a source should be used?

Let us suppose that the scintillation detector comprises a 50 mm × 50 mm crystal of sodium iodide. In a radiation field of 10 microsieverts per hour, such a detector will produce an output of approximately 10^4 counts per second. In fact, assuming a 10-second counting period, a countrate of 5000 counts per second would be perfectly adequate to ensure a result of adequate accuracy (see Appendix). The size of the source should, therefore, be selected so that the radiation field at the detector is approximately 5 μSv per h. We have seen (Chapter 6) that a source of cobalt-60 of strength 37 GBq produces a radiation dose of 13 mSv per h at a distance of one metre. Using the inverse square law relationship, such a source would give a dose at 8 m of $13/64$ mSv h^{-1} $= 200 \, \mu$Sv h^{-1}.

We must also take account of the effect of the vessel wall thickness. The total thickness of steel which the radiation must traverse is 50 mm—which is approximately 2 half-thicknesses for cobalt-60 (Chapter 2). The walls of the vessel will, therefore, further reduce the intensity at the detector by a factor of 4. The dose at the detector from a 37-GBq source will, therefore, be approximately 50 μSv. The source strength we require is, therefore, 3.7 GBq. Armed with this information, the practitioner may now select the basic equipment required to carry out the measurement.

The selection of technique and the choice of the basic measurement equipment was, in the example chosen, fairly straightforward. However, the thought-processes described are typical of those which precede every sealed-source measurement: first, decide upon a technique; then select a suitable source–detector combination—bearing in mind at all times the need to restrict the radiation exposure to an absolute minimum.

The remainder of this book is devoted to plant applications of sealed-source techniques. Though some beta-particle and X-ray applications will be described, the emphasis will be heavily upon neutron and gamma-ray methods since, as we have described, these are more appropriate for direct application to the study of problems on full-scale plant. For detailed descriptions of alpha-particle, beta-particle and X-ray applications, the reader is referred elsewhere[4-8].

References

1. *The Radiochemical Manual.* The Radiochemical Centre, Amersham (1966).
2. Reiffel, L. (1955) *Nucleonics* **13**, 22.
3. Bardell, A. G., Technological applications of neutron sources. *NPLNEWS* **362**, Winter 1984–1985.
4. *Industrial Measurement and Control by Radiation Techniques.* IEE Conference Publication No. 84, The Institution of Electrical Engineers, London (1972).
5. Lorch, E.A. (1979) Industrial and analytical applications of radioisotope radiation sources. *J. Radioanal. Chem.* **48**, 209.

6. Putnam, J.L. (1969) Radioisotope techniques in process control and analysis. In *Nondestructive Testing*, Egerton, H. B. (ed), Oxford University Press.
7. *Radioisotope Instruments in Industry and Geophysics.* IAEA, Vienna (1966).
8. Cameron, J. F. and Clayton, C. G. (1971) *Radioisotope Instruments.* International Series of Monographs on Nuclear Energy, Pergamon, Oxford.

13 Gamma-ray absorption techniques

J. S. CHARLTON

13.1 Introduction

Techniques based upon the absorption of gamma radiation are used very widely in plant and process investigation. Even setting to one side industrial radiography, which is an adjunct to most construction and maintenance operations, this category of technique constitutes a very significant proportion of all investigations involving sealed radioisotope sources (see for example Chapter 1, Table 1.1).

Before going on to describe the more important applications of gamma-ray absorption, it is worthwhile taking a little time to review some of the key aspects of the phenomenon.

The interaction of gamma-radiation with matter was discussed at some length in Chapter 2. It was demonstrated that, quantitatively, the attenuation of a beam of gamma-rays of intensity I_0 by an absorber of thickness x and density ρ is described by the relationship

$$I = I_0 B e^{-\mu \rho x} \tag{13.1}$$

where μ is a constant known as the mass absorption coefficient, and B is the 'build-up factor'. For many purposes it is a reasonable approximation to re-write (13.1) as:

$$I = I_0 e^{-\mu_{\text{eff}} \rho x} \tag{13.2}$$

where μ_{eff} is an effective mass absorption coefficient determined empirically.

There are three main processes responsible for attenuating the beam: the photoelectric effect, Compton scattering and pair production. The photoelectric effect is important only for gamma-rays of low energy, while pair production does not occur at all unless the energy of the gamma-ray exceeds 1.02 MeV. In the important energy range of 0.5 to 2.5 MeV, Compton scattering is the process chiefly responsible for the attenuation. In the majority of applications on industrial plant we are concerned with gamma-rays in this latter category, since radiations of lower energy are in general incapable of penetrating the (usually) substantial thickness of material from which plants are constructed.

For a given gamma-ray energy and for an absorber of specified chemical composition, μ_{eff} is a constant. (Indeed, for high-energy gamma-rays where

Compton scattering predominates it is also approximately independent of the elemental composition of the absorber.) Equation (13.2) then indicates that the attenuation of the gamma-ray beam is an exponential function of the mass per unit area of the absorber. This forms the basis of the gamma-ray absorption techniques used for plant and process investigation. The basic technique involves positioning a gamma-ray source and a radiation detector on opposite sides of the medium of interest and relating changes in the transmitted intensity to changes in the mass per unit area of the material. There are, broadly speaking, three main classes of application.

(a) Material of constant thickness. The attenuation of the beam provides information about the density of the material.
(b) Material of constant density. The attenuation of the beam provides information about the thickness of the material.
(c) Applications in which we are interested in the mass per unit area, or more specifically changes in the mass per unit area *per se*. In this category we are not usually interested in absolute values of mass per unit area but in changes and differences which can provide insights into changes in the status of some plant or process parameter. Applications of this kind are among the most valuable of all uses of radioisotopes in plant investigation.

13.2 Equipment for plant applications

Although the details of experimental arrangement and apparatus vary considerably from one application to another there are several items of equipment which are common to all investigations. Such general equipment requirements will be described here: more specialized items for specific applications will be described in the case histories (section 13.6).

13.2.1 *Radioactive sources*

A large number of radioisotopes are commercially available in the form of sealed sources[1]. However, it has been found that the vast majority of gamma-ray absorption applications can be carried out using one or other of the isotopes ^{60}Co, ^{137}Cs, ^{241}Am and ^{192}Ir. The relevant properties of these isotopes have been described in Chapter 4. Wherever possible, the source used should be in the 'special form' category. In use, the source is housed in a shielded source holder—usually of lead or heavy alloy construction with a 'window' defined by appropriate collimation.

13.2.2 *Radiation detector*

For short-term investigation the NaI scintillation counter has been found to be most useful, combining ruggedness and moderate cost with very high detection efficiency. For certain longer-term applications where reproducibility of data is important, Geiger–Müller detectors may be used. This is particularly true of gauges installed permanently or semi-permanently to continuously monitor a parameter such as density of the process material or the level of material inside a vessel. Detector shielding and collimation is useful for certain applications.

13.2.3 *Associated electronics*

For plant investigations with sealed radioisotope sources it is advantageous to make use of portable electronic equipment. Several such systems are commercially available. Ideally such an instrument should possess the following features:

(a) Small dimensions and low weight. The importance of being able to take the equipment into areas of plant with difficult access cannot be overemphasized.

(b) A detector power unit and high voltage supply compatible with scintillation counters, Geiger–Müller tubes and proportional counters. (The options are usually available via a selector switch). Similarly, the front end of the detection electronics must be capable of handling signals from the different types of detector.

(c) The ability to function either in the 'ratemeter' or the 'scaler' mode.

(d) Electronics stable over the normal range of ambient temperatures.

(e) Good weatherproofing. In particular, water ingress, even under conditions of heavy rainfall, should be zero.

(f) Simplicity of operation. The equipment will often need to be used in very adverse environmental conditions and, for example, the device should be possible to operate with a gloved hand.

(g) Energy discrimination. It is often useful to have the facility to set a counting threshold (perhaps to cut out electronic noise) or to count within an energy 'window'.

(h) Output for a chart recorder. Occasionally it is useful to record the ratemeter output continuously. As portable data loggers become more commonplace, it will become increasingly useful to incorporate a digital output to record scaler output directly.

Such devices are not, in general, intrinsically safe from an electrical point of view—though it goes without saying that they are most unlikely to spark under operating conditions. Nevertheless, if the equipment is to be operated in an area with a potentially flammable or explosive atmosphere, air-sampling devices should be used to ensure that conditions are safe for the duration of the investigation.

In contrast, the detection systems used in association with installed gauges *can* be made intrinsically safe and this should be an important factor in selecting equipment for use in areas with fire potential.

13.3 Thickness measurement

Measurement of thickness may be required for several purposes: to assess the depth of deposits on the inner walls of pipes and process vessels; for the quality control of metal sheet; to check for voids in concrete and similar materials; to measure metal thinning by corrosion or erosion.

In many cases, repeat measurements may be required and, as for example in corrosion measurement, these may extend over a considerable period of time. Clearly accuracy and reproducibility are both important.

The effects of the build-up factor (equation 13.1) can be virtually designed out, and the experimental arrangement set up in such a way that B approaches unity. This can be achieved either by collimating the detector (thereby approximating to narrow-beam geometry) or by electronically discriminating against all but the non-scattered component of the beam. In practice a combination of both methods would probably be used. However, this does not remove the need to calibrate the equipment: it is dangerous to rely upon tabulated values of mass absorption coefficients.

By differentiating equation (13.2) with respect to thickness (x) we can derive the expression

$$\frac{\mathrm{d}I}{I} = -\mu_{\text{eff}}\rho\mathrm{d}x \qquad (13.3)$$

This shows that the fractional changes in transmitted intensity (or recorded count rate) are directly proportional to the *absolute* changes in thickness. The expression also clearly demonstrates that the larger the mass absorption coefficient and the higher the density, the more sensitive the measurement (i.e. a smaller change in thickness can be measured for a given percentage change in count rate).

The density of the medium is, of course, beyond our control but by judicious choice of radioactive source (and, therefore, gamma-ray energy) we can, within limits, ensure that the mass absorption coefficient is of an appropriate magnitude. In practice the choice of source is usually a compromise between the desire for high sensitivity and the need to ensure that the transmitted beam is sufficiently intense to avoid unduly long data acquisition periods. It can, in fact, be readily shown[2] that for a fixed gamma-ray intensity, a good compromise between high sensitivity and small statistical variations in count rate is achieved when:

$$\mu_{\text{eff}} = \frac{2}{\rho x} \qquad (13.4)$$

Table 13.1 Gamma-ray sources for steel thickness measurement

Isotope	Gamma-ray energy	Useful range for iron (mm)
Cobalt-60	1.17, 1.33	Up to 150
Caesium-137	0.66	Up to 100
Americium-241	0.06	Up to 10

13.3.1 *Installed thickness gauges*

Though somewhat outside the scope of this book it is worth briefly noting that instruments based upon the gamma-ray absorption technique have for many years found useful application—chiefly in metallurgical industries. Gauges are used to detect cavities in rolled steel ingots[3], to measure the thickness of hot rolled strip and plate steel[4] and to measure the wall thickness of steel tubes[5]. Useful sources for steel thickness gauging are listed in Table 13.1. The quoted ranges should be regarded as guidelines only.

13.3.2 *Thickness of deposits in pipes and vessels*

The deposition of solids on the walls of pipes and vessels and the build-up of liquid 'pools' in low-lying sections of plant are frequently sources of major plant problems. Some of the difficulties which can be encountered are listed below.

(a) Areas of deposit can act as nucleation centres for blockages in pipelines. Even a partial blockage can seriously impair plant efficiency while a complete blockage may lead to a crash shutdown.

(b) Deposit build-up affects the pressure-drop through vessels (tubular coolers for example) and can both limit throughput and lead to unnecessarily high loads being placed upon compressors, air blowers and other machines[6].

(c) Even a very thin layer of deposit can seriously affect the heat-transfer characteristics of the surface upon which it is laid down. This is particularly worrying in vessels designed specifically for the purpose of heat exchange[7].

(d) A problem particularly relevant to gas transmission lines occurs when liquid condenses out from the stream and 'pools' in the pipe. Here is a danger that at some point a large 'slug' of liquid will be carried forward with the process gas—perhaps causing major damage to downstream apparatus such as rotary compressors.

(e) In certain cases, particularly in tubular reactors, the build-up of deposits can seriously affect the quality and yield of the product since, in diminishing the free volume of the reactor, the residence time of the

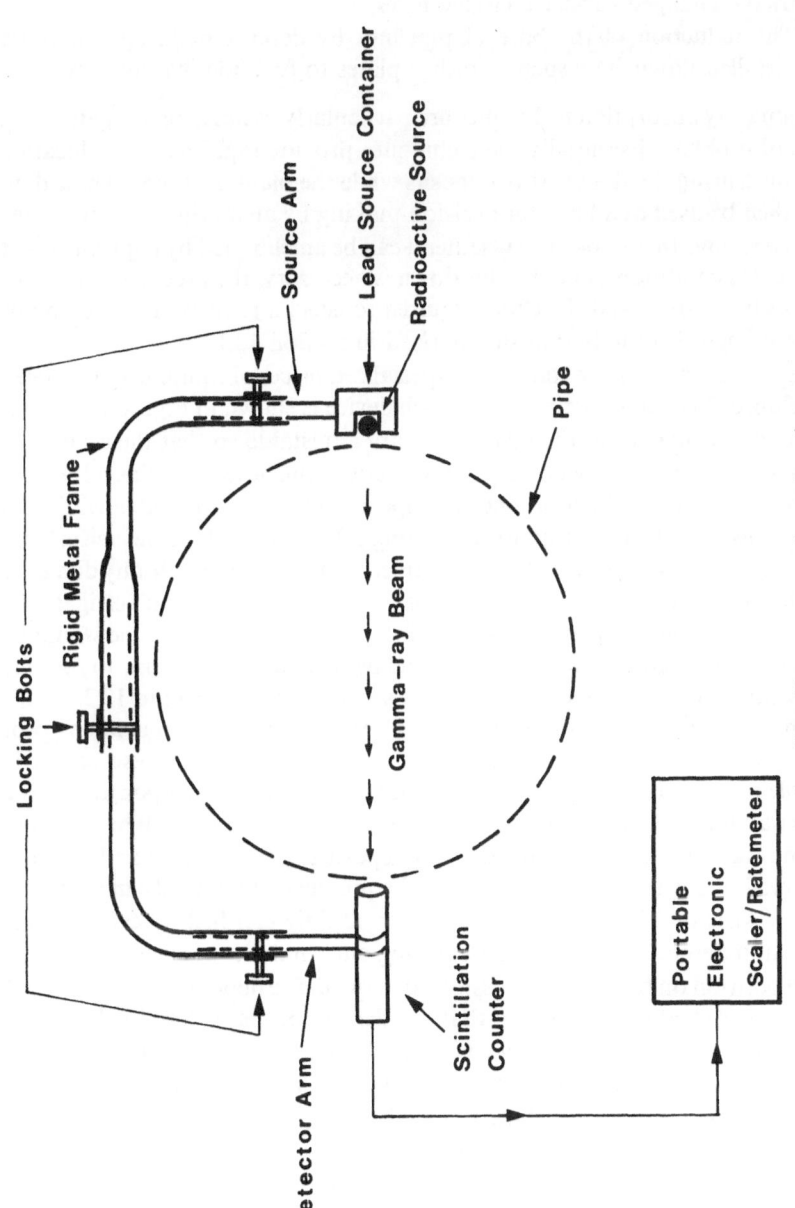

Figure 13.1. An adjustable gamma-ray absorption pipe-scanner.

reactants is reduced below that which is required for the process to function as per design. There may also be adverse catalytic effects caused by the changed surface characteristics.

(f) The reduction of the bore of pipelines by deposit build-up can cause installed flowmeters such as orifice plates to function inaccurately.

Gamma-ray absorption techniques are particularly useful in investigating this type of problem. Essentially the techniques provide rapid means of locating and measuring the thickness of deposits while the plant is on-line. These data can then be used as a basis for decision-making by answering such questions as how serious the problem is; whether it can be ameliorated by adjusting plant or process conditions, and if a shutdown *is* necessary, the precise locations of the problem areas and the effort required to effect a remedy. The method of measurement is exactly that summarized in section 13.1.

For the examination of pipework, specialized pieces of equipment have been developed. The construction of one such device is shown in Figure 13.1. Both span and armlength of the metal yoke are adjustable so that the scanner is suitable for use over a wide range of pipe sizes. The device is adjusted so that source and detector are positioned at opposite sides of a pipe diameter. Figure 13.1 shows the device set up to scan along a horizontal diameter: clearly by rotating the yoke and by making adjustments to the arm-length, any diameter or chord can be examined. For example, if it was required to investigate the depth of material lying in the bottom of the pipe a vertical scan line would be employed, though horizontal scans would be carried out from time to time for calibration purposes. The instrument is shown in use in Figure 13.2.

The presence of deposits inside the pipe results in a reduction in intensity of the gamma-rays reaching the detector. However, a single measurement is unable to give unambiguous information about the precise disposition of the material: it may be evenly distributed around the wall; it may all be adhering to one wall; it is even possible that no deposit is present at all and that the observed intensity change results from the presence of material carried in the process stream. The situation is, however, rapidly clarified by carrying out scans at different orientations. By carrying out a number of such scans at the same position on the pipe it is possible to map out the depth profile of material in the pipe[8,9], always assuming that the density of the material is known.

For many purposes, however, such detail is unnecessary and it is sufficient simply to record that the measurements indicate the presence of 'x centimetres of density one' in the scan line. For high-energy gamma rays where μ_{eff} is approximately constant this statement could properly be replaced by 'x/ρ centimetres of density ρ ($g \, cm^{-3}$)' in the beam.

For vessels other than pipes it is not usual to employ specialized equipment: the source and detector, suitably shielded and collimated, are positioned independently at points marked on the vessel at the commencement of the

Figure 13.2. Inspection of pipework for deposits using a gamma-ray absorption technique. Photograph by courtesy of ICI PLC.

measurements so that the separation of source and detector is accurately known.

Several case histories of actual applications are given later in the chapter.

13.3.3 Detection and measurement of corrosion

Corrosion is one of the most serious and most general problems on industrial plant. It has been estimated that in a single year corrosion costs between one and three per cent of the Gross National Product of industrialized countries[10]. This enormous figure does not take into account the very large sums of money

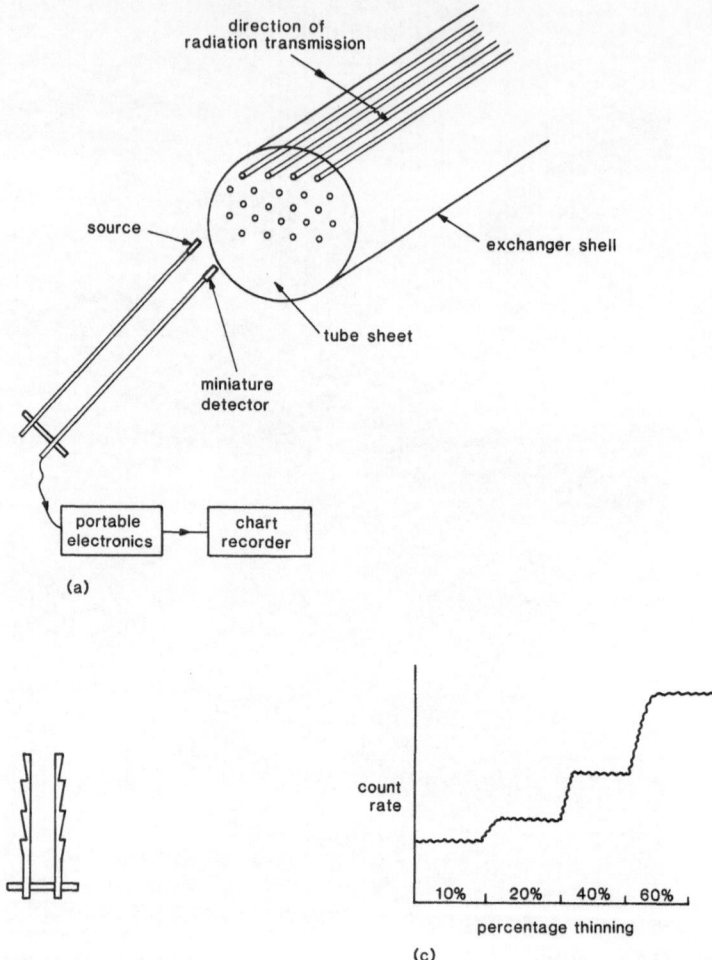

Figure 13.3. Gamma-ray absorption technique for heat exchanger tube inspection. (a)
Experimental arrangement; (b) machined calibration piece; (c) typical calibration curve.

associated with lost production occasioned by breakdown of plant and
unscheduled shutdowns; it simply reflects the replacement cost of plant
rendered inoperable as a result of corrosion. It is obviously essential, both
from economic and from safety considerations, to have appropriate methods
of measuring corrosion both on a continuous and a spot-check basis. Many
such methods exist. Activation methods have already been discussed in
Chapter 11; mention may also be made of the use of ultrasonics, of eddy
currents, corrosion coupons, visual inspection techniques and resistivity

measurements. The subject is extensively treated elsewhere[10]. No single method is universally applicable and the above techniques should be regarded as complementing each other. In this context, techniques based upon gamma-ray absorption have an important part to play.

Gamma-radiography provides one very useful method which is particularly appropriate for 'difficult', irregular regions such as the areas of pipe around welds. Another method involves the use of a 'gamma-ray pipe wall caliper'[11]. This clamps on to the pipe under investigation and a very thin radiation beam traverses the wall and is detected by a NaI scintillator.

A further application is the measurement of wall-thinning in the tube-bundles of heat exchangers[12]. A gamma-ray source (which is usually americium-241 but may occasionally be caesium-137 for tubes with thick walls) and a miniature Geiger counter are inserted simultaneously down adjacent tubes in the bundle. Radiation transmitted through the tube walls is recorded. The transmitted signal is related to the local thinning of the tube wall following calibration using machined test pieces of specification identical to that of the tubes in the exchanger (Figure 13.3). Measurements may be performed on a spot-check basis to gain a very quick impression of the distribution and degree of corrosion throughout the bundle. A more comprehensive picture can be built up by systematically carrying out the measurement procedure for each pair of tubes. This technique is not restricted by the alloy composition of the tube metal, is rapid to apply and has adequate sensitivity (0.1 mm thinning is readily detectable on a typical tube). Heat exchanger scanning using this technique is now routinely incorporated into many plant shutdowns so that the progress of corrosion can be regularly monitored. Alternatively, it is used in emergency situations to rapidly obtain quantitative data about the condition of a tube bundle in which corrosion is suspected. Such information facilitates decision making as to whether a bundle needs to be retubed or whether certain tubes need to be plugged to avoid leakage.

An example of the use of the technique is given in the case histories (section 13.6)

Though gamma-ray absorption techniques have their place in corrosion monitoring, they are inherently limited by the fact that access to both sides of the object under inspection is required. This is not true of gamma-ray backscattering methods, which will be discussed in Chapter 14.

13.3.4 Detection of voids

The detection of voidage is essentially a form of thickness measurement. Gamma-ray absorption techniques have been used to detect voids between stressing-ducts in concrete structures[13]. A similar technique has also been applied to checking the refractory linings of pipes for use at high temperatures. The technique in each case is essentially the same: only the refractory testing will be described here. The source is positioned at the end of a graduated rod, and circular spacers ensure that it is positioned centrally within the pipe. The detector is moved over the outside surface of the pipe to receive the radiation from the source at different axial positions. The presence of voidage is indicated by increases in the transmitted intensity: calibration is effected using test-pieces with known voids incorporated.

A simple transmission geometry can be used to measure voidage in packed columns, provided that the diameter of the column is not such as to cause complete cut-off of the beam.

13.4 Density measurement

The use of gamma-ray absorption techniques for density measurement is widespread throughout industry. In addition to continuous density-gauging applications, the techniques are used extensively in process investigation.

The theory of the technique is very like that underlying thickness gauging: in this case, however, the *thickness* of the system under investigation is constant so that variations in transmission can be related directly to density changes. The equipment and methodology are also closely similar.

13.4.1 *Installed density gauges*

Gamma-ray absorption gauges have been in use for many years in practically all branches of industry. Many reviews of the subject have been compiled[14-17] and it is not proposed to duplicate them here. It is perhaps worth noting the fact that the most recent developments make use of microprocessor technology for data handling. By feeding the detector signal through an appropriate interface device to a microcomputer, it is possible to introduce a high degree of sophistication into the signal processing. For example, composition or temperature of the medium can be allowed for, and empirically-derived calibration curves can be incorporated in the software. There is little doubt that the increasing availability and versatility of microprocessor systems will even further increase the accuracy and range of applicability of installed density gauges.

13.4.2 *Vessel density profiles*

A knowledge of the density distribution of material inside a process vessel can be important to the industrial process operator. The gamma-ray absorption technique provides a means of measuring the density profile while the plant is on line and (usually) without elaborate preparations or modifications to the plant. Some of the more useful applications are described below.

Packing of catalyst beds[18]. This is often critically important in ensuring that a process operates at highest efficiency. It is useful to be able to check that the bed has been correctly packed; to monitor the expansion or contraction of the catalyst as a function of time on line (this provides indication of when re-charging or regeneration is required); and to investigate suspected malfunctions—slumping of the bed, failure of the support basket, excessive attrition of the pellets.

Steam/water mixtures. The quality ('wetness') of steam is very important in determining its heat transference potential. Measuring the density of steam in pipework can also provide a useful insight into the operation of steam generating equipment (steam drums, quench boilers and the like). A single absorption measurement across a pipe diameter will provide a measurement of the mean density in the beam path. Occasionally more detailed information is required and it is necessary to perform a series of measurements at different orientations, as described in 13.3.2. Very sophisticated applications of this technique have been reported[19].

Frothing and foaming. This can be a problem in that it is often a cause of liquid entrainment. The presence of a foam layer above the liquid in a reaction vessel or gas/liquid separator greatly increases the chance that liquid droplets will be carried forwards with the gas stream from the vessel[20], with potentially dire consequences for downstream machines. There are several ways in which gamma-ray absorption measurements can help. Thus, by measuring the 'gas' density above and below a demister pad, the performance of the pad can be checked. At plant start-up the techniques can be applied to rapidly monitor entrainment in gas streams as a function of throughput rate, so that optimum process conditions can be established. Bubbling and gas 'slugging' in liquids can also be quantitatively assessed[21, 22].

Density of packing beds in absorption columns. Loss, breakdown or maldistribution of packing rings can give rise to liquid distribution problems. Liquid channelling or stagnation are just two of the malfunctions which may result. These (and others) can be investigated using the gamma-ray absorption technique with standard source/detector geometry. We have found, however, that such investigations are usually best carried out in conjunction with liquid residence-time studies (Chapter 9).

Liquid–liquid interfaces in vessels and pipelines. Frequently the interface between two liquids is not clearcut, but extends, sometimes over several tens of centimetres, the interface region being characterized by a gradual transition of density from one liquid to the other. To measure and control this interface in plant vessels is important for successful process operation. In pipelines, the problem usually manifests itself when batch transfers are being carried out and one liquid is being used to sweep the other out of the line. It is clearly very important to know when the 'sweeping out' process has been completed so that the new liquid flowing through the line is not contaminated by the old. Gamma-ray absorption provides a fast-response method of registering both the time of arrival and the extent of the interface[16]. However, the gamma-ray absorption technique as applied to density measurement has an inherent limitation. If the total mass per unit area of the system under investigation is such that any more than about eight half-thicknesses are interposed between source and detector, the measurement is rendered very difficult, since, with a source of reasonable activity, insufficient radiation reaches the detector to give adequate counting statistics. (It is perhaps obvious, though worth mentioning, that the material in the beam does, of course, include the vessel walls and lagging or lining material as well as vessel contents.) There are, however, a number of ways in which this problem can be alleviated. Three approaches which have been found to be successful (in the appropriate circumstances) are illustrated in Figure 13.4.

(a) It is sometimes possible to position source and detector on opposite sides of a chord, rather than a diameter, of the vessel, thereby reducing the path length. There is a limit to how far this approach can be taken, however: if the chord is made too shallow, air scattering of the gamma-rays will become a problem by adding to the background radiation at the detector. In addition, the shorter the chord the greater is the relative absorption in the walls and lagging as opposed to the vessel contents. Nevertheless, this technique has been used on numerous occasions to provide useful data on otherwise intractable problems.

(b) The second approach makes use of a dip-pipe installed inside the vessel at a position such that there is an appropriate path length (say 2 or 3 half-thicknesses) between it and the vessel wall. A radioactive source is lowered down the dip-pipe on a graduated tape and a detector positioned outside of the vessel is simultaneously moved to record the transmitted radiation. In this way, the density profile of the vessel contents is measured. It must be noted that this approach assumes that the material between the dip-pipe and the wall of the vessel is representative of the vessel contents as a whole—an assumption which is not always justified. Also, of course, this type of measurement cannot be performed as an afterthought—it must be planned well in advance and an appropriate dip-pipe installed. The technique has however, produced useful results—especially in monitoring the changing density profiles of packed-bed reactors as a function of time on line.

(c) If there is access to the vessel contents—as is the case with an open-topped vessel or, perhaps one supplied with an inspection port—it is possible to lower a yoke comprising a source and detector held in a fixed geometry to selected parts of the vessel. The yoke must be of robust construction and

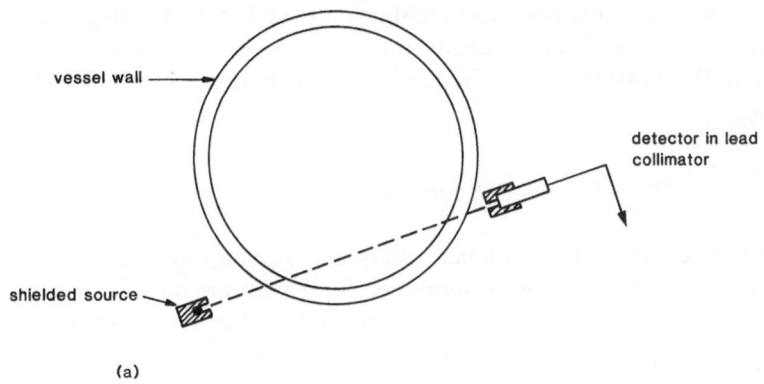

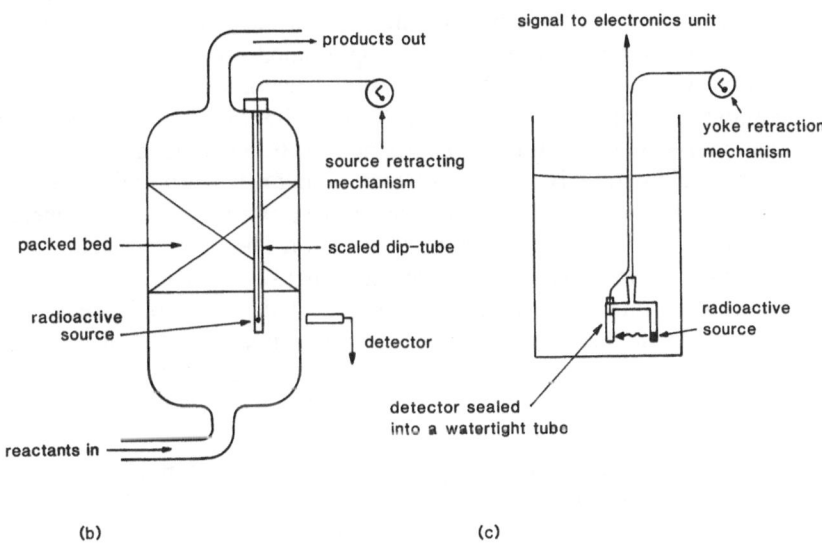

Figure 13.4. Methods of density measurement when the vessel is too large for direct transmission. (a) Reducing the path length by scanning through a chord of the vessel; (b) use of an installed dip-tube; (c) where access is available, source and detector, in a rigid frame, can give density checks at any position in the fluid.

the detector of course watertight. This approach has been successfully applied to the study of deposition in settling tanks and to the investigation of aeration of sewage digestors. On a larger scale still, the method has been applied to the measurement of silt densities at the bottom of lakes[23] and estuaries.

13.4.3 *Density profiles of distillation columns*

The inspection of distillation columns using the gamma-ray absorption technique is essentially a further example of density profile measurement. However, the requirement for such measurements is so frequent[7] and their potential value so great that the subject merits separate consideration. The technique is usually referred to as 'column scanning'.

The size and design of a column is very much a function of the duty the column is meant to perform, but the method of carrying out a scan varies only in points of detail: the basic technique for the inspection of a 9-m diameter primary fractionator on an olefine plant is the same as that employed for scanning a 30-cm diameter column on a semi-technical-scale unit.

A radioactive source of appropriate strength and gamma-ray energy, housed in a lead container, is lowered on a graduated tape down one side of the column, while at the same time a radiation detector (usually a NaI scintillation counter) is similarly lowered down the opposite side. Care is taken to ensure that source and detector are maintained in the same horizontal plane. The source collimation is such as to give a uniform panoramic beam in the horizontal plane. Axial rotation of the source container thus has no effect upon the intensity of the beam reaching the detector. The sources used for this work are almost invariably cobalt-60 or caesium-137. The detector measures the transmitted intensity, which is a function of the mean bulk density of the material in the beam path. In recording the data the transmitted intensity, expressed in detector count rate, is plotted on a logarithmic scale so that the resulting transmission characteristic faithfully reflects the density profile of the column (equation 13.2).

Figure 13.5 is an idealized representation of the information which can be obtained from a column scan. Vapour spaces are represented by the peaks in the profile (the high-count regions) while tray structures and liquid of relatively high density are represented by the troughs. It is, mercifully, rare to encounter a column in such a bad condition as that shown in Figure 13.5, but this serves to illustrate the type of fault-condition which can be readily identified.

(a) The position of the demister pad can be checked. Loss of the demister could give rise to liquid entrainment in the gaseous overheads stream from the column and so the ability to confirm that it is present and undamaged (or otherwise) is an important diagnostic aid.

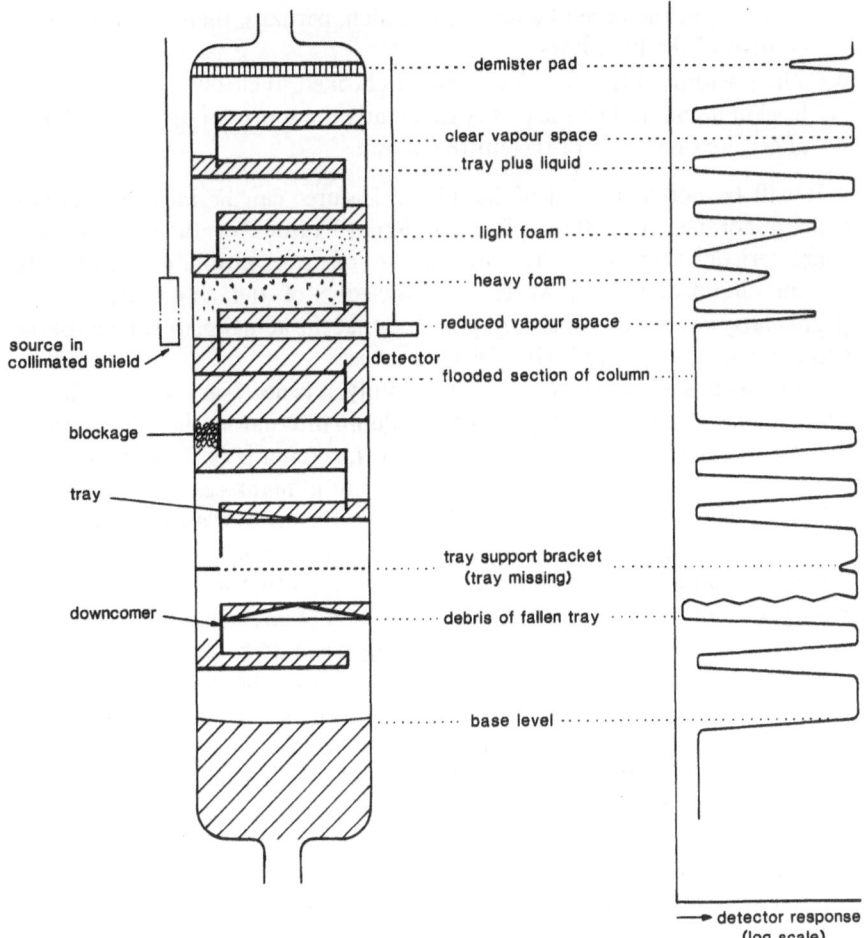

Figure 13.5. Gamma-ray absorption scan of a distillation column.

(b) The presence (and extent) of foaming can be identified. Such a malfunction can, of course, seriously affect the liquid/vapour equilibrium and result in reduced column performance. The ability to perform on-line measurements rapidly enables the efficiency of remedial measures, such as the addition of antifoam, to be assessed.

(c) Tray flooding is one of the more readily identifiable column faults. The condition is often caused by downcomer blockages. This can usually be inferred from a gamma-ray scan but can be identified positively using a neutron backscatter technique (Chapter 15). The two techniques are frequently used together for column inspections, the one providing information complementary to that provided by the other.

(d) Missing or damaged trays are also easily identified. This situation is often observed at the base of columns in which, perhaps, there has been poor control of the base level.

(e) The position of the base level can be checked. Ineffective control of the level of liquid in the base is very frequently the source of upsets which can affect the operation of the entire column.

It will be noted that all of the above features can be identified simply from a comparative study of the different regions of the transmission characteristic: there is no requirement to measure absolute densities. In certain circumstances, however, a knowledge of the actual densities—particularly vapour densities *is* required so that the actual performance of the column can be compared with design criteria.

To perform such measurements successfully, it is necessary to know (a) the effect of walls, lagging or lining and any column internals in the beam, and (b) the effective mass absorption coefficient of the contents of the column.

To investigate (a), it is desirable to carry out a 'blank scan' on the empty column. This, of course, must be planned in advance. Plants which make regular use of radioisotope techniques frequently arrange for such scans to be carried out on all columns (at plant shutdown) so that a kind of 'reference library' is built up to facilitate the interpretation of any measurements which may be required during operation. (In passing, it should be mentioned that it is likewise advantageous to have a reference scan of the column when it is operating under good conditions. This too facilitates subsequent interpretation and the identification of faults). If a blank scan is not available, it may be necessary either to calculate the absorption in the walls from knowledge of the column design or, better, to stimulate part of the column in a test-rig.

The effective mass absorption coefficient is far less amenable to computation, and if high accuracy is required, calibration on a simulated section of column is essential. Alternatively, efforts should be made, either by collimation or by electronic discrimination, to ensure that the measurement geometry approximates to narrow-beam conditions.

It should, of course, be noted that when scans of a column are made on different occasions for comparison purposes, it is necessary to standardize. This is done by checking (by pulse-height analysis) that the same region of the gamma-ray spectrum is being counted, and by using a small reference source to measure overall counting efficiency on each occasion. The source-detector geometry must also be reproducible.

The simple scanning technique described above has the virtue that it is rapid and that it can be applied to practically every column to which access is available. In practice, most columns have convenient platforms or walkways from which source and detector can be suspended and it is rare indeed that additional scaffolding or staging is required. On large columns, small lateral movements of source and/or detector have (because of the inverse square law) negligible effect on the transmitted count-rate.

Occasionally, however, where highly reproducible measurements are required or with columns of small diameter, more accurate positioning is required. Metal guides attached to the column are one way in which accurate positioning has been achieved,[6] though with the panoramic collimation described earlier, plastic drainpipe fixed physically to the columns has been found to be equally effective.

On certain columns which are scanned frequently, or in which conditions change so rapidly that scans must be carried out very quickly, mechanized systems have been installed[24, 25]. Several different designs have been used but the general principle is that a motor drives source and detector synchronously up and down guides attached to the column. The detector output, perhaps from a logarithmic ratemeter, is recorded directly on a chart recorder. With the availability of small portable data loggers, further improvements in methodology are likely.

13.5 Measurement of mass per unit area

Under this heading we will discuss those applications in which we are not especially interested either in the thickness or in the density but rather the total mass of material lying in the path between source and detector. Since the spread of the beam is of finite dimensions this is equivalent to saying that we are interested in the mean mass per unit area in the beam path. Some very important applications come into this category: industrial radiography; level gauging—both spot checks and installed gauges; blockage detection; mass flow of material on conveyor belts. Some of the techniques described under thickness measurement (section 13.3) or density measurement (section 13.4) could perhaps also be placed in this category—distillation column fault-location being a good example. This is largely a question of nomenclature and does not reflect fundamental differences in the principles underlying the measurements.

13.5.1 Industrial radiography with gamma-rays

Radiography of items of industrial plant using radioisotopes as the source of radiation is arguably the most frequently used, and therefore the most familiar, of all techniques involving radioisotopes. The technique has long been part of the standard repertoire of every non-destructive testing (NDT) company and there can be few plants which have not, at some stage, been subject to radiographic examination. Because of the very familiarity of the technique and because, in general, applications are oriented more towards NDT than to process investigation, we will not accord the subject much space here. For detailed discussions of the theory, equipment and practice, the reader is referred to comprehensive tests on the subject[26-28]. The object of this section

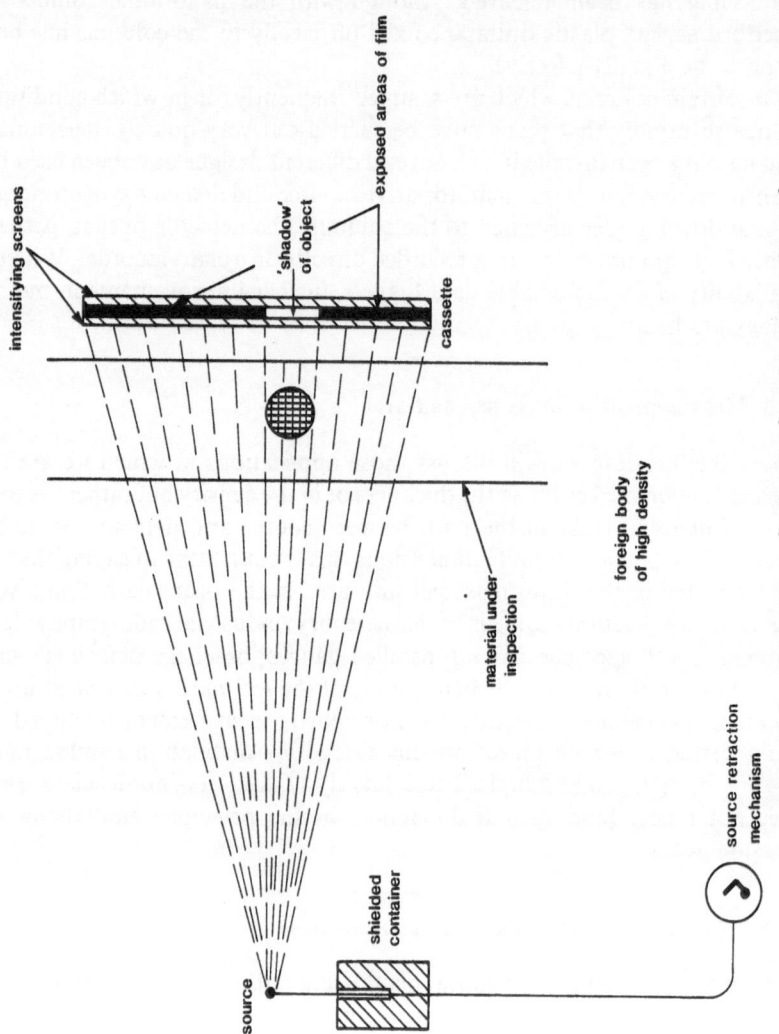

Figure 13.6. Arrangement of apparatus for gamma-ray radiography.

is to state briefly the key considerations in performing radiographic measurements and to indicate how the technique can complement the types of gamma-ray investigation which have already been described.

Radiography, broadly speaking, provides a method of visualizing the internal structure of vessels and plant components. This may be for the purpose of examining integrity of construction—weld radiography for example—or to provide a check that vessel contents are correctly positioned and that no foreign objects are present. Whatever the purpose of the measurement, the principles of the technique are the same (Figure 13.6). A source of gamma-radiation is placed on one side of the material of interest and a photographic X-ray film (housed in an appropriate cassette holder) on the other. Gamma-rays transmitted through the intervening medium impinge upon the photographic film. If the vessel contains an object whose mass per unit area is greater than that of its surroundings, fewer gamma-rays are transmitted and a shadow pattern will be produced on the film. After adequate exposure, the film is developed. Parts of the medium of high mass per unit area will give rise to underexposure and produce light regions on the film; the converse is true of regions of low mass per unit area. The radioisotope sources in most common use are cobalt-60, irridium-192, ytterbium-169 and thulium-170—though various other radioisotopes have been studied[29]. The properties of the more useful gamma-ray sources relevant to their use in radiography are summarized in Table 13.2.

The radioisotope source for any given application should meet two basic criteria: it must be of sufficient activity to give adequate film exposure in a reasonably short time and it should be of small physical dimensions. If the latter criterion is not met, some blurring of the image on the film results. In practice this effect can usually be made very small by ensuring that there is sufficient distance between the source and the object, though the inverse square law limits the extent of this separation. The relative insensitivity of photographic film compared to radiation detectors dictates that source

Table 13.2 Some useful radioisotope sources for gamma radiography*

Optimum working thickness	Cobalt-60	Ytterbium-169	Iridium-192	Thulium-170
steel	50–150 mm	2.5–15 mm	12.5–62.5 mm	2.5–12.5 mm
light alloys	150–450 mm	7.5–45 mm	40–190 mm	7.5–37.5 mm
other materials	40–120 g cm^{-1}	2–12 g cm^{-1}	10–50 g cm^{-1}	2–10 g cm^{-1}
Half-life	5.26 y	32.0 days	74 days	128 days
γ-energies (MeV)	1.17, 1.33	0.008–0.308	0.206–0.613	0.052, 0.084
Exposure rate (for 1 curie equivalent activity at 1 m in Rh^{-1})	1.30	0.12	0.48	0.0025

*By kind permission, Amersham International PLC.

strengths must be correspondingly higher for radiographic applications. The effective exposure of the film is commonly increased (and exposure time and/or source strength thereby reduced) by using image-intensifying screens[30]. Nevertheless, a difference in two or three orders of magnitude in strength is not uncommon and radiographic sources are in general in the multi-curie range.

With sources of this strength, handling becomes something of a problem. (The dose, for example, at 10 cm from a 10 Ci source of cobalt-60 is approximately 13 sieverts per hour.) Various designs of shielded-source container and remote exposure equipment are available. Wherever possible, radiographic exposures are made in properly designed bays, equipped with shielding, physical demarcation barriers, alarms and interlocks to minimize the chances of accidental exposure of personnel. Sometimes, of course, this is not possible and exposures must be made on operating plant. In these circumstances, stringent safety precautions are put into force and personnel excluded from the vicinity of the source, which is operated remotely.

A point worth mentioning in this context is that installed nuclear gauges— as well as personnel—can be affected by exposure to gamma radiation. Level gauges, for example (section 13.5.2) often operate at radiation levels as low as 0.002 mSv per hour, and so stray radiation from a radiography source can easily produce a spurious reading. Since in many cases nuclear gauges control the operation of critical parts of the plant, such an occurrence is potentially disastrous. Fortunately, the problem can be avoided if it is recognized at the outset: in ICI plants, for example, a section of the Permit to Work (the completion of which is mandatory) makes it incumbent upon those responsible for the work to ensure that as far as possible, the source is collimated to avoid interference with installed gauges. This notwithstanding, the reading of any gauge in the vicinity of a radiographic exposure is critically examined for the duration of that work.

Several types of X-ray film of various sensitivities are available[31,32]. Film exposures of between about 10 and 200 milliroentgens are not uncommon.

In making a radiograph, one or two simple rules, if adhered to, can make the process simpler and the results easier to interpret.

(a) As discussed, the source should approximate as closely as possible to a point, and the source-to-object distance should be as large as possible. Considerations of source strength, exposure time and the physical geometry of the object being inspected limit the extent to which the guideline can be followed.

(b) The X-ray film should be placed as closely as possible to the object under inspection for optimum sharpness of image.

(c) To avoid distortion of the image, the film should be angled so that the gamma-rays impinge upon it perpendicularly.

Gamma-radiography has a useful role to play in process investigation and by careful design of radiographic equipment, exposures may be made under

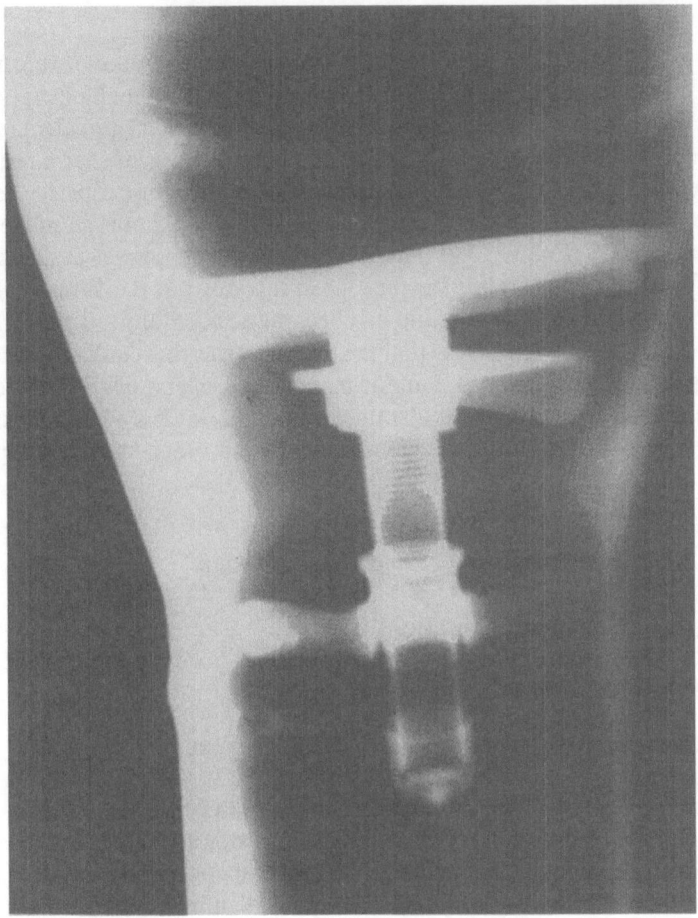

Figure 13.7. Gamma radiograph of a reduction section in a pipeline, showing internal 'pipe-blocker'. Light areas on the radiograph correspond to regions of high density. Photograph courtesy of ICI PLC.

arduous conditions. For example, one type of cassette which has been designed[34] for operation upto 300 °F will probably find extensive applications in inspecting operating plant.

As a general rule, because of the problems attendant upon the use of large sources we tend to exhaust other possibilities before turning to radiography. There are, however, cases in which alternative solutions (including the use of gamma-ray source/detector techniques) are not forthcoming and in such situations gamma-radiography is an invaluable asset.

An example of how gamma-radiography can complement simple gamma-absorption scans occurred recently, when, following a routine shutdown, a high pressure-drop was observed along a gas supply line to a major chemical complex. A line restriction was suspected and gamma-ray checks were made at selected points—bends and changes in pipe diameter where blockages would be likely to occur. At one such point, increased absorption equivalent to about 15 mm of steel was observed. There was no obvious cause for such a restriction and detailed absorption scans in the vicinity could not properly define its nature. A radiograph carried out with an iridium-192 source immediately clarified matters (Figure 13.7). The cause of the blockage was a pipe-stopper (used for temporarily sealing the pipe when carrying out pressure tests) which had been mislaid at shutdown and had become accidentally lodged in the line. The radiograph showed clearly that the blockage was most unlikely to become dislodged and so the line was shut down and the stopper quickly removed. The rapid location and subsequent identification of this fault is a good illustration of how the two techniques can complement each other in saving expensive down-time.

13.5.2 *Level measurement by gamma-ray absorption*

During plant operation, occasions frequently arise in which the operator requires information about the level of material inside a plant vessel. Some typical examples are:

(a) Calibration (or proving) of an installed level gauge
(b) Stocktaking
(c) Monitoring the filling or emptying of road and rail tankers
(d) Checking the packing level in absorbers and monitoring catalyst bed levels
(e) Providing information to diagnose a malfunction—for example a high liquid level in a knock-out pot or separator is often the cause of liquid entrainment problems
(f) Checking that correct operating conditions have been established—base levels in distillation columns, levels of process material in reactors.

Whatever the reason, the operator wants the information quickly and with the minimum of inconvenience. There are, of course, several ways in which level measurement can be achieved and the gamma-ray absorption technique is complementary to other methods. (Other radioisotope level-measuring techniques are discussed in Chapters 14 and 15.) Its strengths lie in the universality of application, the simplicity of operation, the unambiguous nature of the results and the ease of interpretation.

The experimental method (Figure 13.8) is almost identical to that employed in density profiling. Indeed, the only difference is that in level measurement the mass per unit area of the material in the vessel (ρx) is usually such as to cause complete extinction of the gamma-ray beam. Since we accept that no information on density will be obtained, the technique, unlike density profiling, is applicable to vessels of very large diameter.

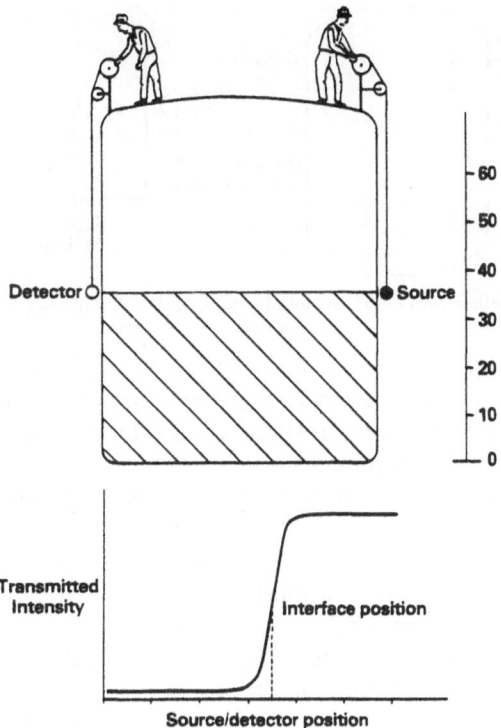

Figure 13.8.   Level measurement by gamma-ray absorption.

The measurement of levels on a spot-check basis, though extremely useful, can become both tedious and expensive if the measurement is frequently required. In such circumstances, a more cost-effective solution is to permanently install a level gauge on the vessel of interest. In this respect gauges operating on the gamma-ray absorption principle have much to offer. Their chief advantages are:

(a) Instruments operate external to the process. They, therefore, suffer from none of the problems usually associated with measurements made on corrosive, toxic, high-temperature or high-viscosity materials.
(b) Instruments can be installed while a plant is on line and this can lead to significant savings.
(c) Instruments can be used with either solids, liquids or slurries.
(d) They are of rugged construction.
(e) Running costs are negligible and little or no maintenance is required.

Item (b) is of crucial importance when one considers the gauges as adjuncts to problem-solving. Gauges can be installed in an emergency situation (perhaps occasioned by the failure of conventional level-measuring equipment) and can thus enable a plant to continue operation in circumstances in

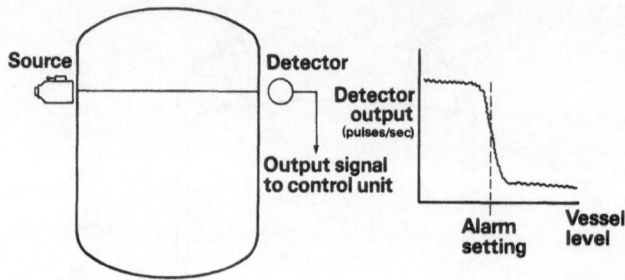

Figure 13.9. Installed level alarm (switch) based upon the gamma-ray absorption principle.

which a shutdown would otherwise be necessary. (It is of interest to note that gauges installed to cover an emergency period usually remain as permanent features of the plant: operators are impressed by their accuracy, high reliability and minimal servicing requirements.)

Gauges fall into two general categories: level alarms and proportional level indication.

Level alarms are in common use. They are used to ensure that the level of process material in a vessel does not rise or fall below some critical position in the vessel. The equipment consists of three units: a control unit, a detector unit and a γ-emitting radioactive source housed in a shielded container.

The principle of operation is illustrated in Figure 13.9, which shows the arrangement of the equipment on a hypothetical gas–liquid reactor. Gamma-radiation from the source is directed through the vessel to the detector which is mounted opposite the source on the plane where the level is to be monitored. When used as a high-level alarm, any material which rises through the gamma-beam reduces the radiation intensity at the detector. This modifies the signal from the detector to the control unit and causes an alarm relay to de-energize. The control unit may also be used as a low-level alarm.

For certain applications, however, it is not sufficient simply to provide a level alarm system, but a continuous indication of level is required over a range of heights in a vessel. Figure 13.10 illustrates the basics of the most simple

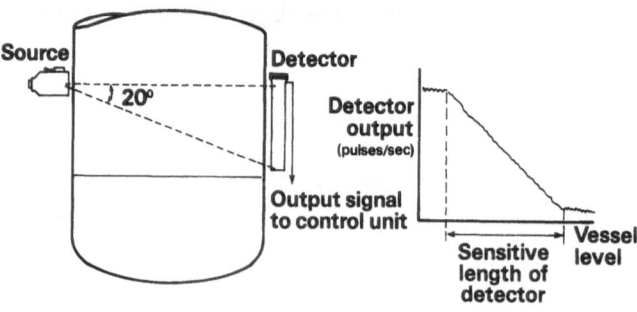

Figure 13.10. Proportional level gauge based upon the gamma-ray absorption principle.

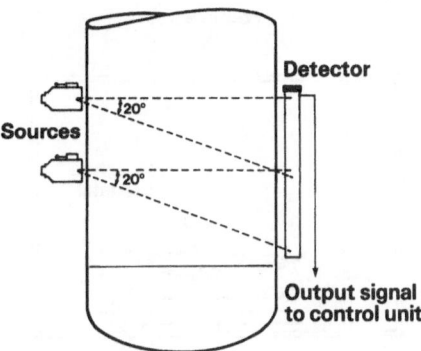

Figure 13.11. Proportional level gauge—multiple external sources to extend the vertical range.

gamma-ray absorption technique. Here the radiation is directed across the vessel in a narrow vertical plane. The angle of dispersion is normally fixed at about 20° so that the path lengths of the various rays which make up the beam are similar and the radiation intensity over the detector length is reasonably uniform (N.B.: inverse square law applies). As the level of material rises, it attenuates the radiation over an increasing length of the detector, and the mean voltage pulse rate from the detector decreases. This change in pulse rate is sensed by the control unit which produces an output signal proportional to level.

A consideration of the inverse square law explains why only a limited measurement range can be obtained from a single source system. For applications where a large vertical range of operation is required, multiple sources are used. One such arrangement is illustrated in Figure 13.11.

13.5.3 *Measurement of mass flow on conveyor belts*

The measurement of the mass flow of solids is an extremely widespread requirement: mining, steelmaking, chemical and construction industries are typical of those in which the problem arises. The gamma-ray absorption technique forms the basis of an instrument widely used for such applications— the so-called 'nucleonic' belt-weigher or weighscale.

In this type of instrument, gamma-radiation from a source is directed through the material on the conveyor towards a radiation detector which is positioned immediately below the belt (shown schematically in Figure 13.12). The instantaneous count rate generated by the detector is an exponential function of (density × depth of material on belt). By means of an electronic linearizer in the control unit, this count rate is converted into a dc voltage signal which is directly proportional to the quantity (density × depth of bed). Multiplication of this voltage by a factor which varies linearly with belt speed (i.e. tachometer output) produces an output signal which is directly pro-

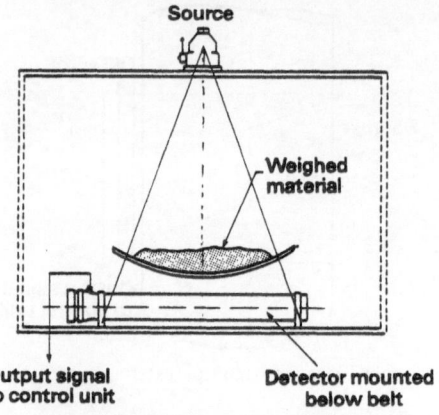

Figure 13.12.   Nucleonic weighscale.

portional to instantaneous mass flowrate. This output can be used to control the addition of other materials to the product on the belt or to optimize the conditions of downstream processing units (product driers etc). If required, the output voltage can be processed to give a digital display of the total tonnage of material which has passed the weighing apparatus.

Unlike their more conventional counterparts, nucleonic weighers make no contact with the belt and require virtually no maintenance. They take up only about 18″ of belt length and, since they are totally unaffected by changes in belt tension, they can be installed with a minimum of inconvenience at any position on the belt. As in the case of level gauges, nucleonic belt-weighers are often installed in an emergency but remain as permanent plant control instruments.

13.6 Case histories

13.6.1 *Detection of corrosion : methylamines interchanger*

After several years' service, leaks developed on the tube bundle of an amines interchanger such that cold reactants entering the interchanger on the tube side bypassed the reactor via tube leaks to the shell (Figure 13.13). This effectively reduced the efficiency of the process and so remedial action was needed. In order to ascertain the extent of the repair work required it was decided to examine the tube bundle of the exchanger at shut-down to determine the location (and degree) of tube thinning. Since the time available for repairs was short it was important to obtain the necessary information as rapidly as possible. The gamma-ray absorption technique described in 13.3.3 was used. Since the tube wall thickness was 2 mm, americium-241 was an ideal gamma-ray source. Because the interchanger was 5 m in length it was decided that the examination could best be carried out by using a probe with an

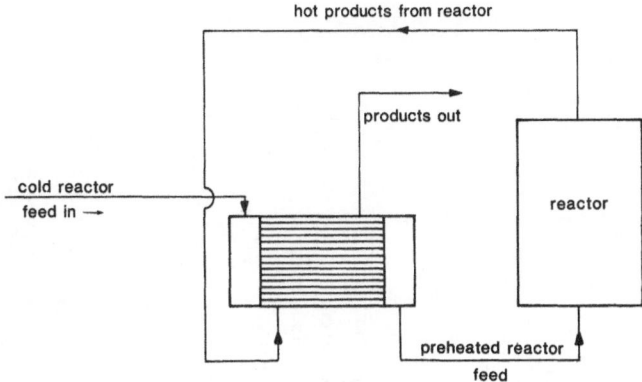

Figure 13.13. Plant configuration: methylamines interchanger leak.

effective length of 2.5 m and making the measurements from each end of the bundle. (A probe of length much in excess of 2.5 m would have been unwieldy.)

The results of the scans were plotted out on a drawing of the tube-plate using a colour code (simulated in Figure 13.14 by line hierarchy—see key) to represent the degree of corrosion inferred from each measurement. It rapidly became apparent that, in a small section of the bundle immediately opposite the inlet line bringing hot products from the reactor (i.e. the hottest part of the bundle) there was thinning in excess of 30%. Elsewhere, the bundle was relatively free from corrosion. On the basis of these observations, the plant engineer decided to plug off those relatively few tube pairs on which corrosion in excess of 30% had been observed. It was also decided to totally re-orient the interchanger (which was of symmetrical construction) so that what had been the hot end became the cold end and vice versa. By these actions it was possible to extend the life of the interchanger and to avoid re-tubing—costly both in materials and time.

13.6.2 Investigation of deposit distribution in a serpentine cooler

On a pigment plant, hot powdered pigment carried in an air stream was cooled to ambient temperature in a serpentine cooler (Figure 13.15) before passing forwards to the product silos. Over a period of time it was noted that pressure-drop limitations restricted the throughput of powder. A partial blockage was suspected and it was decided to carry out gamma-ray absorption measurements after plant shut-down so that the extent of the problem could be determined and appropriate remedial action taken. Since the total length of cooling coil was more than 250 m it was decided initially to carry out measurements at 30 positions only, distributed more or less evenly along the coil. The measurements were carried out using a caesium-137 source and NaI

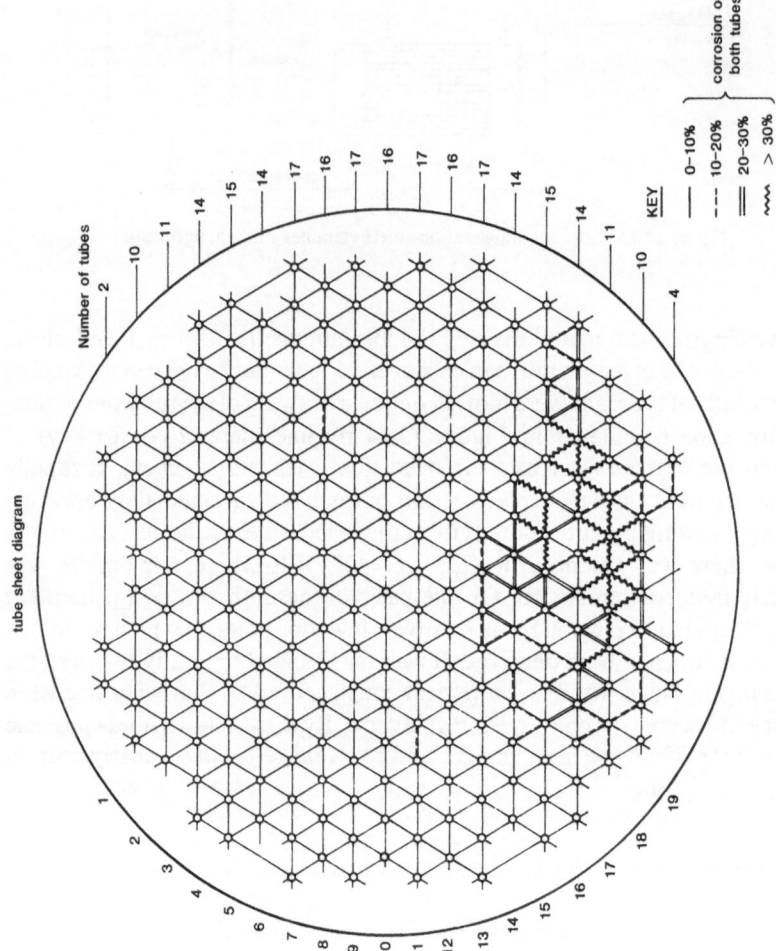

Figure 13.14. Methylamines exchanger. Tube sheet diagram—tube corrosion measurements results of gamma-ray absorption scans.

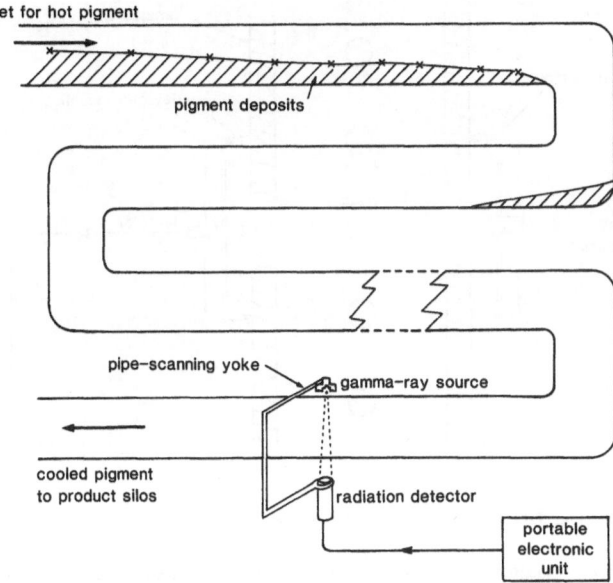

inlet for hot pigment

pigment deposits

pipe–scanning yoke

gamma–ray source

cooled pigment
to product silos

radiation detector

portable
electronic
unit

Figure 13.15. Gamma-ray absorption measurements of deposit thickness in a serpentine cooler.

scintillation counter in the pipe-scanning 'yoke' described in section 13.3.2. Calibration was effected by using a section of empty pipe, of the same specification as the cooler, filled to various depths with powdered pigment.

It very soon became apparent that the problem was almost entirely localized in the first pass of the cooler which in places was over half choked with deposits. Elsewhere in the cooler, deposits were slight or non-existent.

Armed with this knowledge it was possible for maintenance personnel to open a flange in the top pass and to rod out the deposit, which was found to be pigment which had became 'baked' on to the pipework—presumably as a result of the high temperatures in this section of plant. By directing effort towards the precise problem area and eliminating unnecessary break-ins into parts of the cooler which were functioning perfectly adequately, the shutdown was shortened with consequent (significant) savings.

13.6.3 Investigation of coke deposits in the radiant section of a cracker furnace

The laydown of coke on the walls of the tubes in the radiant section of cracker furnaces was a source of problems. Because of the relatively low thermal conductivity of the coke, heat transfer from the furnace to the feedstock in the tubes was impaired. As the deposits built up, a progressive rise in tube-skin temperature was observed. Ultimately, as the temperature limit for the tube

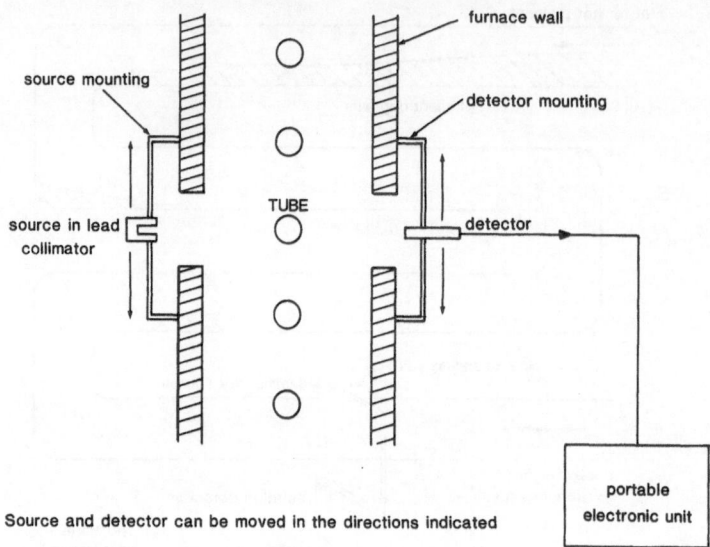

Figure 13.16. Gamma-ray absorption measurement of coke build-up on a furnace tube.

metal was approached plant shutdown was necessitated so that the tubes could be cleaned. In fact, cleaning was usually carried out well before this point was reached since the lowered heat transfer properties of the tube wall caused a reduction in process efficiency.

A gamma-ray absorption technique was used to monitor the rate of coke deposition so that the effects of varying furnace operating characteristics and feedstock additives could be investigated. The method of measurement is illustrated in Figure 13.16. The gamma-ray source (^{137}Cs) and radiation detector (a NaI scintillation counter) were positioned on opposite sides of the furnace at a conveniently-situated observation port. Both were mounted on rigid framework and each could be moved in a direction parallel to the walls of the furnace. Initially the position of the source holder was checked visually and adjusted, if necessary, to ensure that the collimated gamma-ray beam was directed through the centre of the tube. The intensity of the radiation traversing the tube was measured at some arbitrary position of the detector and was recorded. The detector was then moved by 20 mm and the intensity was again recorded. This procedure was continued until the detector had completely traversed the furnace tube and the recorded intensities were then graphed as a function of detector position. A typical transmission characteristic is shown in Figure 13.17. Examination of this curve revealed the presence of two dips in intensity corresponding to the edges of the furnace tube. The point midway between the two extremes corresponded to the position at which the detector, source and centre of the furnace tube were in alignment.

The alignment procedure was carried out on each occasion on which

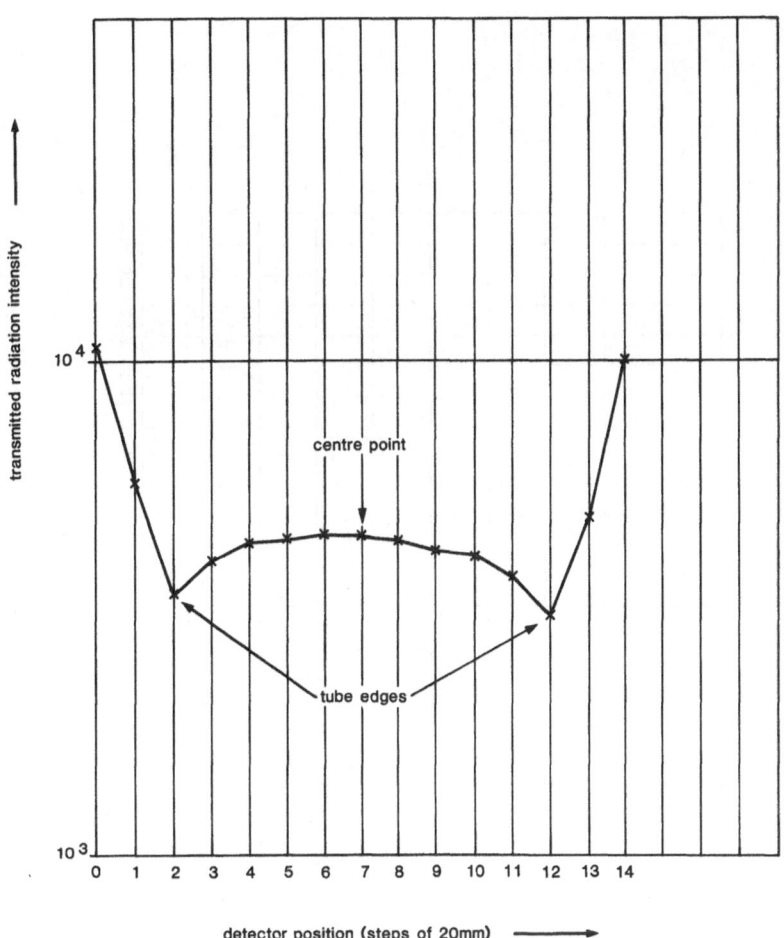

Figure 13.17. Gamma-ray transmission through a furnace tube.

measurements were taken. This was necessary because, over a period of time, the position of the tube changed due to thermal expansion effects. Alignment ensured that all measurements were made on the same (central) point of the tube.

Having set up the source and detector in the correct measurement position, an extended radiation count was taken. This count rate was related to the thickness of carbon in the tube using a calibration procedure in which the measurement geometry was duplicated in a laboratory test-rig using graphite sheets of various thickness to simulate the carbon deposits. The results of the measurements, which were carried out over a period of two months, are shown in Figure 13.18. After an initial 'quiescent' period extending typically over 3 days, deposits rapidly built up to a maximum of approximately 8 mm (i.e.

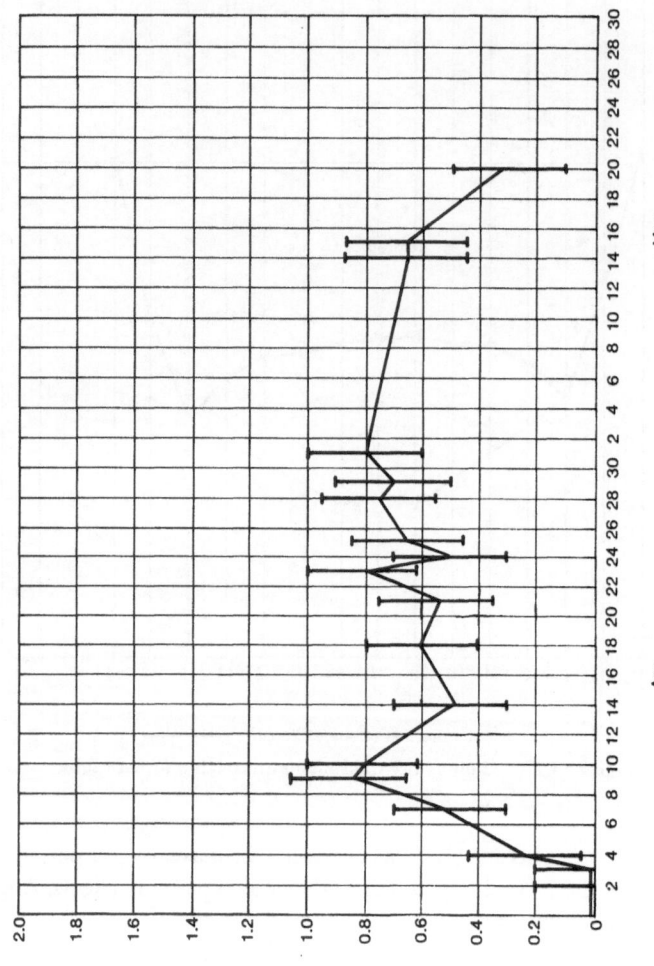

Figure 13.18. Radiant coke build-up measured by the gamma-ray absorption technique.

4 mm evenly distributed around the tube circumference) and thereafter remained constant within the limits of accuracy of the measurements.

These data obtained for several sets of operating conditions led to a better understanding of the factors which were important in influencing the rate of coke laydown and assisted in establishing optimum procedures for furnace operation. It is difficult to compute the value of such work. However, on a plant with a throughput of up to 200 te h^{-1} of naphtha, an efficiency improvement of as little as 1% leads to savings of several millions of pounds per year.

13.6.4 *Detection of water deposits in a hydrogen transmission line*

In cold weather, water vapour present in hydrogen gas passing through an overland pipeline to an aromatics unit condensed out leading to localized 'pooling' at low points in the line. The accumulation of water deposits led to flow restrictions and it was, therefore, important to have a means of identifying the location of the water so that remedial action could be taken. By using the adjustable pipe-scanner (section 13.3.2) it was possible to carry out a rapid examination of the pipeline and to measure both the position and the extent of the condensed water. By strategically positioning 'wrap-around' heaters on the pipeline, process personnel were able to eliminate the problem.

Although this problem was initially solved using a gamma-ray absorption technique it was subsequently found that the position of the water could be more rapidly established using a method based upon neutron backscattering (Chapter 15). However, for accurate measurement of the *depth* of the deposits the gamma-ray absorption technique was more reliable. This is a good example of how the two techniques can often be used to complement each other to provide the most effective solution to many plant problems.

13.6.5 *Detection of the interface between liquids during a batch transfer*

The principles of this technique were described in section 13.4.2: the ability of the gamma-ray absorption technique to provide a measurement of density is used to discriminate between two liquids in a pipeline, thereby enabling detection of the interface between them. In the case in question, a stream of liquid propylene (density approximately 0.5 g cm^{-3}) was being used to sweep out mixed C_5s (density 0.7 g cm^{-3}) from a 100 mm-diameter line several miles in length. Using the arrangement shown in Figure 13.19 the time of arrival and extent of the interface were clearly discernible. Process personnel were alerted and this early warning enabled them to operate the appropriate diverter values to ensure that the two pure materials were segregated and that the volume of mixed material, which would subsequently need to be reprocessed, was kept to a minimum.

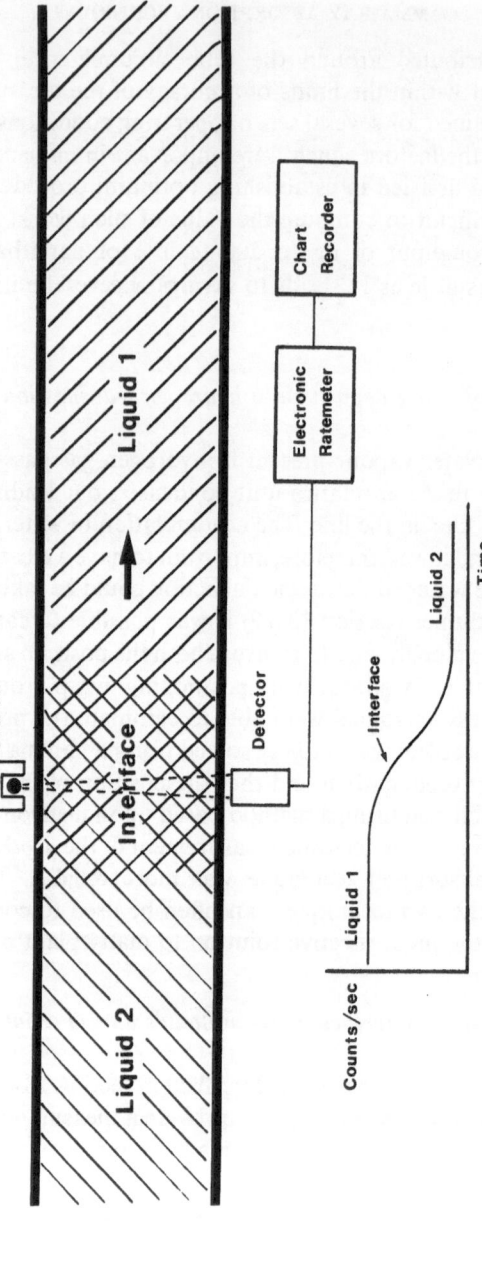

Figure 13.19. Interface detection in a pipeline by density gauging.

This is a further example of a problem which could have been solved using a neutron backscattering technique (Chapter 15). Indeed, had the diameter of the line been so large as to necessitate the use of a very large gamma-ray source, the neutron method would have represented the more acceptable alternative from a radiological point of view.

13.6.6 Density profile of an absorber column to investigate the effectiveness of antifoam addition

An oil absorber column on a refinery was experiencing severe foaming problems resulting in a high level of liquid entrainment in the vapour overhead line. In an attempt to provide sufficient space at the top of the column for liquid to disengage, the top two trays of the column had been removed, but the problem still persisted. It was decided to incorporate an antifoaming additive in the feed to the column. In order to ascertain the effectiveness of this treatment, gamma-ray absorption scans were carried out before and after the addition of the antifoam. The gamma-ray absorption measurements were carried out using the method described in section 13.4.3.

The results of the two sets of measurements are shown in Figure 13.20. Without antifoam, very heavy foaming—approaching flooding—was observed at the top of the column. Further down the column the foaming problem became progressively smaller, though all trays, with the exception of Tray 1, were heavily laden. In contrast, the scan carried out during the antifoam addition indicated that the column was operating normally with well-defined tray levels and vapour spaces. These data provided positive evidence that the antifoam addition was successful. Subsequent scans carried out with different concentrations of antifoam enabled production management to assess the minimum level of additive needed to prevent foaming.

It is, perhaps, worth commenting that experience of similar problems has shown that antifoam addition does not always provide a viable solution. Indeed, situations have been encountered where injecting additives to suppress foaming has, in fact, exacerbated the problem. It is the peculiar virtue of the gamma-ray absorption technique that a rapid on-line assessment can be made of the effectiveness of any measure taken to remedy the column malfunction with consequent savings of effort and materials.

13.6.7 Investigation of tray-fouling in a distillation column

A fractionation column on a phenol plant was producing material of poor specification. It was decided to examine the column using the gamma-ray absorption technique to investigate the source of the problem. The results indicated that in the upper part of the column there was good phase-separation with well-defined liquid levels and vapour spaces. However, in the

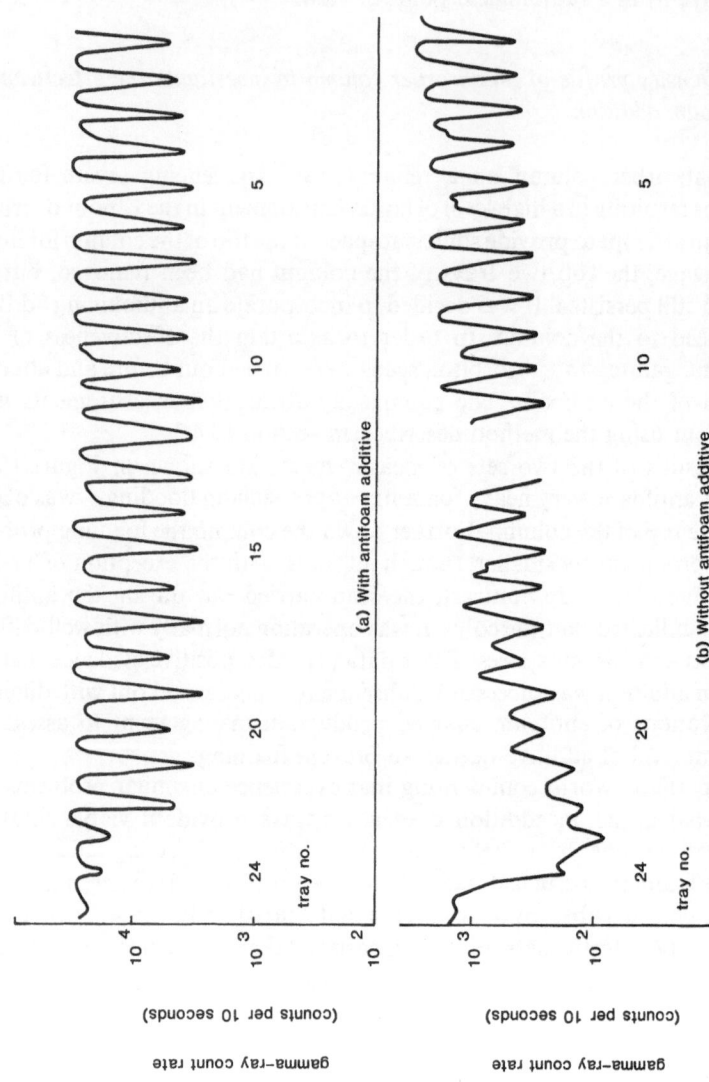

Figure 13.20. Gamma-ray transmission characteristics of an oil absorber column. (a) With antifoam additive; (b) without antifoam additive.

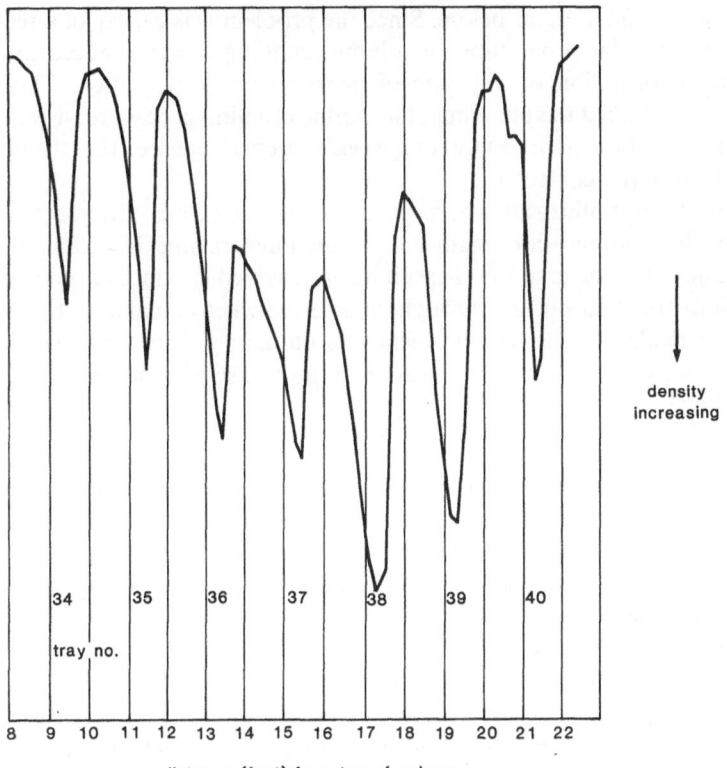

Figure 13.21. Gamma-ray absorption inspection of a fractionation column (note heavy vapour density and high liquid loading of trays 38 and 39).

lower part of the column, progressively high vapour densities and tray loadings were observed (Figure 13.21). Trays 38 and 39 in particular were carrying very large amounts of liquid and were approaching the 'flooded' condition.

On the basis of these results, the column was opened and the suspect region examined. In the lower (hotter) part of the column heavy deposits of an insoluble salt were found on the trays. Clearly, these deposits were impeding the transfer of vapour up the column, leading to inefficient fractionation. The trays were therefore washed with an appropriate solvent and the column brought back on line. A second gamma-ray absorption scan carried out immediately after start-up confirmed that the flooding in the lower part of the column had cleared and this, together with the fact that the product quality had returned to normal, showed unambiguously that the salt deposition had been the cause of the problem.

However, over a period of several weeks the product quality progressively deteriorated and further column scans indicated that the same part of the column was flooding as before. Since the problem was clearly of a recurrent nature, a regular programme of column scanning was introduced to obtain precise information on the rate of performance-deterioration. This information facilitated the planning of a routine cleaning procedure: it was found that a solvent wash carried out at 3-weekly intervals ensured that the problem was kept under control.

This example illustrates the importance of close collaboration between the radioisotope applications team and production personnel in optimizing plant operation. In this case the measurements enabled production personnel to maximize the time on-line throughout a period when demand for the product was particularly high. The problem was ultimately solved by changing the design of the lower part of the column in the course of a routine maintenance shutdown.

13.6.8 *Investigation of catalyst deterioration in a packed-bed converter*

A converter on an amines plant contained five catalyst beds (Figure 13.22). As the catalyst became spent, the packed beds expanded to fill the intervening gaps. A gamma-ray absorption technique was used on a routine basis to provide production management with an up-to-date view of the situation and, in this way, facilitated the decision as to when catalyst regeneration should be carried out. A similar technique has been used on methanol plants to determine the drop in catalyst level resulting from a reduction process.

13.6.9 *Gamma-ray absorption for level measurement and control during plant start-up*

The start-up period of a major chemical plant is a time during which the gamma-ray absorption technique can be used to particular advantage. Production personnel are breaking new ground; if the plant is of new design, establishing the correct operating conditions can be a delicate matter and rapid feedback on the effects of any changes initiated from the control room is particularly welcome. One aspect of start-up operations which frequently gives rise to problems is concerned with establishing the correct levels of liquid in plant vessels—columns, separators, reactors, catch-pots etc. The gamma-ray absorption technique, because of its ease of application and the unambiguous nature of the data it produces, is especially suitable for such applications. The following example illustrates the general usefulness of the technique.

During the start-up of a major ethylene complex, problems were experienced in controlling the operation of a stripper column. The overheads stream from the column, which should have comprised 'light', gaseous hydrocarbons was, in fact, badly contaminated with material of high molecular weight, to the

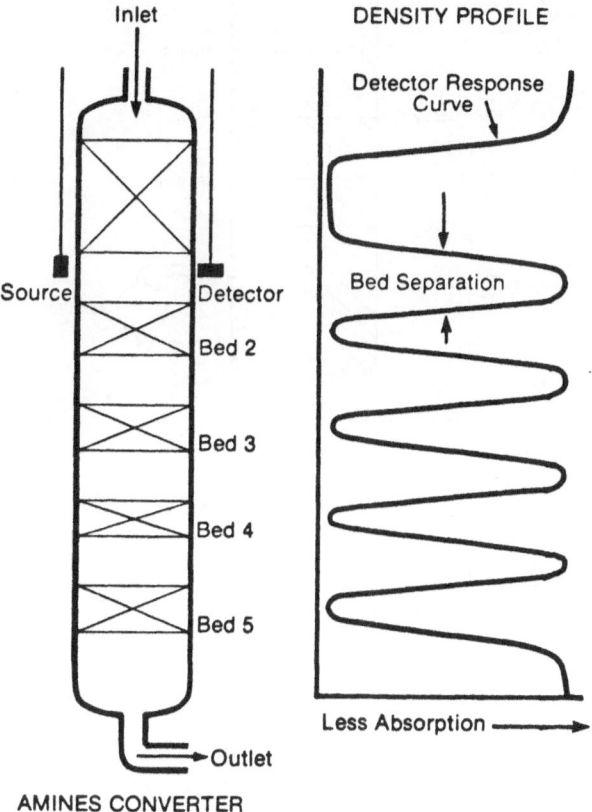

Figure 13.22. Catalyst bed-level measurement.

extent that liquid carry-over into the gas stream was suspected. A radioisotope applications team, which had been seconded to the plant for the duration of the start-up, was able quickly to establish the cause of the problem.

A gamma-ray absorption scan showed that the base-level controller on the column had failed: the level of liquid in the column had steadily risen to the extent that, at the time the scan was performed, the bottom half of the column was completely full of liquid (Figure 13.23a). On the strength of this result, the liquid feeds to the column were severely cut back until the gamma-ray data confirmed that the base level had returned to an appropriate position — approximately mid-way between the take-off and return lines of the 'bottoms' reboiler. The problem still remained, however, of retaining control of the base level in the absence of a reliable level indicator/controller.

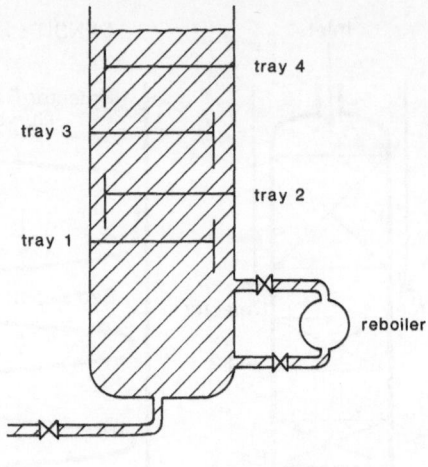

(a)

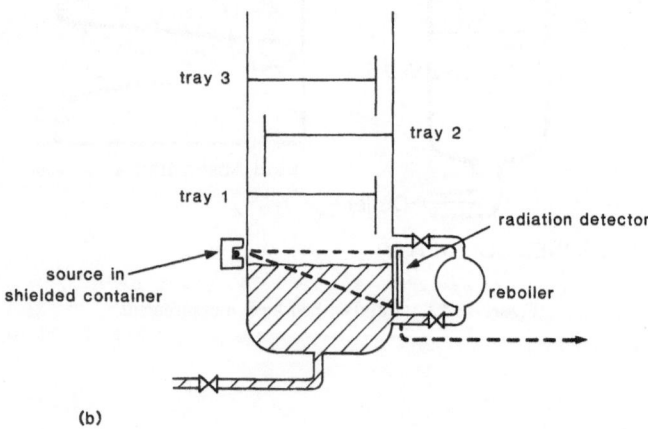

(b)

Figure 13.23. Diagnosis and cure. Flooding in the base of a hydrocarbon stripper. (*a*) Abnormally high base-level diagnosed by gamma-ray absorption scan; (*b*) base level controlled by nucleonic proportional level indicator.

Again, the gamma-ray absorption technique was able to provide the solution. A proportional level indicator of the type described in section 13.5.2 was fitted to the base of the column (Figure 13.23*b*). Since all parts of the gauge were completely external to the process, the installation proceeded in parallel with other start-up activities and delay in bringing the plant up to full production was minimized. The reliability of the gamma-ray gauge was such that the instrument was retained permanently as the preferred means of level control of the column base-level.

The economic benefits of this particular application were not precisely quantified. However, in initially diagnosing the source of the problem and then in providing a cure, the gamma-ray absorption measurements saved several days of unproductive operation and averted a temporary shutdown of the unit. It is clear that the resulting savings were orders of magnitude greater than the costs incurred in carrying out the measurements.

References

1. *The Radiochemical Manual.* The Radiochemical Centre, Amersham (1966).
2. Jordan, G. G. *et al.* (1956) Application of radioisotope to control technological processes. In *Proc. First Int. Conf. on the Peaceful Uses of Atomic Energy,* Geneva, Vol. 15, United Nations, New York, 135.
3. Kohn, A., (1963) Non-destructive testing of pipes in hot blooms, by gammascopic inspection. *Non-Destructive Testing* **21**, 241.
4. Syke, G. (1958) Automatic control in steel strip manufacture. *J. Brit. IRE* **18**, 117.
5. Cameron, J. F. and Clayton, C. G. (1971) *Radioisotope Instruments.* International Monographs in Nuclear Energy, Pergamon Oxford, 207.
6. Johnson, P. *et al.* (1963) Some applications of radioisotopes in the chemical industry. *Chem. Ind.,* 750.
7. Charlton, J. S. *et al.* (1982) Radioisotope techniques for the investigation of process problems in the chemical industry. In *Proc. of IAEA Conf. on Industrial Application of Radioisotopes and Radiation technology,* IAEA, Vienna, 393.
8. Myers, H. G. (1981) Radioisotopes in plant operations. *Chemtech,* 489.
9. Anon. (1975) New meter will 'see' steam, water inside pipe. *AECL Review* **10**, 1.
10. Shrier, L. L. (ed.). (1976) *Corrosion.* Newnes-Butterworth, London.
11. Krolicki, R. P., (1977) Internal corrosion examination and wall thickness measurement of pipe by radiographic method. *Materials Evaluation,* February, 32.
12. Charlton, J. S. and Ross, J. F. (1975) UK Patent Specification 1406489.
13. James, P. A. (1972) Radioactive isotopes solve industrial and processing problems. *Atomic Energy,* July, 20.
14. *Radioisotope Instruments in Industry and Geophysics.* Proc. Symp. Warsaw 1965, IAEA, Vienna (1966).
15. Kohl, J. *et al.* (1961) Measurement of mass per unit area, specific gravity and thickness gauging. In *Radioisotope Applications Engineering,* Van Nostrand, Princeton, 461.
16. Williams, J. (1979) Tips on nuclear gauging. *Instrument and Control Systems,* Jan., 47.
17. *Industrial Applications of Radioisotopes and Radiation Technology.* Proc. Conf., Grenoble 1981, IAEA, Vienna (1982).
18. Hunt, R. H. *et al.* (1957) Find catalyst density with isotopes. *Petroleum Refiner* **36**, 179.
19. Loeffel, R. (1982) Two-phase flow measurements using a 6-beam gamma-densitometer. Reference 17, 459.
20. Coogan, C. H. *et al.* (1975) Measurement of local density by gamma-ray attenuation in a free surface flash evaporator. *Mech. Eng.* **97**, 56.
21. Heywood, N. I. and Richardson, J. F. (1979) Slug flow of air-water mixtures in a horizontal pipe: determination of liquid hold-up by gamma ray absorption. *Chem. Eng. Sci.* **84**, 17.
22. Hewitt, G. F. and Whalley, P. B. (1979) Flow measurement in two-phase (gas-liquid) systems. *IChemE Symp. Ser.* No. 60, 131.
23. Crompton, C. E. (1956) The versatility of radiation applications involving penetration or reflection. *Proc. First Int. Conf. on the Peaceful Uses of Atomic Energy,* Geneva, Vol. 15, United Nations, New York, 124.

24. Fulham, M. J. and Hulbert, V. G. (1975) Gamma scanning of large towers. *Chem. Eng. Progr.* **71**, 73.
25. Severance, W. A. N. (1981) Advances in radiation scanning of distillation columns. *Chem. Eng. Progr.*, September, 38.
26. Basic Radiography (series of articles) *Non-destructive Testing*, Vols. 2 and 3 (1969–1970).
27. Rockley, J. C. (1964) *An Introduction to Industrial Radiology*. Butterworth, London.
28. *Gamma Radiography*. The Radiochemical Centre, Amersham (1971).
29. Birchall, I. *et al.* (1975) Gamma radiography using short half-life radioisotopes. *Int. J. Appl. Rad. Isotopes* **26**, 141.
30. *Radiographic Film, Intensifying Screens, Film Processing and Darkrooms*. Engineering Industry Training Board Publication TE A25. Hills and Lacy, Reading, (1977).
31. *Kodak Data Book of Applied Photography*. Kodak Ltd., Hemel Hempstead.
32. *Industrial Radiography using Ilford Materials*. Ilford Ltd., Basildon.
33. Daggs, R. G. (1956) Portable isotopic X-ray units. In *Proc. First Int. Conf. on the Peaceful Uses of Atomic Energy*, Geneva, Vol. 15, United Nations, New York, 174.
34. Karchnak, G. F. and Naylor, C. A. (1977) Radiography while maintaining 300° F minimum temperature. *Mater. Eval.*, May, 24.
35. Clayton, C. G. and Cameron, J. F. (1966) A review of the design and application of radioisotope instruments in industry. In *Radioisotope Instruments in Industry and Geophysics*, Proc. Symp. Warsaw 1965, Vol. 1, IAEA, Vienna, 15.

14 Radiation Scattering techniques

E. A. EDMONDS

14.1 Introduction

The previous chapter dealt in detail with industrial applications of measurements of gamma-ray absorption for the purposes of determining levels in vessels, thicknesses of materials, density profiles and so on. In the main, these gamma-ray absorption techniques are the simplest and most flexible to employ on industrial plant. The basic principle is quite straightforward, and usually so is the interpretation of the results, especially since much analysis can be done quasi-theoretically relying on versions of the simple formula $I = I_0 \exp(-\mu dx)$ first given in Chapter 2, with values for μ or half-thickness being determined on site for the usually simple geometries of measurement employed. Furthermore, the design of the equipment is frequently fairly obvious, being governed merely by the need to straddle the item under inspection with a suitable source of radiation and a detector: and indeed the same equipment can often be used in a variety of situations without modification.

There are occasions, however, when a different approach is useful: instead of measuring the amount of radiation transmitted through an item under inspection, it is sometimes worth looking at the amount of radiation which is returned from it, or *scattered back*, to a point close to the primary source of radiation. Methods which rely on this kind of approach are called *radiation backscatter* techniques. The need for this kind of technique arises most obviously when access to the item under inspection is restricted to one side, so that it is not possible to straddle it with a source and detector. The detector is then placed alongside, or at least close to the radioactive source and usually shielded from direct radiation in some way. In this simple configuration, level measurement and interface detection can be accomplished but that is not all. Additional information can be derived about the composition and geometrical arrangement of materials in the item under inspection by looking at low energy secondary radiations stimulated in the test item by the primary source.

Backscatter techniques suffer from a major complication and that is their acute sensitivity to geometrical effects. The amount of radiation which reaches a detector looking particularly at scattered radiation, instead of transmitted (essentially unscattered or lightly scattered) radiation, is critically dependent on the exact juxtaposition of source, detector and scattering materials. Not only that, but the different kinds of radiation are scattered in very different ways. This sensitivity to geometrical effects and type of radiation can be a drawback. It means that it is difficult to interpret the results of anything but the simplest measurements without carefully calibrating the experimental arrangement under laboratory conditions. Usually the equipment to be used must be thoughtfully designed and tested in advance of application on plant. On the other hand, the strong dependence on geometrical effects can confer great sensitivity of measurement in many circumstances and this can be particularly valuable when looking at properties of materials which vary over short distances.

14.2 Radiation scattering processes

14.2.1 Gamma-rays and X-rays

Gamma-rays and X-rays are scattered when they interact with electrons and nuclei in matter through which they pass. In Chapter 2 the main processes were described, these being photoelectric absorption, Compton scattering and electron–positron pair production. At low energies, below 0.5 MeV, the most important process is photoelectric absorption. At higher energies, Compton scattering becomes predominant and then, above a threshold of 1.022 MeV, pair production takes effect.

Compton scattering, in which a photon bounces off an electron, is clearly a simple scattering process. If the interaction of the photon is with an electron loosely bound to a nucleus, at the outside of an atom, the products of the interaction are an energetic electron and a photon of diminished energy travelling in a different direction. This photon could be scattered in any direction—it could be returned in the direction from which it came, that is, backscattered. Essentially, the amount of scattering will be proportional to the electron density in the material when Compton scattering is the dominant process occurring.

Photoelectric absorption does not look like a scattering process at all, at first sight. After all, the photon imparts all its energy to one of the inner electrons of an atom and so, as stated in Chapter 2, ceases to exist. The product of this interaction would seem to be merely a high-energy electron ejected from the atom. There is, however, the ionized atom to consider. This atom is in a highly excited state, with a vacancy in an inner electron shell. This is the precursor condition to X-ray emission. When a free electron, or perhaps an electron from an outer shell, falls into this vacancy, an X-ray is emitted. This phenomenon is called *X-ray fluorescence* (XRF) and it is dealt with in more detail in the next section. Suffice it to say, here, that this stimulated X-ray emission can occur in all directions and X-rays can emerge from irradiated materials travelling towards the source of primary radiation. This X-radiation is, in effect, also backscattered radiation.

In addition to backscattered gamma-ray photons and the X-ray photons mentioned, any detector placed to observe these radiations will inevitably detect a component due to bremsstrahlung from the slowing down of the high-energy electrons, the products of ionization.

14.2.2 X-ray fluorescence

When energetic photons stimulate the emission of X-rays from atoms by creating vacancies for electrons in the inner orbital shells, this process is called X-ray fluorescence. In effect, the photons are doing the same job as the high-energy electrons fired into a target in a medical X-ray machine. The X-rays emitted are characteristic X-rays, that is, characteristic of the electron level structure of the atom.

In the nomenclature of X-rays, the inner electron shell is called the K-shell, the next shell out is called the L-shell, and so on. When an X-ray is emitted because an electron falls into a vacancy in the K-shell, this characteristic X-ray is said to be a 'K' X-ray of the particular atom or element under consideration. Depending where the electron that fell into the vacancy came from, there is a further subdivision of the nomenclature using the Greek letters α, β, γ.... For instance, if the electron came from the next shell out from the K-shell, it would be the 'Kα' X-ray. Figure 14.1 illustrates the point and gives some hint of the complexity which arises, both in the number of possible X-rays which can be produced and in the nomenclature itself. In Chapter 2 a fairly simple treatment was adopted, dealing essentially with decay of excited states by the capture of

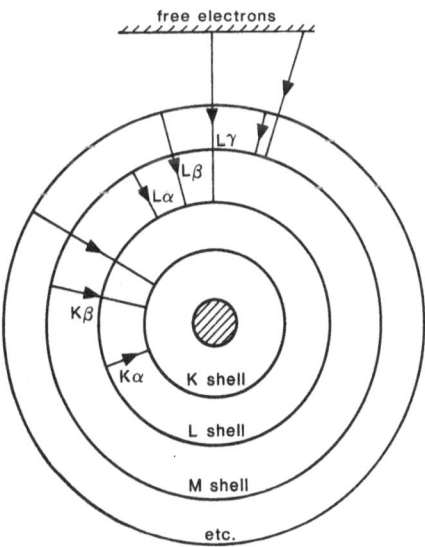

Figure 14.1. Characteristic X-rays.

free electrons. That process generates the most energetic X-rays which an atom can emit. In reality, there are often internal transitions in the electronic level structure of the atom which yield characteristic X-rays of lesser energies. There can be complex cascading of electrons within the orbital shells so that the yield of X-rays is not one simple characteristic 'line' as shown in Figure 2.7 but a multiplicity of lines, the low-energy, 'soft' and 'mushy' X-rays which were almost dismissed from consideration in Chapter 2.

(Were one looking at photon transmission at these low energies one would notice a sudden decrease in half-thickness for each element at the energy when a photon is just capable of ejecting one of the tightly-bound K-shell electrons. This shift in the half-thickness is quite substantial—a factor of about six—and it is referred to as the *K-absorption edge* for the element.)

14.2.3 *Beta-rays*

Beta-particles are energetic electrons and they interact very strongly with the electrons in matter, being scattered and deflected very severely. The amount of scattering, including backscattering, is roughly proportional to the atomic number of the scattering material providing bulk density does not vary too rapidly. This is merely another way of saying that the amount of scattering depends on the number of electrons per unit volume. Again, geometrical effects are important, both for detection of backscattered electrons and for their production. Electrons have a fairly short range in solids and liquids, and providing the thickness of scattering material is above some threshold the backscattered signal does not change much as thickness increases. But below this threshold the amount of backscattering is a strong function of thickness because the higher energy component of a beta-ray flux is able to penetrate the material and so will not contribute to backscatter signals. The higher the energy of a beta-particle, the less likely it is to be scattered through large angles.

14.3 Industrial applications

Radiation backscattering is used as a tool in industry in a variety of guises. The case studies in the next section describe some specific kinds of application — and it is in relation to the case studies that some of the important points of detail are amplified. What follows in this section is a general summary of the range of applications.

Gamma-ray backscattering is primarily used for short-range density determinations, such as detection of voidage in vessels or behind containments of one kind or another. Level gauging can be important where gamma-ray absorption techniques or neutron backscattering techniques (discussed in detail in Chapters 13 and 15 respectively) fail or are impracticable for one reason or another. Level gauging using backscattered gamma-rays is usually based on relative density determination, looking for density-related signal changes under conditions of simple and constant geometry. Gamma-ray backscatter level-measuring devices are usually portable, prototype instruments put together to match the exact requirements of the job in hand. Although technically feasible, permanently installed gauges for level determination are rarely based on backscattering of gamma-rays because the point of measurement must always be local to the source/detector assembly and spanning an interface which is liable to move significantly up or down would require the detector head to be driven to track or hunt it. It is usually easier to separate the source and detector so that they straddle the vessel under inspection and use the gamma-ray absorption technique, taking advantage of the span conferred by spreading the beam over the distance traversed. In one special case, however, the most convenient method of interface detection is based on gamma-ray backscattering, and this is in downhole well logging operations, which are discussed in section 14.4.1 (iii). In this application, the probe containing the radioactive source is indeed driven up and down to hunt the interface, under the control of an operator on the surface.

X-ray fluorescence is an important analytical technique of wide application in industry. Sophisticated laboratory tools have been developed to analyse the elemental composition of materials by looking at characteristic X-rays emitted by samples. These are often based on stimulation of X-rays by electron beams, just as in medical X-ray machines. The use of electron beam stimulation of X-rays is convenient for the specialist laboratory because it is easy to vary the energy of an electron beam merely by altering the driving voltage—simply turning a knob, in other words—to seek or scan for particular X-ray lines of interest. Outside the laboratory, however, on operating process plants, devices which incorporate electron accelerators are not favoured, for a variety of reasons to do with convenience and electrical safety. Fortunately, the fact that these X-ray emissions can also be stimulated by photon capture provides a means of developing instruments containing sealed radioactive sources which emit radiations of suitable energy to excite the X-rays of interest. The major field applications on operating plant arise in two main areas. The first is in applications which require that a device be carried to points of inspection on the plant. Portability of the apparatus is an important criterion in this case. For example portable XRF instruments may be used for metals identification by establishing the presence and concentrations of trace elements in different grades of steel. The second major area of application of XRF instruments is in installed, on-line analysers which return process information to plant operators for the purpose of control and optimization.

Beta-ray backscattering is well-established in the field of thickness gauging, particularly for the measurement of thin layers on substrates, like the thicknesses of plastic coatings, varnishes and paints on metals, coatings on optical lenses and so on. In addition, quantitative analysis can be performed on mixtures of materials, taking advantage of the dependence of beta-ray scattering on electron density or atomic number. For example, weight fractions of different hydrocarbons in mixtures can be analysed on-line, again for the purposes of control and optimization of industrial processes.

14.4 Case studies

14.4.1 *Gamma-ray backscattering*

(*i*) *Portable level gauges.* A portable gamma-ray backscatter level gauge is illustrated in Figure 14.2. This apparatus incorporates a sodium iodide scintillation detector, a radioactive source and lead shielding to prevent direct radiation from the source from penetrating to the detector. The base-plate carries a small handle, but the casing of the detector provides the main point of purchase. The initial reason for building such a level gauge was to enable one operative to measure levels in process vessels and tanks which had previously been gauged with portable gamma-ray absorption equipment, that approach requiring at least two operatives—one to control the detector and electronics and one to control the radioactive source. Successful trials have been conducted in the laboratory and on plant using this equipment with cobalt-60 and caesium-137 sources, but the system does have limitations. The main problem is that the 'range' of the instrument is low. Figure 14.3 shows the results of trials of this apparatus set up for optimized geometry using a caesium-137 source of 1.5×10^8 Bq. (Clearly, the exact spatial and angular relationship of source and detector are variables.) On an unlagged vessel, the sensitivity of the instrument to changes in electron density in the contained liquid diminishes with increasing wall thickness of the vessel until, at 32 mm of steel wall, the sensitivity is too low to allow accurate gauging of interfaces between different liquids. Fortunately, the majority of process vessels have walls which are thinner than 32 mm: about 15 mm is not untypical. Unfortunately, most process vessels are clad in thermal insulation to at least 50 mm. Once the detector head is removed from the surface by a distance of 50 mm the device becomes insensitive. So, even though thermal insulation itself is fairly light, that is of low density compared with steel, its very presence as a 'spacer' severely compromises the performance of the instrument. Some improvement can be sought by changing the caesium-137 source for one which emits gamma-rays of higher energy, that is more penetrating gamma-rays, for example by using a cobalt-60 source, but as the penetrating power of the gamma-rays increases so too does the difficulty of shielding the detector from photons which travel directly through the lead shielding. Increasing the lead shielding soon makes the instrument too unwieldy for one person to use, whereupon much of the justification for using gamma-ray backscattering as a means of level gauging becomes lost.

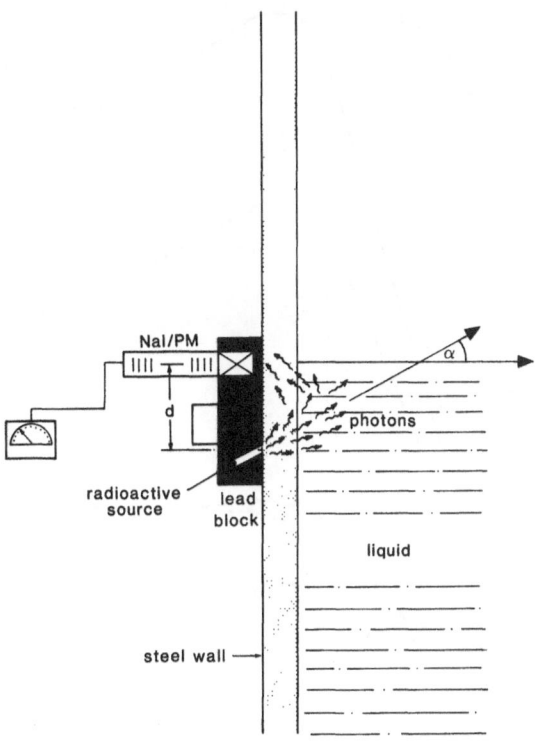

Figure 14.2. Gamma-ray backscatter level gauge.

Ultimately, portable level gauges exploiting backscattered neutrons supersede gamma-ray backscattering techniques. Neutron backscattering techniques are discussed in detail in Chapter 15. These techniques are superior provided the materials being gauged are hydrogenous, as are most process materials in refineries and chemical plants. For non-hydrogenous process materials like titanium tetrachloride ($TiCl_4$) or chlorine, gamma-ray backscattering is still used as a gauging technique, an alternative to gamma-ray absorption measurements.

(ii) Voidage determination and thickness measurement. The very sensitivity of the instrument illustrated in Figure 14.2 to short-range geometrical effects has made it successful in a number of measurements conducted on plant to locate and determine the extent of 'voidage' in steel walls and in thermal insulation itself.

In one application, a spherical vessel containing titanium tetrachloride was due for routine inspection. A previous inspection using ultrasonic techniques to measure the thickness of the steel walls had revealed suspect points of apparent thinning which had tentatively been ascribed to the presence of sub-surface 'inclusions' in the steel. These inclusions were expected to reflect

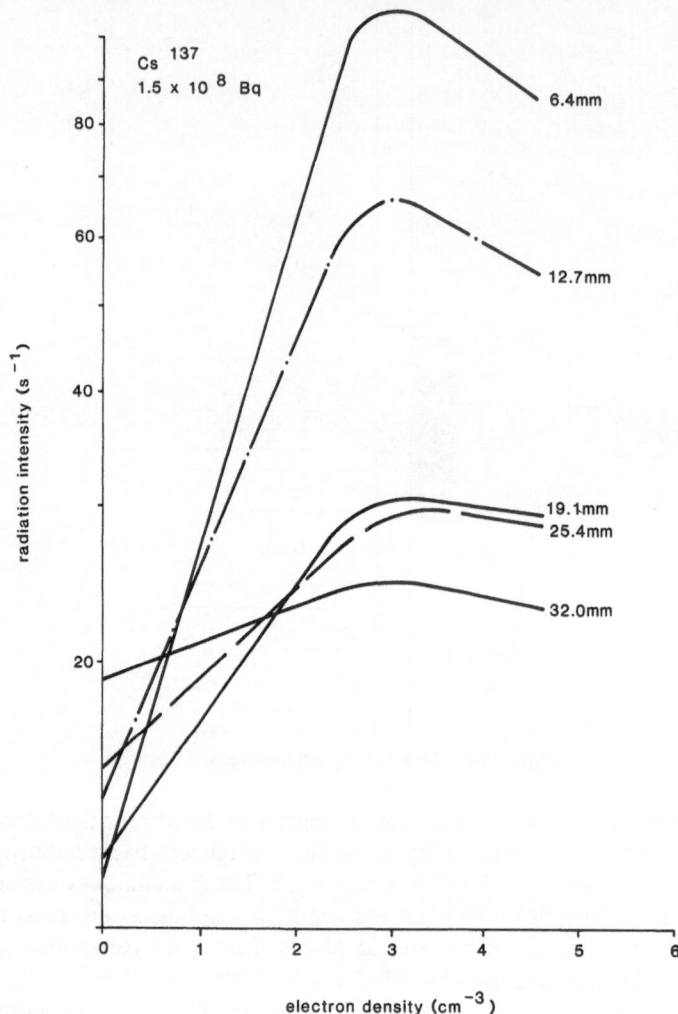

Figure 14.3. Gamma-ray backscatter gauge variation with electron density and vessel wall thickness.

ultrasound from some intermediate point before the inner steel/contained-liquid interface, and thus cause an erroneously low reading for wall thickness. Such inclusions might, in reality, be no more than internal crystalline boundaries or regions of impurity in the steel which would not necessarily compromise the integrity of the vessel. Gamma-ray backscatter measurements were able to confirm the distance between the outer wall of the sphere and the contained liquid; in other words the real separation of the liquid and the radiation detector placed on the surface of the sphere. The results showed that inclusions did exist which were reflecting ultrasound and causing the wall thickness to be under-evaluated.

A similar application was when gamma-ray backscattering was used to determine the extent of 'graphitization' of cast-iron pipes carrying river water for cooling in a power station. The river water was tending to leach out iron from the castings, leaving behind an insubstantial, graphitized material. The graphitized layer extended from the inside of the pipe towards the surface with no external indication. This graphitization behaviour was first found by visual inspection of certain parts of the system during a routine shutdown and internal inspection. There was a need systematically to evaluate the extent of the problem over all the pipework, and gamma-ray backscattering provided a simple means of monitoring the cooling-water system while it was on line, using a cobalt-60 source of 4.5×10^7 Bq. Local access to the surface of the pipe was all that was required. As the pipework became graphitized, water permeated the damaged layer, moving closer to the external surface of the pipe. This made the pipe wall appear to be effectively thinner to gamma-rays than in its original, non-graphitized state. The effect on backscattering was very clear, as the calibration curves in Figure 14.4 show. The sensitivity obtained was much better for gamma-ray backscattering than for absorption, because the percentage change in density, to which gamma-ray absorption measurements are sensitive, was quite small over the total diameter of the water-filled pipes.

Gamma-ray backscattering measurements have been used to search for voidage in *in-situ* foamed thermal insulation. The thermal insulation of pipework in low-temperature service is sometimes accomplished by encircling the pipe with aluminium alloy cladding in sections or spools, leaving an air gap into which a foam mixture can be introduced by pumping. An internal reaction in the foam mixture causes it to expand, filling the gap and displacing the air. The foam then sets hard. This procedure was introduced as a cost-effective alternative to cladding the pipework by hand with solid insulation. The only problem was that there was no test procedure whereby quality control could be maintained. There was no way of telling whether the procedure was fully immersing the pipework in insulation or if voids and holes, gaps and faults were being left behind. The testing procedure in force was actually to strip the cladding off items selected at random and visually inspect them. This was, of course, a destructive test, since the inspected section then had to be repaired. The non-destructive testing which was proving most successful before gamma-ray backscattering was introduced was judicious tapping of the cladding with a coin. A keen ear could locate a substantial void of the kind that occasionally occurred when, due to some failure in communication or record-keeping, a section was deemed to have been insulated when, in fact, no such thing had been done. This 'wheel-tapping' test was swiftly superseded by gamma-ray backscattering measurements. It was found in laboratory trials that, using an americium-241 source of 1.5×10^{10} Bq, cubical voids of 25 mm side could be located with ease through

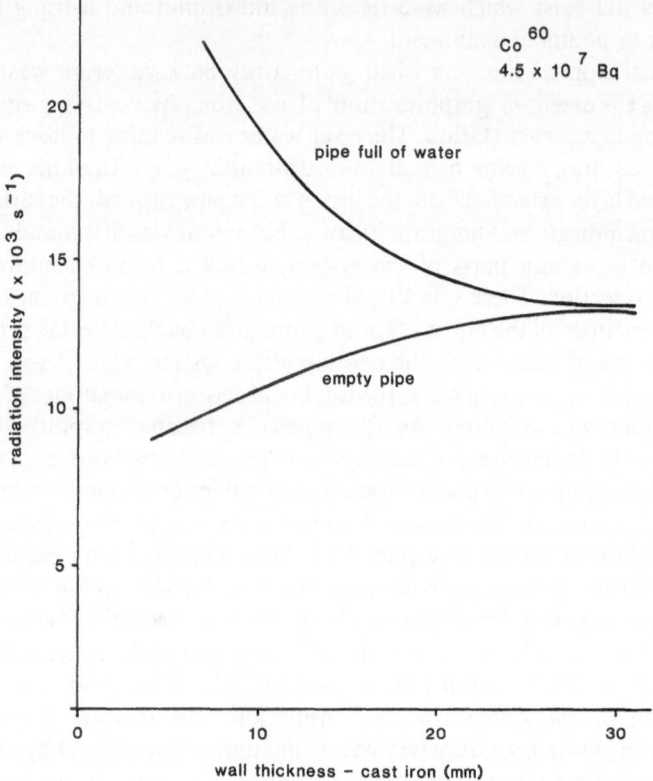

Figure 14.4. Gamma-ray backscatter gauge response through cast iron.

the cladding. It was possible to determine the position of a void, and whether
it was adjoining the cladding or the pipe wall. In fact, when detailed scanning of
the surface was undertaken very minor faults were located in the test spool in
the laboratory, amounting to no more than folds or ripples in the insulation,
but these were judged to be of no practical consequence for the thermal
insulation properties of the insulation. Accordingly, very detailed scanning on
site was abandoned and a fairly rapid screening technique adopted whereby
the source-detector assembly was planed over the surface of the alloy cladding.
A problem was encountered on site, however, which was not originally
expected. There were large variations in backscatter signal due to different
thicknesses of cladding. The cladding tended to vary in gauge from sheet to
sheet as it was drawn from stores and it was quite common to come across
adjacent sections of pipeline which had been clad in alloy of different
thicknesses. This had the effect of shifting the zero point (no void) for the
apparatus quite significantly from spool to spool. Fortunately, the cladding
showed no tendency to sudden, localized changes in thickness and so

variations in the cladding did not preclude seeking local voids in the thermal insulation under it.

The measurements of graphitization could probably equally well have been conducted using neutron backscattering. In Chapter 15 an example is given of voidage location using backscattered neutrons. In that case, the hydrogenous liquid was in fact moving away from the vessel wall due to gas formation. Voidage determination in the *in-situ* foamed insulation was attempted by neutron backscattering and in that case the technique did not work. It had been hoped that there would be sufficient residual solvent in the essentially inorganic foam when it had set to give a reasonable slow neutron signal. In practice, this was not so and gamma-ray backscattering with all its associated problems became the favoured technique. Certainly, had the use of neutron backscattering been possible then variations in thickness of the alloy cladding would have been of no consequence at all. On the other hand, the actual resolution of voids—the minimum detectable void size—would probably have been poorer because of the greater penetrating power of the fast neutrons and their consequently wider spread in space.

(iii) Well-logging. Gamma-ray backscatter probes are used to evaluate the contents of underground storage cavities containing a variety of different materials. Chemicals like ethylene, propylene, natural gas and nitrogen are all stored underground in wide, flat cavities at depths of around 500 m. Usually, but not always, the cavities are formed in underground salt layers by 'solution mining'—that is, brine extraction. Material for storage is then pumped down and kept over brine. As the inventory of stored product alters—when material is produced and sent down for storage or when stored product is removed from the cavity and sold—the position of the interface between the brine and the product moves in the cavity. Brine is pumped in or out to 'balance' the well.

Such wells are 'logged' by lowering down into them a gamma-ray backscatter probe on a cable by which it is connected to recording instruments and power supplies at the surface. The probe consists of a steel casing which houses a radiation detector and a gamma-ray emitting radioactive source, scandium-46 as scandium oxide, contained in a removable, threaded, nose-piece. When not in use, the radioactive component can be removed from the probe for safe storage. When the probe is lowered into the cavity the amount of radiation scattered back to the detector depends on the nature and density of the material surrounding it. As it descends there is returned to the surface a characteristic profile of the kind shown schematically in Figure 14.5. The position of the brine/product interface can be located with relative ease and referred to standard reference points. It can usually be located to within 15 mm with little difficulty.

The radioactive component scandium oxide is 'activated' in a nuclear reactor and installed in the probe in the form of an unsealed source. The half-life of scandium-46 is quite short (84 days) but a 'fresh' source of about

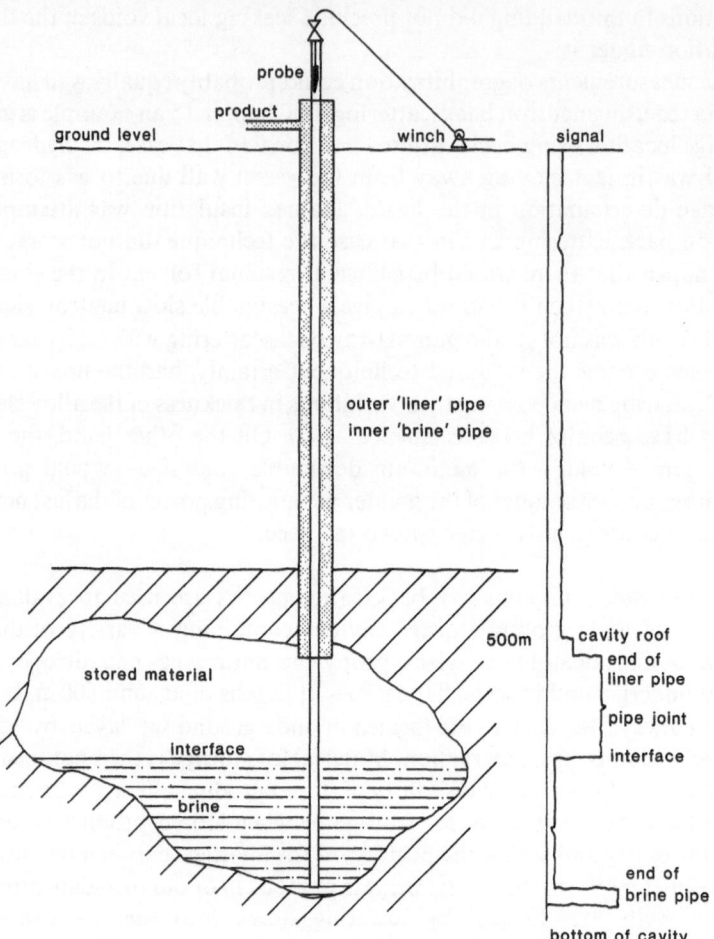

Figure 14.5. Gamma-ray backscatter well logging.

4.5×10^8 Bq has a useful lifetime of about one year. The advantage of using a short-lived unsealed source is that should a probe be lost down a well, the radioactive material can be 'written off' as safely disposed of. It poses no radiological hazard. Probes are occasionally lost, when cables are cut accidentally by the internals of the valves sealing the wells. The probe then plunges to the bottom of the well and the operator on the surface has a few moments of silent contemplation looking at the sad end of his cable. (These stories sometimes have a happy ending when probes are retrieved or 'rescued'. The record to date is twelve years. A lost probe was hooked back to the surface during the course of routine pipe maintenance after being missing for a dozen years and it was found to be in working order, when a new radioactive source was attached—the original source was long since dead.)

This kind of logging is also used during routine inspection and maintenance of the downhole pipework associated with the well. Notice on Figure 14.5 that pipe joints and the ends of pipes can be located on the logging profile. This well logging tool is used during cavity development to ensure that the pipework is correctly installed in the well.

It is worth mentioning that these wells are usually sonar-scanned in a series of horizontal planes to establish their shape and lateral extent. No two wells are the same, and the shape of a cavity should be well known so that calibration graphs showing interface position against volumetric capacity can be prepared. Otherwise the interface measurement itself would not be very useful for the purposes of routine stocktaking. Routine stocktaking aside, well-logging by gamma-ray backscattering is extremely valuable for monitoring the position of the interface at critical points in the cavity, such as when the interface is approaching the end of the brine pipe—that is, when the well is nearly full. It is important to be able to gauge swiftly and accurately the position of the interface so that stored chemicals can be prevented from getting into the brine system, with potentially disastrous consequences.

The probe also contains temperature-sensitive oscillators so that the temperature at different points in the cavity can be measured to allow corrections to be made when volumetric capacity is converted to stored mass. This is very important for compressible fluids like the ones usually stored.

14.4.2 X-ray fluorescence

(i) *Portable analysers.* Figure 14.6 shows a portable XRF probe, consisting of a radioactive source, a scintillating crystal/photomultiplier tube type detector, and *balanced filters.* The source is selected for the particular analysis required, and when the unit is not in use, it is shielded by the movable shutter. When the shutter is opened, radiation from the source falls on the sample and generates fluorescent X-rays from the elements present. The X-rays are passed in turn through each of the balanced filters, which are selected for the particular analysis required. The intensity of the radiation falling on the scintillating crystal is measured.

The balanced filters are selected for the energies of their K-absorption edges, which were mentioned in section 14.2.2. One filter is selected to transmit the X-rays from the element of interest while the other is selected to absorb them strongly. By comparing the intensity of the radiation passed by one filter with that passed by the other, a good measurement of the intensity of the X-ray line of interest can be obtained, irrespective of any other X-rays and scattered radiation present in the spectrum. The filters provide a specific pass-band for the X-rays of interest. The intensity of the X-rays is related to the concentration of the element of interest in the sample. The transmission characteristic of a pair of filters is shown in Figure 14.7. This example

K

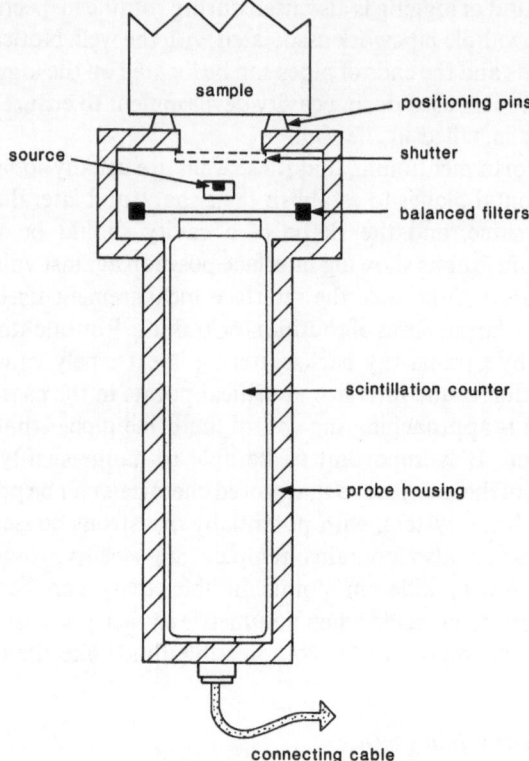

Figure 14.6. A portable XRF analyser.

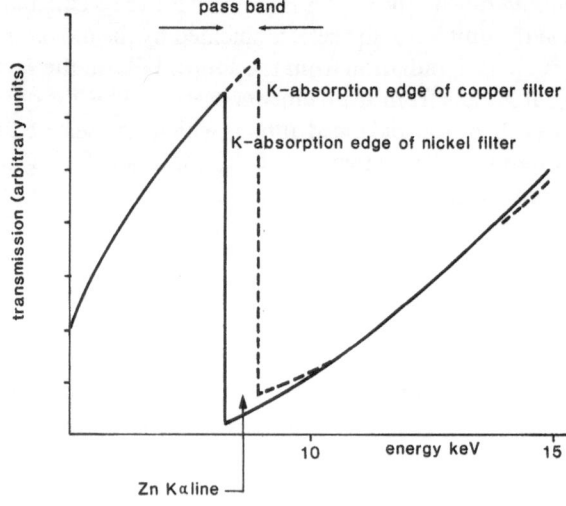

Figure 14.7. Balanced filters for Kα line of zinc.

illustrates the use of filters to select the Kα X-ray from zinc. The thicknesses of the nickel and copper filters are chosen so that they each absorb the same amount of radiation outside but near the well-defined, narrow pass-band. Some sources suitable for stimulating XRF emission are listed in Table 14.1. Balanced filters appropriate for analysing a range of elements using K-absorption edges are listed in Table 14.2. Table 14.3 lists some filter pairs for analysing Lα X-rays. For heavy elements it is usually preferred to excite Lα

Table 14.1 Some radioactive sources suitable for XRF in portable analysers

Source	Half-life (years)	Emission energy (keV)	K X-rays usefully excited	L X-rays usefully excited
Fe^{55}	2.7	5.9	Al—Cr	Br—Xe
Pu^{238}	87.8	12—17	Mn—Y	Eu—Bi
Cd^{109}	1.2	22	Fe—Mo	Nd—Th
Am^{241}	433	59.5	I—Lu	
Co^{57}	0.7	122 and 136	Hg—U	

Table 14.2 Balanced filters for Kα X-rays

Element	Kα energy (keV)	Filter pair
Ti	4.51	Sc/Ti
V	4.95	Sc/Ti
Cr	5.41	Ti/V
Mn	5.90	V/Cr
Fe	6.40	Cr/Mn
Co	6.93	Mn/Fe
Ni	7.48	Fe/Co
Cu	8.05	Co/Ni
Zn	8.54	Ni/Cu
Nb	16.61	Sr/Y
Mo	17.48	Y/Zr
Cd	23.17	Ru/Rh
Sn	25.27	Pd/Ag
I	28.61	In/Sn
Ta	57.52	Ho/Tm
W	59.31	Er/Tm
Hg	70.82	W/Re
Pb	74.96	Re/Ir

Table 14.3 Balanced filters for Lα X-rays

Element	Lα energy (keV)	Filter pair
W	8.40	Ni/Cu
Pt	9.44	Cu/Zn
Hg	9.99	Zn/Ga
Pb	10.55	Ga/Ge
Bi	10.84	Ga/Ge

X-rays, chiefly because these are of lower energy than the Kα X-rays of the same element; and for heavy elements (that is elements of high atomic number) quite high energy photons may be needed to excite, Kα X-rays; which would require special sources, possible redesign of the probe and increased radiological protection problems. (Compare the energy of the Kα X-ray energy of lead with the energy of the Lα X-ray.)

The probe is simply placed up against the sample under consideration and the source shutter is opened by movement of a trigger. The total radiation count accumulated in a preset time is recorded for each filter in turn, the filters being moved by a simple actuator. The absolute accuracy of any particular analysis depends on the method of calibration. Highest accuracy is obtained by using standard samples of known elemental concentration which have densities and matrix compositions similar to those of the material under test.

The probe illustrated is capable of taking a wide range of radioactive sources and filter pairs. It can be dismantled and reassembled in configurations suitable for analysing all elements from titanium to uranium. Stripping down such a device, however, is a job for a classified radiation worker. There are now available XRF probes with more limited and specific applications, which are sold as sealed systems. These can be used and operated by relatively inexperienced operators. They usually contain a small set of filter pairs which can be placed in position by means of external switches, slides or levers. Generally they permanently contain, or can be 'armed' simply with, two or three sources. One of these sources is usually a long-lived standard which allows the efficiency of the radiation detector to be checked before each application. Such devices are often sold for the specific analysis of a handful of elements in a particular kind of matrix. The identification of trace elements in different kinds of steel is a favourite application which has already been mentioned. With the advent of cheap microprocessors, these instruments often incorporate LED or LCD displays which announce the type of steel to the operator without bothering him with any details of the overall elemental composition. The microprocessor already contains information about the manufacturing 'mix' of the steels it may be set to identify. Ultimately, these instruments can be made very sophisticated 'black boxes' indeed, with complete microprocessor control of source selection from an internal set, automatic selection of filters both to identify X-rays and to fine-tune the spectrum falling on the sample from the source, and automatic counting and logging.

(ii) *On-line analysers.* Figure 14.8 shows an XRF instrument which was installed to measure the concentration of bromine in a liquid catalyst being used as part of the process on a petrochemical plant. The bromine was an integral part of the catalyst and was present in a fixed ratio with the other components. It was used as a 'marker' by which to measure the throughput of catalyst. This particular analyser allowed the process operators on the plant to control their

Figure 14.8. On-line bromine analyser

catalyst inventory efficiently, enabling them to minimize the stock in circulation in the process.

Bromine $K\alpha$ X-rays were stimulated in a catalyst sidestream diverted from the main flow through a sample cell. The radioactive source was an annular source-target assembly of the type already described in Chapter 12. Gamma-rays of 59.6 keV from americium-241 were used to excite X-rays of 22.1 keV in a silver target. The X-rays from silver efficiently stimulated the 11.9 keV $K\alpha$ X-rays from bromine but were sufficiently far away in energy to be resolved. The detection system in this case was a gas-filled proportional counter, which gives output pulses the amplitude of which is proportional to detected photon energy and this allows for electronic discrimination of photon energy.

The long half-life of americium-241 ensured that the intensity of the source was effectively constant which obviated the need for frequent recalibration of the analyser. The use of X-rays from silver provided a further, less obvious, benefit because the mass absorption coefficients of carbon and hydrogen are identical at 22 keV. By choosing to exploit X-rays from silver, problems associated with variations in the composition of the hydrocarbon matrix were eliminated.

The sampling system consisted of a stainless-steel cell through which a representative fraction of the catalyst stream could be diverted. The cell was fitted with a window made of thin beryllium which minimized the attenuation of the low energy X-rays. The instrument was also provided with a means of diverting the catalyst stream at any time so that samples of known bromine concentration could be flowed through for the purposes of calibration.

14.4.3 Beta-ray backscatter

(i) *Thickness measurement.* Beta-ray backscattering is useful for measuring the depth of thin coatings laid on substrates which differ in atomic number from the added surface layer. This is because the amount of scattering is roughly proportional to atomic number, and beta-particles have a fairly short range in solid materials. The best applications for beta-ray backscatter are therefore ones where one material of high atomic number is laid down on another of low atomic number, or vice versa. In the former case, very thin layers can be measured whereas in the latter case thicker layers can be looked at. The simplest geometry occurs where the substrate is being produced in flat sheets. This is quite a common occurrence in industrial processes and gauges have been used to measure the thickness of tin and zinc coatings on steel for instance[1-3], an application for which beta-particle backscattering is ideally suited. (Interestingly, XRF has also been used in the same kind of application[4].) The measurement of the thickness of plastic coatings on metals and optical lenses and of the thickness of coatings of precious metals on electrodes, printed circuitry and jewellery have all been studied in detail and devices have been installed in some processes[5].

A recent, very successful industrial application of beta-ray backscattering has been in the monitoring of thin oil films in compressors. Lubricating oil breaking through the seals in the compressors was passing into the process streams. In some very sensitive processes this can be both expensive and potentially hazardous. The brief was to produce a device which was small enough to be introduced into the limited space under the cylinders of the compressor, below the oil seals, and look at the piston rod moving up and down, giving an alarm signal should a droplet of oil run down the rod, smearing itself over the surface. A flat beta-ray backscattering gauge was produced which fitted into the 15 mm space available and which gave alarm signals for oil films a few molecules thick.

Thicknesses of the order of one or two microns (1×10^{-6} m) can be measured to a few per cent by beta-ray backscattering in circumstances of reasonable geometry[1,5]. This would seem to cover the accurate measurement of surface layers on rolled sheets and drawn or extruded wires admirably— and these are quite common industrial situations. Additionally, beta-ray backscatter gauges are usually small, self-contained and require quite small sources of low energy emissions. Despite this, beta-ray backscattering gauges have not been widely introduced into industrial processes. Mainly, this is because of geometrical effects which are hard to control and which prevent a good laboratory gauge, simple to set up by hand, being incorporated in the feedback loop of an industrial process. An additional problem is that these gauges are sensitive to the effects of extraneous materials deposited directly on the source or detector window, which is not an infrequent occurrence in 'dirty' industrial environments. This has meant that the technique has largely been confined to smaller-scale processes with low-volume, high-value products, such as the production of optical lenses and mirrors.

Nevertheless, the prospects for wider adoption of beta-ray backscattering gauges in industry now seem more promising. There is a large new industry involved in producing very thin polymer films for use as videotape substrates. Beta-ray backscattering thickness gauges are being tried in part of the control loop of these processes. There are still geometrical problems to be overcome if beta-ray backscattering is to be used to look at flat sheeting coming off an industrial roller at process speeds—say a few metres per second. This material invariably flexes and moves rhythmically as it goes past the detector assembly. This has the effect of introducing a 'wobble' into the backscattered signal because of the changing effective air gap between the source and detector. An additional problem is that the sheet coming off the roller may be a metre or more wide. The backscatter gauge, containing a sealed, point-like source, is looking at only a small part of the width. It might be very important for quality control that the entire width be monitored continuously. This requires an elaborate array of gauges across the width. The alternative, driving the entire gauge back and forth to scan the width, might be appropriate in some circumstances but it requires careful engineering to avoid introducing unacceptable changes in the effective air gap because of mechanical movement in the drive gear. Furthermore, if the gauge cannot scan quickly enough across the width of the product as it moves forward, local variations might be missed.

It should be said that there are few obvious alternatives to beta-ray backscattering in this example. In the past, when beta-ray backscattering gauges have been abandoned they have usually been replaced by strict adherence to the 'cookery school' of industrial process control where a recipe that works is diligently followed and 'off-specification' product is thrown in the bin. The higher the value of the product the more wasteful and hence unacceptable this approach becomes.

(ii) *Analysis of mixtures.* Beta-ray backscatter has been used in industry to determine the composition of mixtures, particularly where the components are molecules which are quite similar. For example, this kind of application arises from refinery operations and petrochemicals processes when mixtures of hydrocarbons have to be analysed. The components of the mixture are essentially merely carbon and hydrogen. (If mixtures contain trace elements which are to be determined then XRF is generally a better approach.) Instruments have been developed which are capable of measuring the hydrogen/carbon ratio in samples of hydrocarbon taken from industrial process streams. These instruments provide a rapid result in a non-destructive way, allowing the exact sample which yielded one set of results to be stored for future reference or passed on for further laboratory analysis.

A sample of hydrocarbon exposed to beta-particles scatters them depending on its mean atomic number and its density. The effect of density on backscattering is smaller than on direct transmission, so backscattering is more sensitive for determining effective atomic number than is transmission.

One particular instrument used a strontium-90 source of 7.5×10^8 Bq suspended on a wire in front of an ion chamber detector which faced a sample cell made of aluminium with a thin 'Melinex' window. The source had a plastic collimator around it to focus the beta-ray beam, and it was mounted on a lead backing to prevent beta-rays and bremsstrahlung from the source from directly entering the ion chamber. Figure 14.9 shows the arrangement. In practice, the current from the ion chamber, which is a measurement of the intensity of backscattered radiation, was compared with a reference current

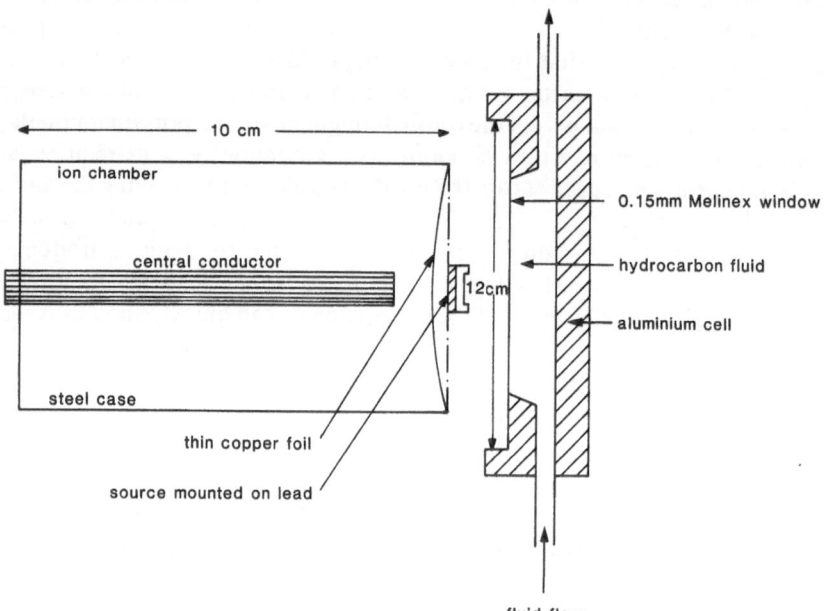

Figure 14.9. Beta-ray backscattering analyser for hydrocarbon H/C ratio.

from an identical chamber exposed to another strontium-90 source, this time of 7.5×10^7 Bq. This had the advantage that long-term drift due to the decay of the source was eliminated. Short-term instabilities due to temperature effects on the ion chamber filling gas were also suppressed.

The instrument was calibrated over the range of H/C ratio from 1 to 2.5 which covered light fuel oil product and naphtha feedstock. Figure 14.10 shows results for standard samples of known H/C ratio corrected for density. This equipment was used for several years as a process investigation tool in the laboratory. It gave rapid feedback of process performance under conditions of changing feedstock which enabled sensitive fine tuning of the complex multiproduct stream to be accomplished. The laboratory device, had it been necessary, could easily have been introduced as on-line equipment out on the plant, allowing continuous monitoring of process streams.

This kind of analytical use of beta-ray backscattering is, in many ways, complementary to gamma-ray backscattering, XRF and even neutron backscattering, which is discussed in the next chapter. In many instances, some particular measurement could be accomplished with more than one technique. Backscattering and emission measurements complement and can be used in conjunction with measurements of radiation absorption. Every different approach offers its own particular advantages and carries with it some difficulties. The examples discussed should give some idea of the wide range of possibilities. When one considers the vast number of radionuclides available and the wide spectrum of radiations they emit, varying in type and energy

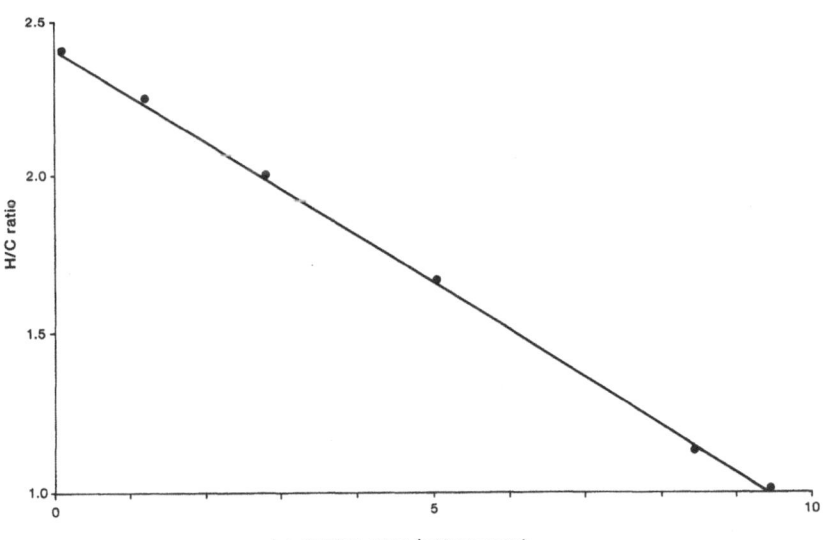

Figure 14.10.   Response of H/C analyser.

over an immense range, and the number of ingenious ways of detecting these radiations that exist, it becomes clear that the possibilities are almost unlimited.

References

1. Fremlin, J. H. (1965) *Applications of Nuclear Physics*. English Universities Press Ltd., [Hodder and Stoughton, London], 274.
2. Hayes, D. M. (1954) Continuous measurement of zinc coatings applied in a continuous coating process. In *Proc. 31st Meeting of Galvanizers Committee, Amer. Zinc Institute* 1.
3. Reider, J. E. (1962) A noncontacting measurement system for coating weights on continuous galvanizing lines. *Iron and Steel Engineer* 39, 73.
4. Cameron, J. F. and Florkowski, T. (1964) Radioisotope sources of low energy electromagnetic radiation and their use in analysis and measurement of coating thickness. In *Proc. Symp. on low energy X-Ray and Gamma Sources and Applications*, Chicago,: ORNL-11C-5, UC-23: Isotopes-Industrial Technology.
5. Clayton, C. G. and Cameron, J. F. (1971) *Radioisotope Instruments*. Part 1. Pergamon, Oxford, 216–229.

15 Neutron techniques

J. S. CHARLTON

15.1 Introduction

The neutron, as we saw in Chapter 2, is one of Nature's building blocks and is present in the nuclei of all isotopes, with the single exception of hydrogen-1. The availability of compact and portable neutron sources, together with the development of appropriate methods of neutron detection (Chapters 3, 12) has led to a continuous growth in the use of neutron techniques as investigational tools on process plant. To understand how and why neutron techniques are so useful, it is necessary first to remind ourselves of the ways in which neutrons interact with matter.

15.2 Neutron interactions

Neutrons emitted from radioisotope sources of the types described in Chapter 12 are energetic particles, with energies up to several MeV, the precise energy distribution depending upon the nature of the source. Collectively, such energetic neutrons are referred to as 'fast neutrons'. Being uncharged, fast neutrons, unlike alpha- and beta-particles, do not interact with the electric fields of atoms and molecules. In addition, because of the large mass of the neutron compared with that of the electron, neutrons are insignificantly affected by electron collision. The only way in which a fast neutron passing through matter can lose its energy is by direct collision with an atomic nucleus. Since, in matter, the spacings of the nuclei are relatively large it may be readily appreciated that fast neutrons are penetrating particles, capable of passing through substantial thicknesses of material.

15.2.1 Neutron slowing down

For fast neutrons, in the range 0.5–11 MeV, it is scattering processes which are responsible for energy loss and the slowing down of the neutrons.[1] Scattering processes fall into two categories, inelastic and elastic. *Inelastic* scattering could be regarded as a special type of absorption process in which the neutron is captured by a nucleus and re-emitted with reduced energy. The energy transferred to the nucleus is usually emitted as gamma-radiation. For light nuclei, inelastic scattering is far less probable than elastic scattering. *Elastic* scattering occurs when the total kinetic energy of the nucleus and colliding neutron remains constant. The neutron is slowed down in the collision and its direction of motion is changed. In the energy range 30 eV–0.5 MeV, elastic scattering is essentially the only process by which a neutron can be slowed down.

If the neutron energy before collision is denoted by E_1, and after collision by E_2, it is possible to show[2] that in a head-on collision, the energy transferred to the nucleus is

$$E_2/E_1 = \left[\frac{A-1}{A+1}\right]^2 \tag{15.1}$$

where A is the mass number of the nucleus.

From equation (15.1) it can be seen that it is possible for a nucleus to lose all of its kinetic energy in a head-on collision with a hydrogen nucleus. It is clear that, for this reason, the presence of hydrogen is a major factor in the slowing down of fast neutrons.

15.2.2 *Neutron thermalization*

Elastic scattering processes of the type described above bring about successive reductions in the energies of the fast neutrons until with energies of about 1 eV they are known loosely as 'epithermal' neutrons. The process by which the neutrons are further slowed down until they reach thermal equilibrium with the medium is called thermalization. Neutrons in thermal equilibrium with the medium are known as 'thermal neutrons' and on average possess a kinetic energy of 0.25 eV at room temperature.

The thermalization process is complicated by the fact that, since the neutron energies are now comparable with the binding energies of molecules, the nuclei of the medium can no longer be considered free. The effect of chemical bonding is to increase the effective masses of light nuclei such as hydrogen. Their effectiveness in slowing down neutrons by elastic collision is thus reduced (equation 15.1). This is partially compensated by the fact that the probability of an elastic collision occurring increases with the effective mass of the nucleus. The net result is, nevertheless, that the thermalization process is, to some degree, dependent upon the chemical composition of the medium.

However, the thermalization process constitutes only a very small part of the total slowing down of neutrons. This is because the epithermal region (0.025–1 eV) is narrow compared with the total range of neutron energies and most neutrons pass quickly through it to attain thermal energies. Thus, chemical effects, although not entirely absent are not, in general, very important in the slowing down of neutrons.

The importance of hydrogen in the neutron slowing-down process, and the relative insignificance of chemical effects, may perhaps be amplified by reference to some laboratory studies conducted by the author.

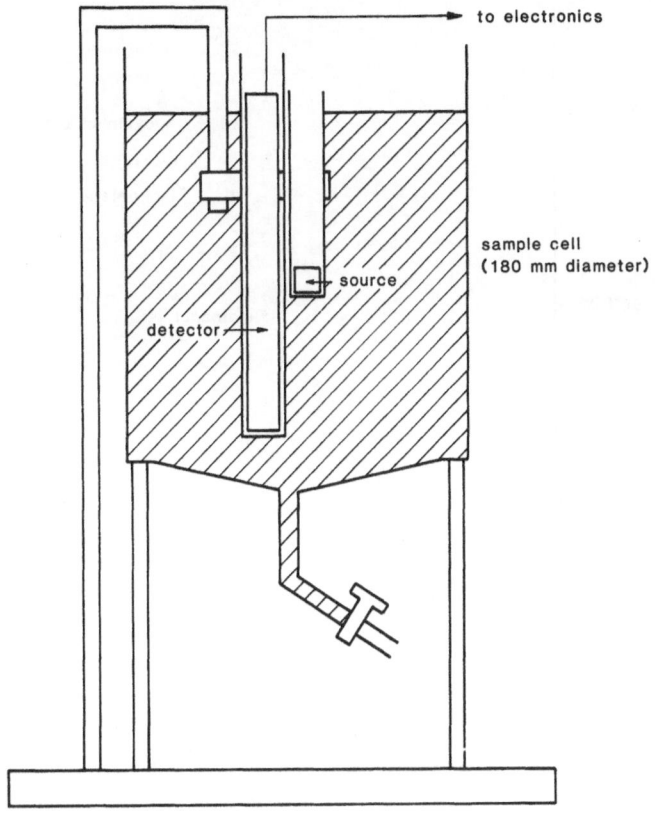

to electronics

sample cell
(180 mm diameter)

source

detector

Figure 15.1. Experimental arrangement for hydrogen concentration studies.

The basic experimental arrangement is shown in Figure 15.1[3]. The fast neutron source ([241]americium–beryllium) and the slow-neutron detector (a BF_3 proportional counter) which was insensitive to fast and epithermal neutrons, both contained in thin-walled steel tubes, were suspended inside the sample vessel in a central position. The cell was filled successively with a number of liquid samples of varying hydrogen content and chemical composition and in each case the number of slow neutrons detected in a fixed time period was recorded. The results are presented in Figure 15.2. It will be noted that there is a very definite upward trend in slow-neutron count rate with increasing hydrogen concentration, confirming our theoretical prediction that hydrogen concentration is the key factor in the neutron slowing-down process. The relationship between hydrogen concentration and slow-neutron count rate is not, however, perfectly linear. The deviations from linearity can be ascribed to the variation in concentration of the other elements present in the samples and to differences in chemical bonding.

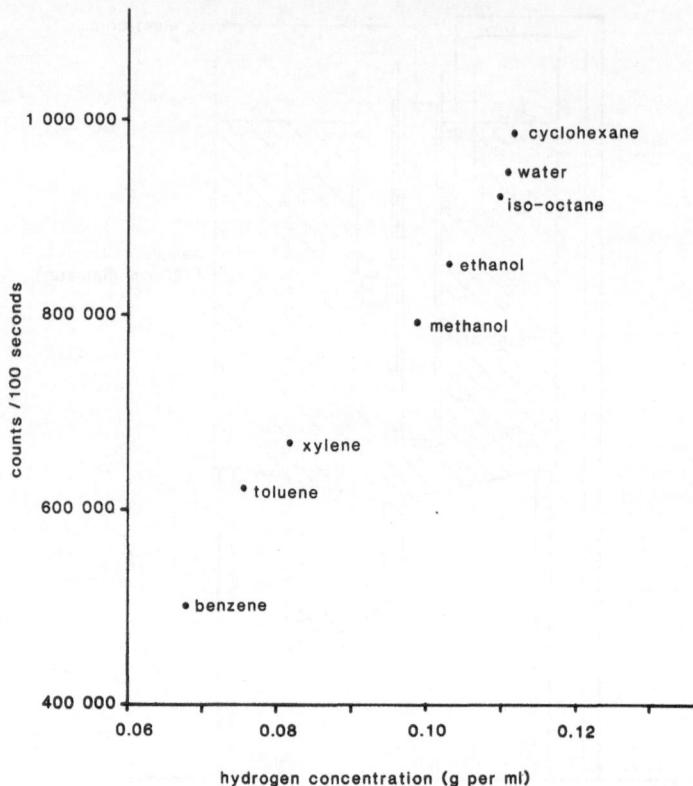

Figure 15.2. Slow-neutron count rate *v.* hydrogen concentration: immersion geometry.

15.2.3 *Neutron diffusion and absorption*

Once the neutrons have reached thermal equilibrium with their surroundings, they diffuse through the thermalizing medium until they are either captured or escape from its boundaries. The concentration of thermal neutrons in the vicinity of a fast-neutron source is increased by the presence of hydrogen. It is also increased by the presence of non-absorbing elements which impede the motion of the neutrons. The higher the concentration of these elements, the shorter are the distances migrated by the thermal neutrons and the higher is their density in the vicinity of the source.

In this energy regime, the neutron absorption process assumes major importance. Different elements differ widely in their capacity to absorb slow neutrons. Unlike the situation found with low-energy gamma-rays where broadly speaking there is a progressive increase of absorption coefficient with atomic number, there is no simple relationship between the ability of an element to absorb neutrons and its position in the periodic table. Absorption is

Table 15.1 Effective slow-neutron absorbers

Element	Absorption cross-section (barns)
Sm	8250
Eu	4370
Gd	3900
Cd	3315
Dy	1100
B	755
Ir	430
Hg	360
In	190
Er	166
Rh	150
Tm	118
Lu	108
Hf	105
Au	98
Re	84
Li	71
Ho	64
Ag	62

dependent upon the detailed nuclear structure of each isotope. Some effective absorbers of slow neutrons are listed in Table 15.1. The neutron-absorption probability is usually expressed as the 'absorption cross-section' in units of barns (1 barn = 10^{-22} mm^2). The presence of any of these elements in significant concentration is clearly a factor which can have a major effect upon the concentration of slow neutrons diffusing through a medium.

The absorption of a neutron is accompanied by the emission of radiation in the form of a gamma-ray. (The process is often referred to as 'radiative capture'.) Because the gamma-ray is emitted immediately after the neutron is absorbed by the nucleus (typically, within 10^{-14} seconds) it is known as a 'prompt' gamma-ray. The energy of the gamma-ray is characteristic of the particular neutron capture process which has taken place, and detection and identification of the prompt gammas can, therefore, be used as a basis for elemental analysis.

15.2.4 Neutron activation

When a slow neutron is absorbed by an atomic nucleus, the mass number of that nucleus increases by one unit. In some cases, the new nucleus will be stable: more often, however, slow-neutron absorption results in the production of a radioactive isotope. This, of course, is precisely the process which takes place inside a nuclear reactor in the course of radioisotope production (Chapter 4).

Each radioactive product of neutron absorption will decay with its own specific half-life and will be accompanied by the emission of its own characteristic spectrum of ionizing radiations. The detection and identification of radiation, and in particular, the gamma-rays emitted from a sample of material which has been exposed to a flux of slow neutrons, thus allows us to determine the elemental make-up of the sample—the so-called Neutron Activation Analysis technique, which has been touched upon in Chapters 10 and 11. It should be noted that these gamma-rays are quite separate and distinct from the prompt gamma-radiation mentioned in the preceding section.

15.2.5 Summary of relevant features of neutron interactions

The foregoing sections have focused upon those neutron interactions which are of relevance to plant investigation. Certainly, no attempt has been made to present a comprehensive coverage, which is far beyond the scope of this book. A number of excellent reference works are available for those who wish to pursue the subject more deeply.[4,5,6] There are three important phenomena to which attention should be drawn.

(a) The slowing down and thermalization of fast neutrons by a medium is primarily determined by the concentration of hydrogen in that medium, more or less independent of chemical form. (As shorthand, let us refer to the slowing down and thermalization processes as 'neutron moderation').

(b) Certain elements are extremely effective absorbers of slow neutrons. Their presence, even at low levels, can bring about significant depressions in slow-neutron flux.

(c) The absorption of a neutron by a nucleus usually produces a radioactive isotope which, upon decay, can be 'fingerprinted' by its characteristic radiation spectrum.

Each of these three phenomena gives rise to a class of techniques which are of importance in the study of problems in industrial plant and in the control of certain features of plant operation.

15.3 Techniques based upon neutron moderation

A high proportion of chemical process streams contain hydrogen in one form or another. (This is especially true of the petrochemical industry where hydrogen is present in practically *every* process material.) In neutron moderation we have a phenomenon which is sensitive to hydrogen concentration and it is, therefore, not surprising that techniques in this category are widely used.

15.3.1 *Basic equipment*

Though precise equipment requirements vary from one application to another there are certain features common to all measurements. The essential components of a measuring system are shown in Figure 15.3.

The measuring head comprises a fast-neutron source positioned adjacent to a slow-neutron detector. Many different source–detector combinations are possible: for plant measurements an americium-241/beryllium neutron source used in conjunction with a helium-3 proportional counter has been found to be particularly appropriate.

In use, the measuring head is held adjacent to, or suspended in, the material of interest. Fast neutrons from the source are slowed down in the material mainly by collisions with hydrogen nuclei, and the resulting flux of slow neutrons is sampled by the detector. The detector count rate, displayed on the electronic unit, is thus principally determined by the hydrogen concentration in the medium. In addition to displaying neutron count rate—either in the scaler or ratemeter mode (Chapter 3)—the electronic unit provides the high-voltage supply necessary for the operation of the detector. The whole system, measuring head and electronics, can be manufactured to be easily portable for field use.

The hydrogen concentration measurement, used either in its own right or in conjunction with other process data, can give us valuable information both

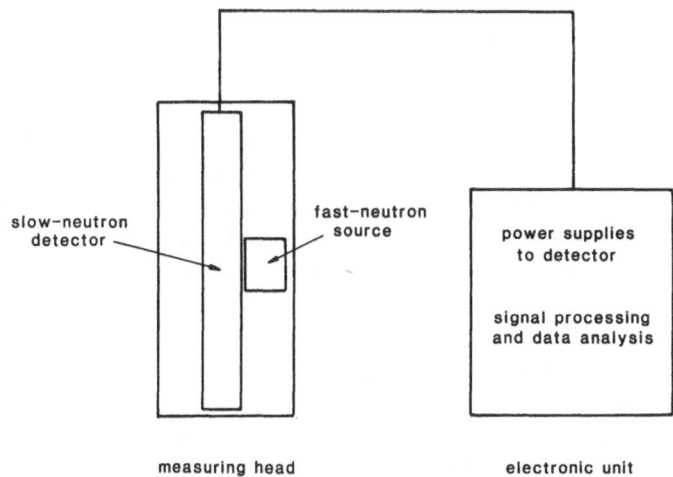

Figure 15.3. Instrumental arrangement of a typical neutron moisture gauge.

about the hydrogenous material itself and about the operation of the plant in which it is being processed.

15.3.2 *Neutron moisture gauges*

The principle of operation of a neutron moisture gauge is exactly as described above. When the hydrogen in the medium surrounding or adjacent to the measuring head is present in the form of water, the detector count rate can be related to the moisture content of the medium, once the instrument has been suitably calibrated. Neutron moisture gauges were originally developed for the *in-situ* measurement of soil moisture content, and instruments of this type are now in routine use in the fields of hydrology[7], agriculture[8] and civil engineering[9]. Used correctly, the gauges can provide field measurements of soil-water content that are both accurate and rapid. However caution must always be exercised as the gauge reading can be significantly influenced by a number of soil parameters—bulk density, presence of elements with high neutron absorption cross-section, bound water (perhaps present as water of crystallization) and organic content. Appropriate calibration procedures need to be adopted, and the more variable the nature of the soil, the more complex the calibration.

There are several potential advantages associated with the use of neutron moisture gauges in industry.

(a) The method is non-destructive
(b) It is rapid
(c) The measurement integrates over a large volume of sample (very important for inhomogenous materials)
(d) Because of the penetrating power of the neutron, gauges can be mounted external to process vessels, with all the associated advantages of a non-invasive instrument.

On the negative side, the technique suffers from some serious disadvantages which are largely responsible for the limited use which the process industry makes of this type of instrument. First and foremost is the fact that neutron gauges are sensitive to total hydrogen concentration. The composition of a hydrogenous process material thus strongly influences the response of the gauge and can lead to large inaccuracies in the assessment of moisture content. Secondly, to obtain a usable signal level, a significant amount of water must be present in the vicinity of the measuring head. Neutron moisture gauges are, therefore, unsuitable for small samples of material and for materials of low moisture content (less than 0.1% water). Thirdly, as with the soil-moisture gauge, response may be seriously affected by variations in the elemental composition of the process material. The presence of neutron absorbers is a particular problem.

There are, nevertheless, several important industrial applications of neutron moisture gauges. The steel industry is probably the major user, and measurement of moisture content of coke and sinter-mix charged to blast furnaces is an important application[10]. Neutron gauges are also used to control the water/cement ratio in ready-mix concrete[11].

15.3.3 Level and interface measurement

Figure 15.5 illustrates how the neutron moderation technique (often referred to as 'neutron backscatter') is applied to the detection of levels of, and

Figure 15.4. Interface measurement in a storage tank using a portable, 'neutron backscatter' gauge. Photograph courtesy of ICI PLC.

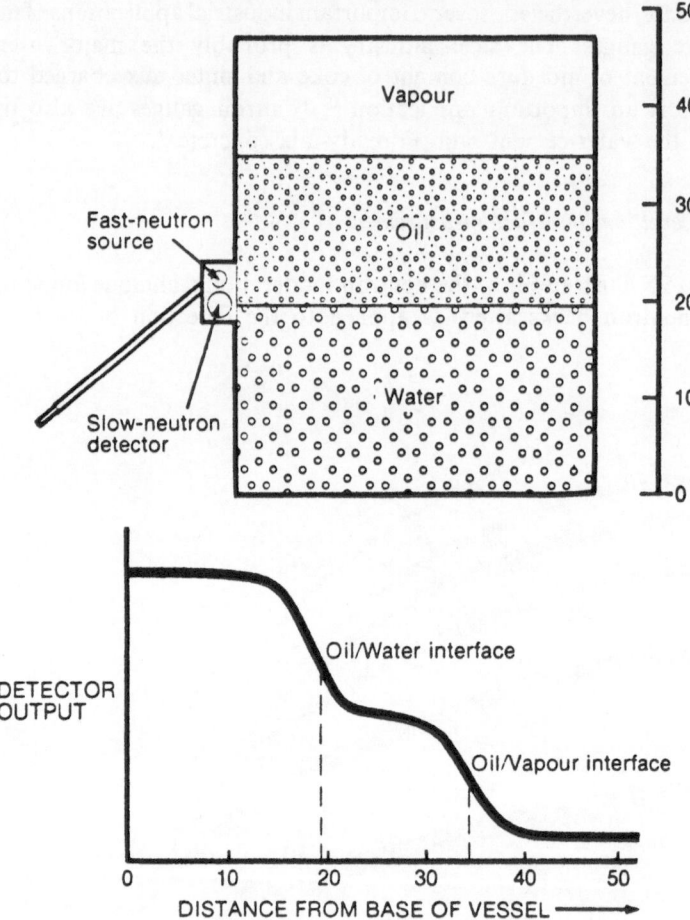

Figure 15.5. Portable neutron backscatter system for interface detection.

interfaces between, process fluids inside of plant vessels. The device illustrated is portable and can be used by an operator working alone (Figure 15.4). The measuring head, comprising fast-neutron source and slow-neutron detector, is placed in close proximity to the surface of the process vessel. Fast neutrons from the source pass readily through the metal walls of the vessel and are moderated, principally by collisions with hydrogen nuclei in the process material adjacent to the measuring head. Slow neutrons diffusing back out through the walls of the vessel are detected. The response of the slow-neutron detector, recorded as a count rate on the portable electronics, is thus determined by the neutron-moderating (and absorbing) properties of the process material.

Thus, as the measuring head is moved up and down the vessel wall the recorded count rate will vary in a manner which reflects changes (as far as neutrons are concerned) in the adjacent material. In the example shown in Figure 15.5, the vessel contains a lower layer of water, an upper layer of light oil and above that a vapour space. The oil (in this case) has a hydrogen concentration some 50% that of the water. The marked change in detector output as the measuring head moves from one level to another provides clear indication of the positions both of the water/oil interface and the oil/vapour level.

Though we have so far discussed a portable instrument for level and interface measurement it will, of course, be clear that it is perfectly possible to devise an installed system based upon the same principles. Instruments of this type have, in fact, been installed—both as interface alarms and as proportional indicators of interface position (see case histories, section 15.7). Even so, their use is extremely restricted as compared with installed gamma-ray absorption gauges (Chapter 13).

To understand the reason, it is necessary to look critically at the advantages and disadvantages of the 'neutron backscatter' technique.

The main advantages are:

(a) The neutron backscatter technique is completely non-invasive—an advantage shared with the gamma-ray absorption technique, in so far as level measurement is concerned.
(b) The neutron technique is not dependent upon the diameter of the vessel. Thus, liquid/liquid interfaces can be detected in vessels of large diameter. In such cases, the gamma-ray absorption technique is ineffective, unless modifications are made to the construction of the vessel to shorten the gamma-ray path length (section 13.4.2).
(c) Measurements are made from one side of the vessel only. This is useful in situations where access to both sides of the vessel is restricted.
(d) Where portable equipment is being used, measurements can in general be carried out by one operator.
(e) The neutron technique distinguishes between materials which have different neutron moderation and absorption characteristics. It can, therefore, be used to measure the interface between two liquids of closely similar density—a circumstance in which gamma-ray absorption density gauging would be ineffective.

Set against these advantages are the following shortcomings:

(a) It is difficult to apply the neutron method to thick-walled vessels. Though, under ideal conditions it is possible to identify a liquid level through steel walls up to 150 mm thick[12], in the field the method is rarely useful in situations where the wall thickness is greater than about 40 mm.
(b) The presence of insulation or lagging on the outside of the vessel can, in some cases, render the neutron technique very insensitive (effectively, the

measuring head is moved back from the hydrogenous process material by the thickness of the lagging). In other cases, if the lagging has hydrogenous inclusions, or has become waterlogged, the results can be misleading. If there is any suggestion of ambiguity, a strip of lagging must be removed to give direct access to the vessel wall.

(c) The neutron technique does not give a representative picture of the interface across the vessel. The response of the detector is very much determined by the properties of the material in the immediate vicinity of the measuring head. Put simply, the instrument 'sees' only 100–150 mm into the process material. If the level in the vessel is not uniform, (if, for example there is 'vortexing' in a liquid medium or pile-up in the middle of a powder or polymeric medium) the results of a neutron measurement can give an erroneous impression of the overall level position. A combination of neutron-backscatter and gamma-ray absorption techniques can sometimes be effective in these circumstances. (In fact, the comparatively small thickness interrogated by the neutron technique is sometimes a positive advantage, as the case histories will show.)

(d) The technique is only really useful for measurements on hydrogenous materials.

(e) Items of neutron equipment (sources and detectors) are expensive compared with their gamma-ray counterparts.

It is this last disadvantage which chiefly accounts for the infrequent use of installed neutron level and interface gauges.

Portable gauges are, however, used very frequently (Table 1.1) and, in experienced hands can provide data of real relevance to process operations.

15.3.4 Analysis

The phenomenon of neutron moderation, as such, is not widely applied to problems of elemental analysis. There are, however one or two classes of application where the method shows promise.

One such is the measurement of the hydrogen: carbon ratio of naphtha and other hydrocarbon process streams. Reference to Figure 15.2 gives a clear indication of the potential of this technique. The data presented there were obtained by immersion of the source-detector measuring head in the medium of interest. However, results obtained with the measuring head external to the containment vessel demonstrate a very similar trend, though the count rates obtained are lower because of the less-than-ideal geometry. It is thus possible to provide a continuous, on-line measurement of hydrogen: carbon ratio by mounting a measuring head on the pipeline through which the material is flowing. The device is non-invasive, sampling is eliminated and analysis is rapid. The method is comparable in accuracy to the beta-particle backscattering technique described in section 14.3.

However the method has not, so far gained wide acceptance: hydrogen: carbon ratio measurements are generally carried out off-line on samples taken from the process stream or by diverting part of the stream through a sample loop. It remains a possibility for the future. The technique *has* been successfully applied to the analysis of hydrogen in small samples of fissile materials[13].

15.3.5 *Detection of blockages and deposits*

The neutron moderation technique provides a rapid and convenient means of locating the position of hydrogenous blockages in pipework and is often used

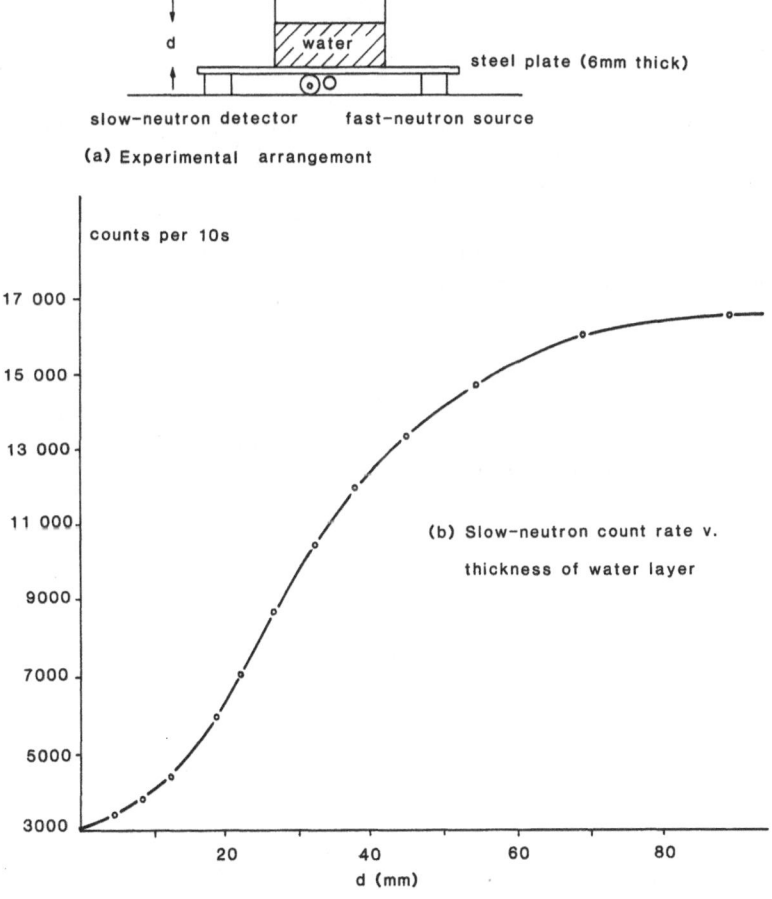

Figure 15.6. Neutron moderation for deposit thickness measurement.

to complement gamma-ray absorption studies (section 13.6.4). The technique is also capable of providing *some* information about the thickness of deposit on the walls of a vessel or about the depth of liquid lying along the bottom of a pipeline.

Figure 15.6a illustrates an experiment designed to examine the limitations of the method. A fast-neutron source and slow-neutron detector were attached to the bottom of a steel tank which was slowly filled with water while continuously recording the slow-neutron count rate. The results (Figure 15.5b) show that useful information can be obtained provided that the depth of water does not exceed about 75 mm. Beyond that point, the technique is insensitive to changes in water depth.

The results of neutron moderation measurements of material thickness should, therefore, be critically scrutinized and, if the possibility of ambiguity exists, complementary measurements (perhaps gamma-ray absorption) should be carried out.

15.4 Neutron absorption techniques

Neutron absorptiometry is used widely as an analytical tool. The likelihood of neutron absorption occurring is a function both of the neutron energy and of the strucure of the absorbing nucleus. A full treatment of the subject should, therefore, include *fast-neutron absorption*, where the neutron–nucleus interaction involves particle emissions, perhaps protons or alpha particles; *epithermal neutrons*, where strong resonances in the neutron capture cross-sections occur in certain nuclei; and *slow neutrons*—radiative capture processes.

Analytical techniques have been based on all of these phenomena but it is the capture of slow neutrons which has found widest application in the industrial environment. We will, therefore, restrict ourselves to applications based upon slow-neutron absorption. For a wider treatment of the subject, the reader is referred elsewhere[14].

15.4.1 *Analysis by slow-neutron absorption*

The method is based upon the fact that certain chemical elements (Table 15.1) have high cross-sections for slow-neutron absorption. If such a neutron-absorber is present in a sample of material then the sample, exposed to a flux of slow neutrons, should bring about a flux-depression which to a good approximation depends only upon the concentration of the neutron absorber.

A schematic instrumental configuration for a slow-neutron absorption analyser is shown in Figure 15.7. A fast-neutron source and slow-neutron detector are embedded in a moderating medium (perhaps polythene or paraffin-wax). The detector samples the flux of the slow neutrons resulting

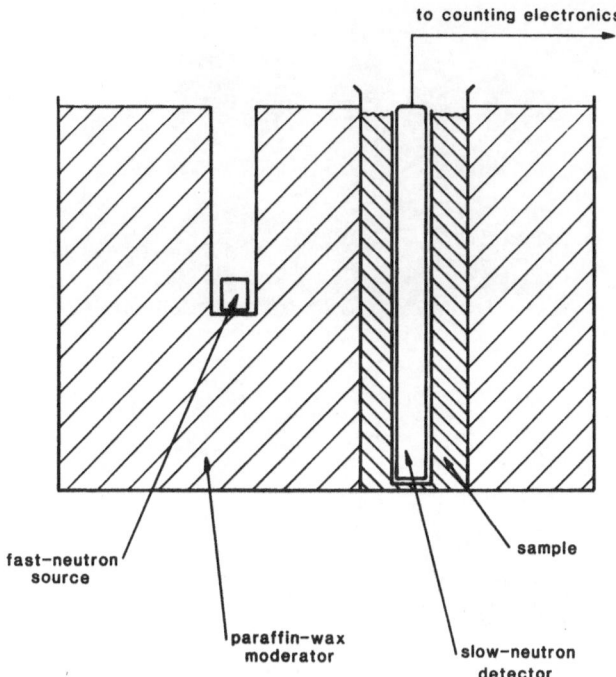

to counting electronics

fast–neutron
source

paraffin–wax
moderator

sample

slow–neutron
detector

Figure 15.7. Basic features of a slow-neutron absorption analyser.

from the moderation process. Samples of the material to be analysed are lowered into position adjacent to the detector, the output signal of which is lowered by an amount dependent upon the concentration of neutron absorber present. Calibration is effected using a range of samples of known absorber concentration. Practical applications of this type of instrument are described in the case histories (section 15.7).

An alternative approach is to make use of the 'prompt gammas' which accompany the radiative capture of slow neutrons (section 15.2.3). Such gamma-rays are generally of high energy (several MeV) and have energies characteristic of the absorbing element. The detection and identification of these gamma-rays (NaI scintillation counter or lithium-drifted germanium detector) forms the basis of a method of analysis. On-line systems suitable for the analysis of metal ore slurries have been described in the literature[15].

15.4.2 Neutron radiography

No résumé of the industrial applications of neutrons would be complete without some mention of the technique of neutron radiography. The technique has many points in common with X- and gamma-ray radiography (section 13.5.1). A uniform beam of neutrons is directed at the object of interest,

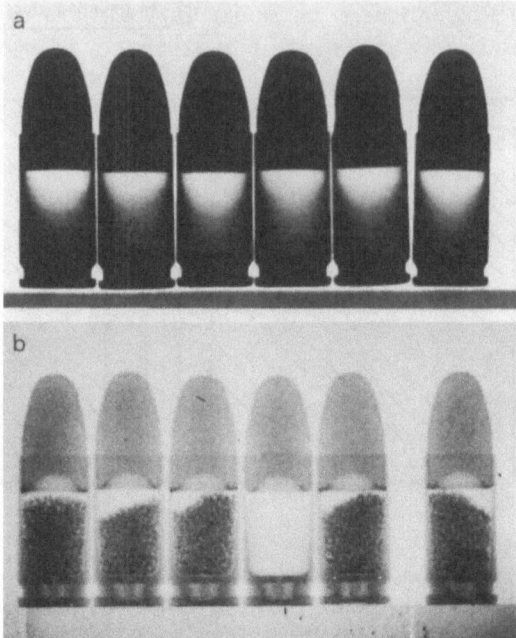

Figure 15.8. Comparison of X-ray and neutron radiography. (*a*) X-radiograph of ammunition cartridges; (*b*) neutron radiograph of ammunition cartridges. Negatives by kind permission of Dr M. R. Hawkesworth, University of Birmingham.

and variations in the transmitted intensity, recorded on a photographic plate, provide information on the spatial distribution of neutron-absorbing elements within the object. (For information on the detailed methodology of neutron radiography, the reader is referred to the literature[16,17].) There are, however, several important points of difference between the two techniques and the information they can provide.

In X- and gamma-radiography, the 'opaque' regions of the object are determined by the presence of material of relatively high atomic number. In neutron radiography it is the presence of elements having high neutron absorption or scattering cross-sections which remove neutrons from the beam and which, therefore, show up as under-exposed areas on the photographic plate.

Since, as we have seen (section 15.2.3) the cross-sections for neutron interactions vary widely from one element to another, neutron radiography can reveal features undetectable by the X- and gamma-ray techniques. This point is well illustrated by Figure 15.8 which compares radiographs of ammunition obtained on the one hand with X-rays and on the other with neutrons. The explosive charge is clearly visible only in the neutron radiograph.

On first sight, then, neutron radiography would appear to offer an excellent complementary technique for the examination of process plant since it is clearly suited to the examination of organic materials inside heavy metal structures. There is, however, a major drawback. To produce a useful radiograph in a realistic time-scale requires a slow-neutron beam of reasonably high intensity (10^3–10^4 neutrons per cm^2 per second). Such beams can readily be extracted from nuclear reactors, but reactors cannot be taken into the field. Transportable neutron sources, often based upon the radioisotope californium-252, are available, but the large mass of moderating material necessary to produce the slow neutrons and to provide radiological shielding makes the source cumbersome and unsuitable for use in the often confined conditions of process plant.

The technique is not, therefore, appropriate for use as a routine tool of plant investigation, though in special applications, such as the inspection of critical electronic components, the examination of aircraft for corrosion and for quality control within the nuclear industry, it makes a valuable and unique contribution.

15.5 Neutron activation techniques

Identification and measurement of the radioactive isotopes produced by neutron absorption is the basis of the powerful analytical technique known as neutron activation analysis (NAA).

In most cases, NAA relies upon a nuclear reactor as a neutron source. Clearly, the neutron flux depends both upon the type of reactor and the design of the NAA irradiation facility, but 10^{12} slow neutrons cm^{-2}s^{-1} would be fairly typical. At this flux the sensitivity of analysis which can be achieved for each element is as indicated in Table 15.2. It will be noted that the sensitivity can vary by several orders of magnitude from one element to its neighbour in the periodic table. If a radioisotope neutron source is substituted for the nuclear reactor, on-line neutron activation analysis for the measurement and control of the element composition of industrial process streams becomes possible. Since the maximum slow-neutron flux obtainable from a moderated neutron source is several orders of magnitude below that from a nuclear reactor, the sensitivities quoted in Table 15.2 are correspondingly reduced.

Figure 15.9 shows schematically the type of arrangement which might be used for this type of measurement. A radioisotope neutron source—possibly californium-252—is positioned at the centre of an irradiation chamber, surrounded by paraffin wax which acts both as a neutron moderator and as a radiation shield. A side-stream from the main process flow is pumped into this chamber where the various elemental constituents are exposed to the flux of neutrons—both slow and fast—from the source. Radioactivity builds up in the sample as a result of the various neutron absorption processes taking place.

Table 15.2 Activation analysis sensitivity in micrograms assuming interference-free limits

Element	Limit of detection (μg)*	Element	Limit of detection (μg)*
Actinium	†	Neodymium	0.02
Aluminium	0.002	Neon	2
Antimony	0.001	Nickel	0.5
Argon	0.001	Niobium	1
Arsenic	0.001	Nitrogen	†
Astatine	†	Osmium	1
Barium	0.05	Oxygen	2000
Beryllium	†	Palladium	0.01
Bismuth	†	Phosphorus	†
Boron	†	Platinum	0.1
Bromine	0.002	Polonium	†
Cadmium	0.005	Potassium	0.1
Caesium	0.001	Praseodymium	0.021
Calcium	0.05	Promethium	†
Carbon	†	Protactinium	†
Cerium	0.2	Radium	†
Chlorine	0.02	Radon	†
Chromium	0.01	Rhenium	0.001
Cobalt	0.005	Rhodium	0.0001
Copper	0.001	Rubidium	0.02
Dysprosium	0.00001	Ruthenium	0.02
Erbium	0.002	Samarium	0.0001
Europium	0.00001	Scandium	0.0001
Fluorine	3	Selenium	0.01
Francium	†	Silicon	100
Gadolinium	0.002	Silver	0.001
Gallium	0.002	Sodium	0.001
Germanium	0.1	Strontium	0.002
Gold	0.0002	Sulphur	200
Hafnium	0.0005	Tantalum	0.1
Helium	†	Technetium	†
Holmium	0.001	Tellurium	0.01
Hydrogen	†	Terbium	0.01
Indium	0.00005	Thallium	†
Iodine	0.001	Thorium	0.1
Iridium	0.0002	Thulium	0.1
Iron	20	Tin	0.02
Krypton	0.01	Titanium	0.1
Lanthanum	0.005	Tungsten	0.002
Lead	100	Uranium	0.004
Lithium	†	Vanadium	0.0001
Lutecium	0.0001	Xenon	0.1
Magnesium	0.5	Ytterbium	0.02
Manganese	0.00005	Yttrium	0.4
Mercury	0.002	Zinc	0.1
Molybdenum	0.1	Zirconium	0.2

*Interference-free limit   †Analysis not normally performed.

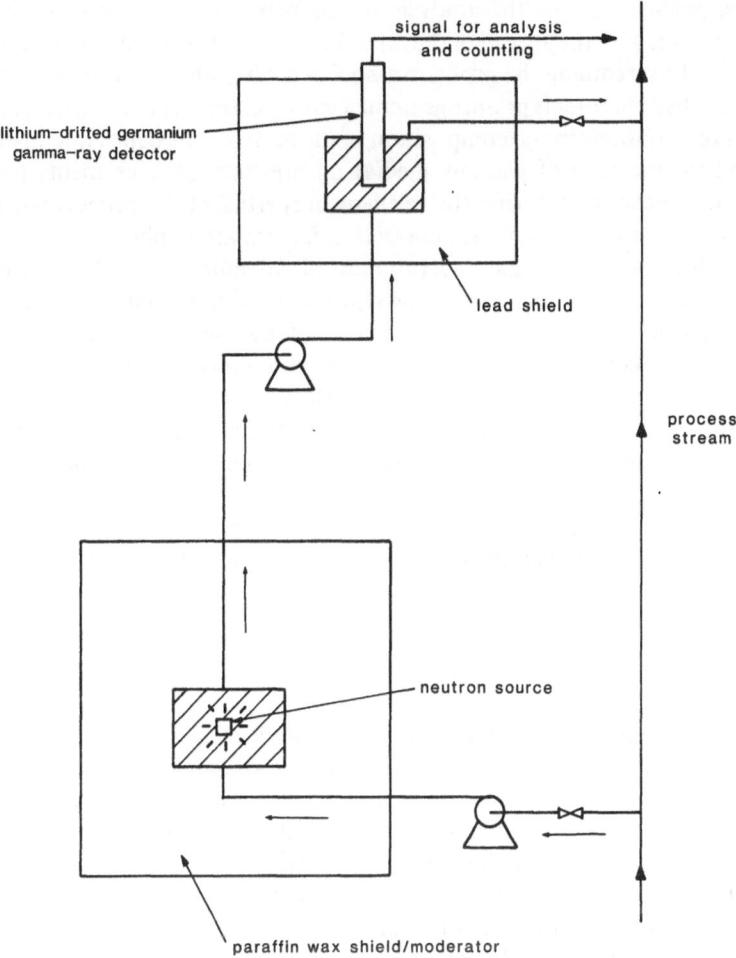

signal for analysis
and counting

lithium–drifted germanium
gamma–ray detector

lead shield

process
stream

neutron source

paraffin wax shield/moderator

Figure 15.9. Schematic of an on line neutron activation analysis system.

The irradiated sample is then pumped to a lead-shielded counting-chamber where it is exposed to a lithium-drifted germanium gamma-ray detector which identifies and counts the characteristic gamma-rays from the sample. From this information the concentration of the neutron-absorbing elements present in the process stream may be calculated using empirically-derived calibration curves. Among other applications, systems based upon the technique have been evaluated for the analysis of fluorine[18] and sodium[19] and for the field analysis of concrete[20].

Calibration may be a difficult procedure if the matrix material—the process stream—varies markedly in its composition. Neutron absorption and scattering interactions with elements other than the ones of interest can introduce significant errors if some compensation is not made. Arguably, it is this factor, more than any other, which restricts the use of on-line neutron activation analysis, particularly for the analysis of complex materials such as metal ores. However, with the advent of the microprocessor, there appear to be good prospects of overcoming the problem associated with interference from other elements. The computer is pre-programmed with correction factors which take account of variations in the composition of the matrix—its density, temperature and (in the case of gaseous media) its pressure. Measurements from subsidiary sensors monitoring the relevant properties of the process stream then ensure that the appropriate correction factors are applied.

With this innovation the several natural advantages of the neutron activation technique in analysis can be realized. It examines large volumes of sample (because of the penetrating power of the neutron) and there is, therefore, less likelihood of errors associated with non-representative sampling of an inhomogenous material. Several elements can be analysed simultaneously. The equipment can be made physically robust and non-contacting—useful for analysing potentially damaging materials such as ore slurries, corrosive or abrasive process streams. It is predicted that, with microprocessor control, the use of on-line neutron activation analysis, and, indeed of the other applications of neutrons to on-line measurement, will be much more widespread in the future.

15.6 Radiological protection aspects

In concluding our brief survey of neutron techniques (and before going on to discuss specific case histories) it is appropriate that we should make mention of the precautions which must be observed to protect against the potentially harmful effects of neutron radiation. Practical means of protection have been described in Chapter 6. In this section, we will explore in more detail the particular problems associated with neutron sources and the measures which are adopted to reduce the hazard to an acceptable level.

What, then, *are* the hazards from a radioisotope neutron source? They may be considered in three groups:

(a) Neutron radiation dose
(b) Gamma radiation dose—both from gamma-rays direct from the source and from neutron interactions in the surrounding medium
(c) Neutron activation of the components of process plant leading to a possible contamination problem.

Considering first category (a) it is instructive to examine Figure 15.10 which compares neutron dose-distance curves for a 37 MBq source of americium-241/beryllium measured firstly in air, secondly in water. The effects of distance and of hydrogenous shielding upon neutron dose are immediately apparent.

Protecting against the gamma-rays does not, in general, present a problem.

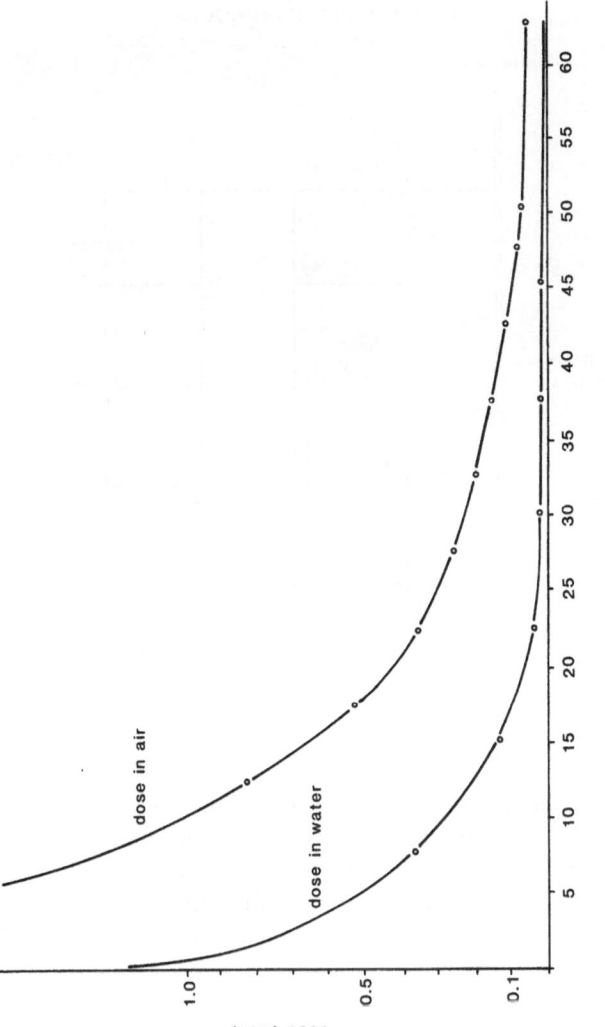

Figure 15.10. Dose-distance relationship for a 37 MBq americium-241/beryllium neutron source.

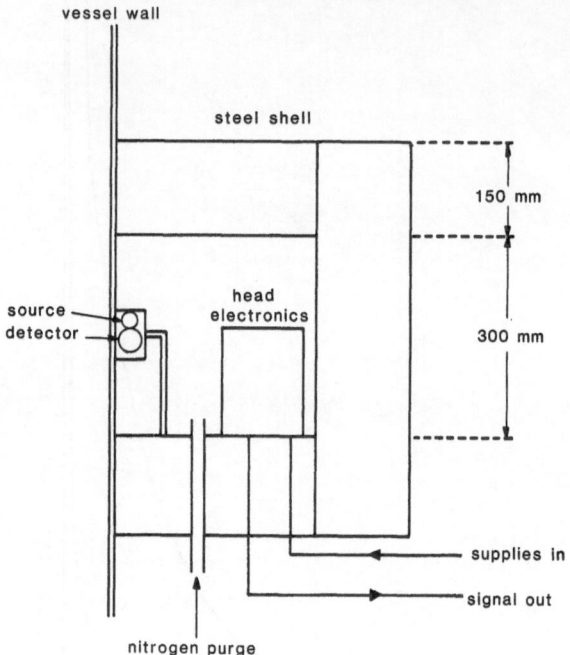

Figure 15.11. A typical neutron shield for a plant installation.

The primary gamma-rays from the americium-241 are of low energy (60 keV) and are, therefore, easy to screen out by encapsulating the source in a small lead container having a wall-thickness of 2 or 3 millimetres. The effect of capture gamma-rays is small compared with that of the neutrons and, therefore, neutron-source storage containers (Chapter 6) and installed neutron instruments are designed primarily to protect against the neutron irradiation hazard.

A typical neutron shielding arrangement for use with a plant level-gauging installation is shown in figure 15.11. In this case, a combination of distance and hydrogenous shielding is used to achieve the required dose-reduction. The dose rate at the surface of this system was, at all points, less than 7.5 μSv per hour.

However, there still remains to be considered the potential problem of neutron activation of plant components, particularly the steel walls of vessels and pipes. With neutron sources used in portable equipment this is not a problem. The neutron flux from the source is comparatively low and the exposure-time short. Such activation that does occur, either of the vessel or of its contents, is of a truly negligible level.

The case of neutron sources which are to be installed on plant for long periods of time, perhaps as part of an instrument system, needs to be examined in rather more detail.

It can be shown[21] that if 1 g of material is exposed to a neutron flux F (neutrons cm^{-2} sec^{-1}) for an irradiation time t seconds, then the specific activity, S, induced in the target material will be

$$S = \frac{0.6\,Fb\,(1 - e^{-0.693t/T})}{W}\ \text{Bq per g.} \tag{15.2}$$

where b is the activation cross-section in barns, T is the half-life of the product isotope and W is the atomic weight of the material.

Several points should be made in connection with this relationship.

(a) The specific activity follows a '1 − e' curve reaching a saturation level for long irradiation times. The saturation value of the specific activity is clearly

$$S = \frac{0.6\,Fb}{W}\ \text{Bq per g.} \tag{15.3}$$

(b) The shorter the half-life of the product radioisotope, the faster the build-up of activity to the saturation level. By the same token, of course, should the source of neutron irradiation be removed a short half-life implies that the product radioisotope will decay rapidly.

(c) The amount of each radioisotope produced depends upon the activation cross-section of its parent isotope. In a material such as steel, several potential parent isotopes are present. Each will yield a daughter radio-isotope upon irradiation, in a quantity which may be calculated using equation (15.2). Thus, in assessing the effects of long-term exposure of a plant component to neutron irradiation, the precise elemental composition of that component needs to be established.

Let us consider, as an example, a mild-steel vessel on which has been installed a neutron system of the type shown in Figure 15.10. Let us further suppose that the system comprises a 37-MBq source of americium-241/beryllium and has been installed for long enough to ensure that all induced radioactivity is at its saturation level. Considering the isotopes present in mild steel, only iron-54 and iron-58 produce active daughters by the neutron capture process.

The slow-neutron capture cross-sections for iron-54 and iron-58 are 2.3 barns and 1.01 barns respectively[22]. With the 37-MBq source, the maximum slow neutron flux to which the steel will be exposed will be approximately 10^4 neutrons cm^{-2} s^{-1}.

Given the isotopic abundances of iron-54 and iron-58, 5.8% and 0.33% respectively, it is a simple matter to show, using equation (15.3), that the specific activities of the daughter isotopes, iron-55 and iron-59, induced in the steel will be approximately $14\,\text{Bq g}^{-1}$ and $0.3\,\text{Bq g}^{-1}$ respectively.

These are extremely low levels of radioactivity. Furthermore, since the radioactivity is distributed throughout the volume of the steel it is 'fixed' in character and is most unlikely to constitute removable contamination. The health hazards from such low levels of radioactivity are negligible. However, under certain systems of legislation, the steel may be considered to be radioactive however low the level of activity, and must be treated accordingly. The handling and disposal of such material does not normally constitute a problem.

15.7 Case histories

15.7.1 *Liquid ammonia level measurement in road and rail tankers by neutron moderation*

One of the first uses of the neutron moderation or 'backscatter' technique was to monitor transport vehicles during the loading and offloading of liquid ammonia. These apparently simple transfer operations are often fraught with difficulty. Vapour 'locking' can occur leading to underfilling. Similarly, when the ammonia is being transferred from the tanker it is quite possible to leave a significant volume behind as a shallow pool in the bottom of the tanker. In theory, of course, these problems might be avoided by using a weighbridge during the transfer operations, but this is not always available, particularly where several tankers are being filled simultaneously from the same storage vessel.

Liquid ammonia is, of course, rich in hydrogen and it was therefore possible to measure the liquid levels in tankers both accurately and rapidly using the neutron backscatter equipment shown in Figure 15.5. This method has proved to be so reliable that its use is now incorporated into a National Emergency Scheme for dealing with accidents involving ammonia tankers and other vehicles conveying hazardous chemicals. Should a transport accident occur (suppose, for example, that an ammonia tanker is derailed and overturned) it is crucially important first to ascertain the volume of material it contains, and secondly to accurately monitor the transfer of all of this material to other vehicles, ensuring none remains. In this way, hazards associated with subsequent handling of the damaged vehicle are minimized. The neutron backscatter technique is used for both of these purposes. The instrument, together with a trained operator, travels with the emergency team to the sites of such incidents and, by providing the Emergency Controller with un-ambiguous, rapid information helps to ensure that the repair crews can carry out remedial action with the minimum of risk.

Though the neutron backscatter technique is invaluable in the emergency situation it is perhaps worth pointing out, as an illustration of the way in which radioisotope methods can complement one another, that installed instruments based upon gamma-ray absorption (section 13.5.2) are being used increasingly to monitor the filling of rail tankers. A shielded gamma-ray source and a radiation detector are mounted on a yoke which spans the rail tanker. When the level of liquid in the tanker rises to intersect the radiation beam the detector initiates an alarm signal which terminates the liquid transfer.

15.7.2 *Installed instruments for interface measurement and control*

Though, as we have stated (section 15.3.2), installed neutron systems for interface measurement are comparatively rare, their use in certain circumstances has proved to be of very great value to process plant operators. Two examples of installed systems will therefore be described: the first a safety system, the second a means of process control.

Figure 15.12 illustrates the effluent holding-tank associated with the manufacture of a plastics intermediate. Effluent from the process, comprising both organic and aqueous liquids, was fed to the tank where phase separation occurred. The tank at any time thus contained:

(a) A lower aqueous phase, predominantly water with traces of organic material. This was transferred to the plant effluent system at appropriate intervals.
(b) An upper organic layer comprising highly toxic material. This was recycled to the plant for further processing.

It was crucially important to ensure that the interface between the two phases never fell to the level of the aqueous effluent take-off line, as this would have

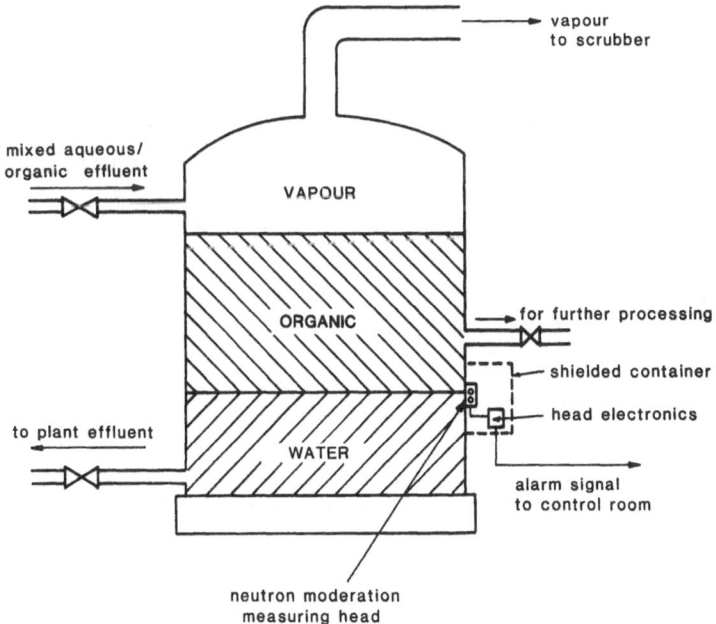

Figure 15.12. Neutron interface alarm on an effluent holding tank.

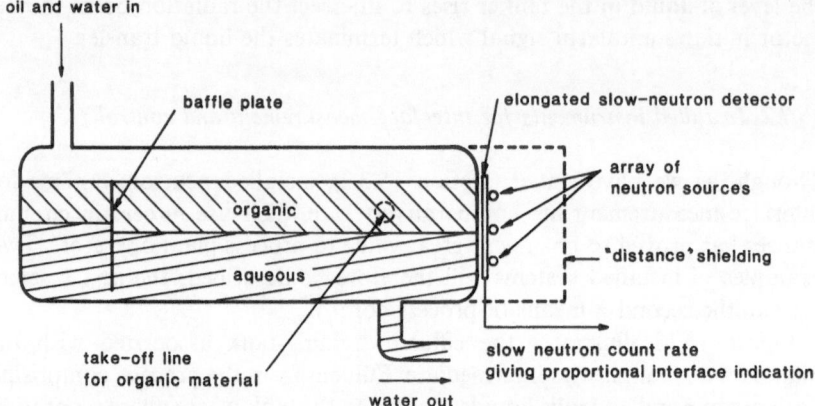

Figure 15.13.   Neutron-based interference controller on a separator vessel.

resulted in the release of toxic organic material. The problem was resolved by installing a neutron backscattering alarm as shown. The difference between the hydrogen concentrations of aqueous and organic phases was such that the neutron detector gave very clear indication of the approach, to the alarm level, of the interface—an invaluable contribution to plant safety.

Figure 15.13 is a schematic representation of an oil-water separator on the same plant. Control of the interface between the oil and water was important if good separation was to be obtained. The installed system, a float arrangement, had failed, necessitating manual control of the interface and leading to consequent inefficiency of operation of the vessel.

An installed neutron system was again able to provide a solution. In this case, because the separator was comparatively small and the interface position varied rapidly, a proportional indication of interface position was necessary for control purposes. As shown in Figure 15.13, this was achieved by using a multiple array of neutron sources (simulating, effectively, an elongated neutron source) and three neutron detectors arranged to span approximately 450 mm about the optimum control position. The summed slow-neutron count from the detectors provided a signal which was dependent upon the interface position and which could be used successfully for control purposes.

15.7.3 Inspection of a flarestack line for ice blockage

Condensation from flare gases in very cold weather may freeze, building up an ice deposit which gradually blocks the flare stack (Figure 15.14). Clearly, this is a potentially dangerous situation and early warning of such a build-up is desirable. The portable neutron backscatter equipment provides a convenient means of detecting flarestack fouling. Because of the greater density (and hydrogen concentration) of the ice as compared with any flare gas which may

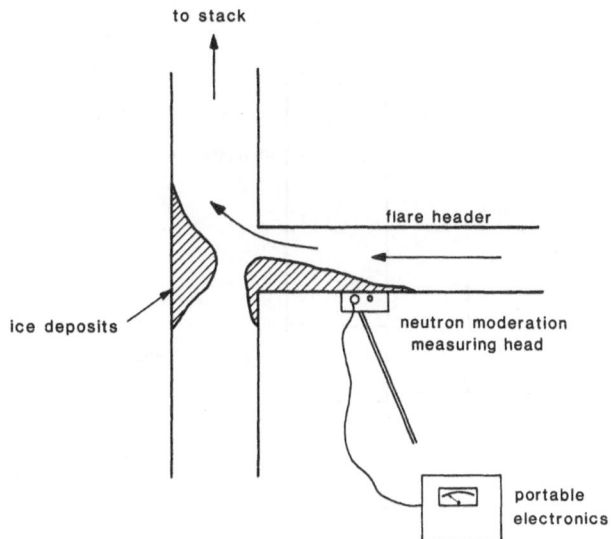

Figure 15.14. Detection of ice in a flarestack line by neutron moderation.

be present in the line, the method is very sensitive and is capable of detecting incipient blockages long before the stage at which they present serious restrictions to flow.

This technique is used for the routine inspection of flarestack lines in winter months. It has also been applied during plant start-up when particularly high flaring rates produced a cooling effect in the pipework—potentially a situation in which ice build-up may have occurred.

15.7.4 Shell-tube heat exchangers: detection of liquid levels in end-boxes and in the shell

The neutron backscatter technique can be extremely useful in diagnosing faults on heat exchanger equipment. As an example, consider Figure 15.15. Here a reboiler is used to heat liquid from the base of a distillation column and return it to the column in a vaporized state. The heating is supplied by steam, which is on the tube side of the exchanger, while the column liquid passes through the shell side. If the supply of steam is too low, insufficient heat will be supplied to the shell side liquid which may then build up to an appreciable level in the exchanger shell. Clearly, in this condition the efficiency of the exchanger will be impaired.

The neutron backscatter technique, because it responds effectively to material in the immediate vicinity of the measuring head, can be used to measure the level of the shell-side liquid and it can thus provide useful information about how well, or how badly, the reboiler is performing. The

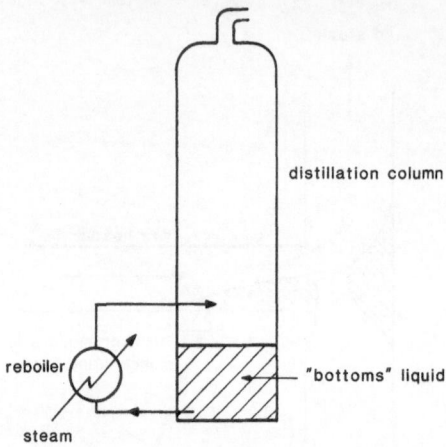

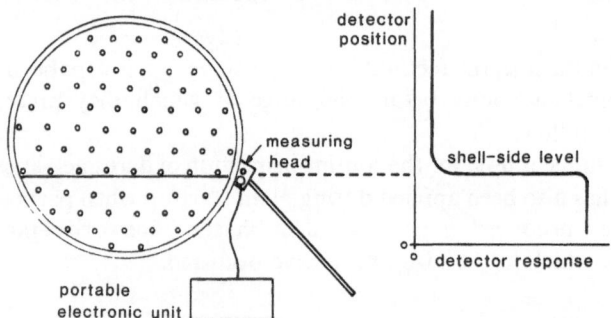

Figure 15.15. Detection of liquid in a reboiler shell using neutron moderation.

gamma-ray absorption technique could not be used for this type of measurement: transmission of gamma-rays through a tube-bundle of normal plant dimensions is not feasible.

The neutron backscatter technique can also provide clues about possible tube-side malfunctions. In the above mentioned case, for example, it would be possible to check the cold end-box of the reboiler for the presence of liquid build-up. This measurement, coupled perhaps with a steam-flow measurement, might suggest that inadequate heat was being supplied to the boiler to allow it to function effectively. Alternatively, an excessive liquid level in the end-box might suggest a restriction in the condensate run-off line.

In either event, production personnel would receive valuable information which would allow them firstly to recognize and understand the nature of the problem and, secondly, to undertake remedial action.

15.7.5 *Boron analysis by slow-neutron absorption*

A producer of a refractory material had a requirement for an instrument to provide a rapid and sensitive analysis of boron in the ore feedstock to the plant. Boron is an undesirable impurity in refractory: being a small molecule it is able to migrate through the matrix material under the influence of heat and can bring about a significant lowering in the useful life of the refractory material. For quality-control purposes, it was clearly important to identify and to reject feedstock of high boron content before that ore entered the production process.

Conventional chemical techniques were long and tedious and led to unacceptable delays. It was therefore decided to provide a control-room analyser, based upon the slow-neutron absorption technique, which would allow assessments of boron concentration to be made rapidly and accurately

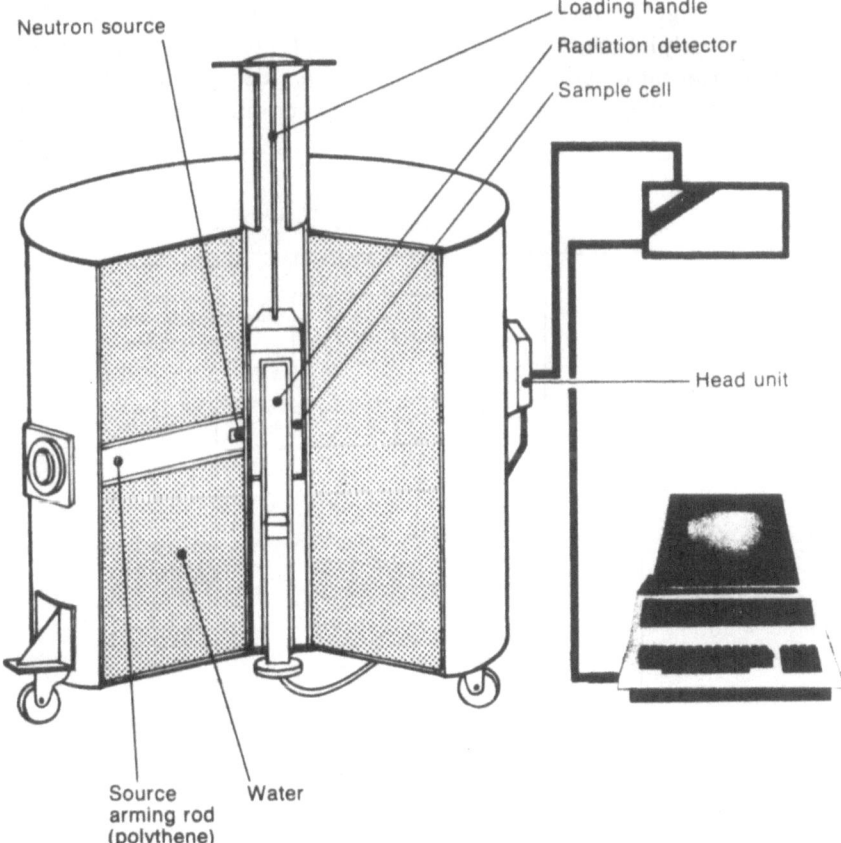

Figure 15.16. Control-room instrument for boron analysis by slow-neutron absorption.

by production personnel. The instrument is shown diagrammatically in Figure 15.16. The fast neutron source (americium-241/beryllium) and slow-neutron detector (BF_3-filled proportional counter) were positioned at the centre of a water-filled sample cell such that the detector sampled the flux of slow neutrons resulting from moderation of the fast neutrons from the source. Weighed samples of feedstock were loaded into cylindrical, watertight sample vessels which were recessed at the base. Each was lowered down the guide-tube into the counting position and the number of slow neutrons detected in a 5-minute count period recorded. Boron has a very high slow-neutron absorption cross-section (755 barns)—orders of magnitude higher than other materials present in the feedstock. Samples of feedstock exposed to the slow neutrons in the counting position, therefore, brought about a reduction in count rate which, to a good approximation, depended only upon the boron content of the sample. The instrument was calibrated using samples of known boron content.

It was found that in the 5-minute analysis time boron levels of approximately 0.001% (by weight) could be measured. Furthermore, provided that the samples were not grossly inhomogenous, no detailed sample preparation was required.

The instrument described above was, of course, designed for off-line use. However, the same technique has been applied to on-line measurement: boric

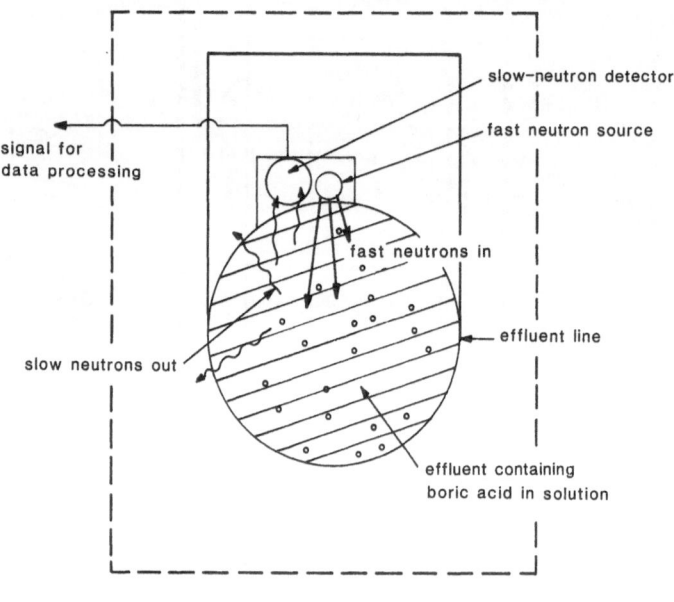

Figure 15.17. On-line analysis of boric acid by slow-neutron absorption.

acid is used in a nylon intermediates process. The cost of the boric acid represents a significant part of the materials cost of the process and it is, therefore, a matter of economic as well as environmental concern that losses of boric acid to the plant effluent system should be kept as small as possible.

Neutron absorption provides the solution to the problem. The instrument configuration is as shown in Figure 15.17. Fast neutrons from the source are slowed down by collisions with hydrogen nuclei in the effluent and the detector continuously monitors slow neutrons which diffuse out through the pipe wall. The reduction in the detector count rate resulting from slow-neutron absorption by boron nuclei can be directly related to the concentration of boric acid once the instrument has been calibrated.

This instrument has operated successfully for several years and is used both for routine calculation of (small) losses of boric acid and as a means of providing alarm indication that an acid carry-over has occurred. This has meant that production personnel have been able to identify and concentrate upon those operations likely to give rise to a carry-over. They have thus been able to achieve much tighter control and greatly reduce boric acid loss.

Arising from the success of this instrument there are now several other installations on the same plant, controlling the catalyst concentration of boric acid at several stages of the process[23].

15.7.6 Neutron moderation technique for voidage measurement in a degasser

An inflammable hydrocarbon gas was entrained in an aqueous effluent from a plastics intermediates plant. Before the effluent could be disposed of to drain, it was therefore necessary to pass it through a degassing unit of a type shown schematically in Figure 15.18a. Essentially, the liquid was constrained by baffles to follow a tortuous path through a horizontal vessel of cylindrical shape. During the process, entrained gas disengaged itself from the liquid, flowed up the degasser stack and into the flare system for eventual combustion.

Because the correct operation of the vessel depended crucially upon establishing a properly-balanced flow through the vessel, a recurring possibility following each start-up of the unit was that localized pockets of gas were forming below the top surface of the vessel. This was undesirable from a safety viewpoint because of the inflammable nature of the gas, and a method of locating and measuring the extent of such gas pockets was required. The problem was solved by the application of the neutron moderation technique. The measuring head, consisting of fast-neutron source and slow-neutron detector, was used to carry out systematic measurements covering the whole upper surface of the vessel. The measured slow-neutron count rates were related to gas voidage below the surface using a calibration curve, derived in the laboratory using the arrangement shown in Figure 15.18b. Large gas

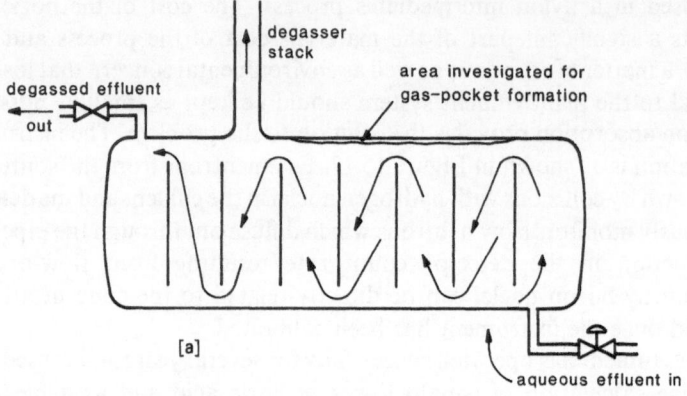

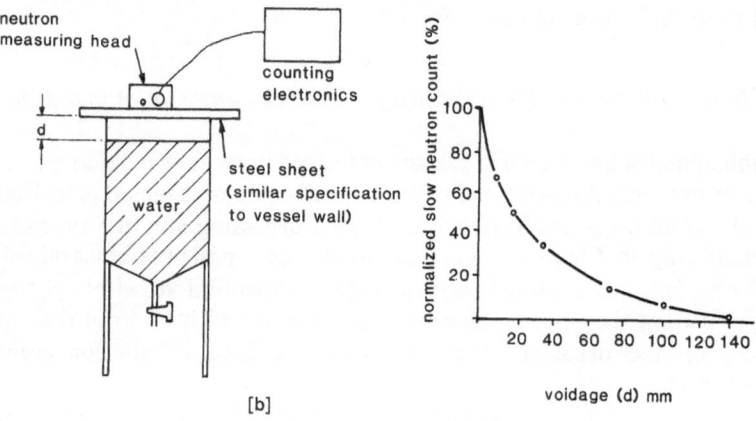

Figure 15.18. (a) Configuration of degassing unit; (b) laboratory calibration of detector response.

pockets were detected on several occasions and the flow through the vessel was subsequently adjusted until further measurements showed that the voidage had become minimal.

This is yet another example of the usefulness of the neutron moderation technique in investigating features occurring near to the surface of a vessel: the calibration curve of Figure 15.18b illustrates the high sensitivity which can be achieved.

15.7.7 *Examination of liquid levels in distillation column downcomers*

The investigation of problems associated with distillation columns by the gamma-ray absorption technique has already been discussed (section 15.4.3). Increasingly, however, the neutron moderation technique is being used to provide complementary data, information which, in some cases, can be crucial in making the correct diagnosis.

A recent application illustrates the usefulness of the method. A de-ethanizer column was operating inefficiently. Pressure-drop measurements suggested that flooding might be a possible cause, but the mechanism by which this occurred was not known. It was decided to carry out a programme of gamma-ray absorption measurements to investigate the problem.

Initially, the column was scanned at low process rates. The column appeared to be functioning correctly with well-defined liquid levels and vapour spaces. At the same time, the neutron moderation technique was used to investigate the liquid levels in the column downcomers (Figure 15.19). These measurements revealed an unusual feature: typically, the downcomers contained liquid of depth approximately 200 mm. However, the downcomer from tray 5 to tray 4 was holding liquid to a depth of 440 mm. This indicated that the downflow of liquid was being impeded, presumably by a restriction in the downcomer, and that this was the cause of the flooding which occurred at higher process rates. To test this hypothesis the process rates were gradually increased while using the gamma-ray absorption technique to examine the column in the region of tray 5. As the rate was increased, the column was, indeed, observed to flood from tray 5 upwards.

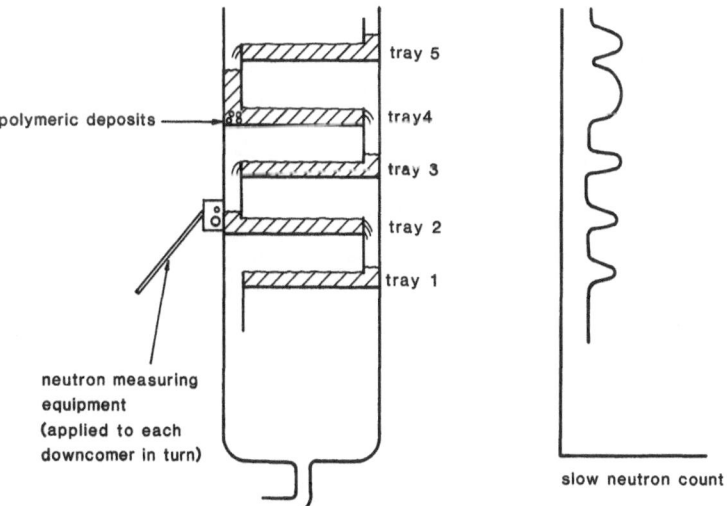

Figure 15.19. Inspection of distillation column downcomers using a neutron moderation technique.

a

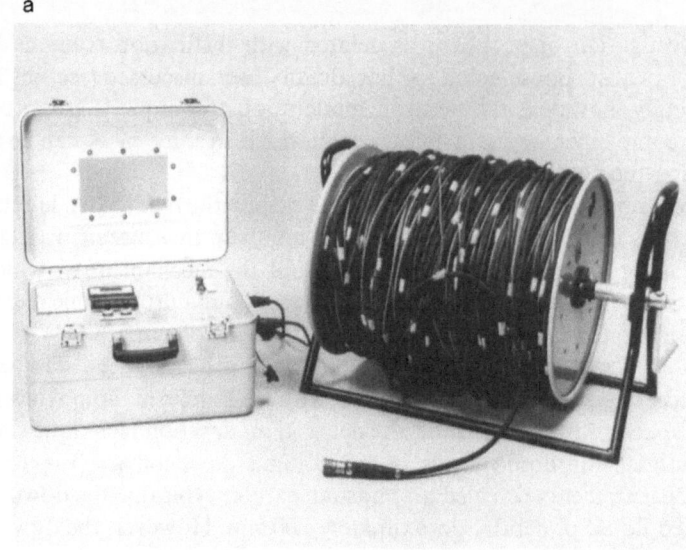

b

Figure 15.20. 'Gammatrol' neutron backscatter system for subsea use. (*a*) Portable 'topsides' electronic unit with winching system and subsea connector; (*b*) diver-operated neutron measuring head.

The source of the problem having been identified, the column was shut down and the offending downcomer examined. A polymeric blockage was found and removed and the column was put back on-line. The problem was found to have been eradicated and the column operated efficiently at full process rates.

The neutron moderation technique facilitated the rapid identification of this problem and obviated the need for detailed and lengthy gamma-ray scanning. The resulting savings were significant since production was at a premium and lost time was, in a very real sense, lost money.

15.7.8 Detection of water flooding in subsea structures

Water ingress into the tubular members of offshore platforms is a particularly serious problem since it indicates serious flaws in the structure. Inspection for water flooding can be complex. Though ultrasonic techniques may be applied they can be time-consuming and do not always give unambiguous results.

To provide a rapid and reliable means of inspection, a neutron moderation measurement system which has been used for several years to investigate similar problems on land-based structures[24] has recently been modified for subsea use. The device is shown in Figure 15.20. The measuring head contains the fast-neutron source/slow-neutron detector combination.

In use, it is placed against the wall of the suspect tubular member, and the slow-neutron flux resulting from neutron moderation by the material inside the structure is measured. Clearly, if the structure is full of water a high neutron count rate is observed, clearly differentiated from the signal obtained from an empty (intact) tubular. Successful subsea measurements have already been undertaken and further applications of the technique have been identified:

(a) Fluid leaks past 'line-pigs' used to isolate subsea pipelines for hyperbaric welding
(b) Voidages in grouting and cement casings
(c) Loss of concrete liner from pipes
(d) Interface determination in subsea vessels (between water, hydrocarbons and heavy-end deposits).

This list is not exhaustive and, no doubt, further applications will be identified. With the incorporation of the neutron system into Remotely Operated Vehicles (ROVs) a rapid expansion of the use of this versatile technique subsea is predicted.

References

1. *Neutron Moisture Gauges*, IAEA Technical Report Series No 112, Chapter 3.
2. *Reactor Handbook Physics*, USAEC, 75 (1955).
3. *Nucleonics* Data Sheet No 23.

4. Glasstone, S. and Edlund, C. M. (1955) *Nuclear Reactor Theory*. MacMillan, London.
5. Price, B. T., Horton, C. C. and Spinney, K. T. (1957) *Radiation Shielding*. International Series of Monographs on Nuclear Energy, Pergamon, Oxford.
6. Bacon, G. E. (1975) *Neutron Diffraction*, Clarendon Press, Oxford.
7. Bell, J. P. and McCulloch, J. S. G. (1969) *J. Hydrol.* **7**, 415.
8. Phillips, R. E. *et al.* (1960) *Soil Science* **89**, 2.
9. Pawliw, J. and Spinks, J. W. T. (1957) *Canad. J. Tech.* **34**, 503.
10. Williams, R. B. (1979) *J. Radioanal. Chem.* **48**, 49.
11. Somer, E. (1969) The BASC combined soil moisture/density gauge and other moisture/density equipment. In *Int. Symp. on Radiometric Methods and Instruments for the Determination of Density and Moisture of Building Materials and Soils*, Brno.
12. Taylor, J. B. (1970) MSc Thesis, University of Birmingham.
13. Close, D. A. *et al.* (1976) 252-Cf-Based Hydrogen Analyser. *Nuc. Inst. Meth.* **136**.
14. Taylor, T. I. *et al.* (1951) Chemical analysis by neutron spectroscopy. *Science* **114**, 341.
15. 252-CP Prompt neutron activation of process control. *Californium-252 Progr.* **17**, 20 (1974).
16. Hawkesworth, M. R. (1968) An introduction to practical neutron radiography. *X-Ray Focus* **8**, 23.
17. Barton, J. P. (1983) Neutron radiography—accomplishments and potential. *Inst. Physics C*, 447.
18. On-line slurry analysis with 252-Cf. *Californium-252 Progr.* **20**, 9 (1976).
19. Kelly, K. J. and Kamp A. J. (1982) *Anal. Chem.-Eng.* **54**, 5.
20. Field analysis technique for plastic concrete. *Californium-252 Progr.* **17**, 17 (1974).
21. *The Radiochemical Manual*. The Radiochemical Centre (1966).
22. Friedlander, G. *et al.* (1964) *Nuclear and Radiochemistry*. 2nd edn., Wiley, New York.
23. Charlton, J. S. (1984) *The Chemical Engineer* No 406, 49.
24. Charlton, J. S. British Patent No 1474395, June 1974.

16 Appendix: Radiation measurement-statistical considerations

K. JAMES

A.1 Introduction

Radioisotopes decay in a random manner and the laws of probability apply to the behaviour of any given radioactive nucleus. Consequently, radiation measurement is always subject to some degree of statistical fluctuation. These inherent fluctuations represent an unavoidable source of uncertainty in all measurements and are often the predominant source of error.

The purpose of this Appendix is to illustrate the use of counting statistics in determining whether any abnormalities exist in the counting system and in estimating the precision that should be associated with various measurements.

A.2 Counting statistics

Suppose we collect a set of counts from a radiation counter, each count reading being of the same duration and having been taken under identical conditions. The count readings are:

$$x_1, x_2, x_3 \ldots x_i \ldots x_N.$$

Two elementary properties of this set of data are

(i) 'sum':
$$\Sigma \equiv \sum_{i=1}^{N} x_i. \tag{A.1}$$

(ii) 'experimental mean':
$$\bar{x}_e \equiv \frac{\Sigma}{N}. \tag{A.2}$$

The data can be represented by a corresponding frequency distribution function $F(x)$. The value of $F(x)$ is the relative frequency with which a particular count reading appears in the collection of data.

$$F(x) = \frac{\text{number of occurrences of the value } x}{\text{number of measurements } (= N)} \tag{A.3}$$

Note that the distribution is automatically normalized, i.e.

$$\sum_{x=0}^{\infty} F(x) = 1. \tag{A.4}$$

Note also that a plot of $F(x)$ against x will be centred about the experimental mean $\bar{x}_e$ and that the width of the distribution function is a relative measure of the amount of scatter about the mean which is inherent for our given set of data. It is usual to characterise the width of the distribution by defining its standard deviation σ :

$$\sigma = \sqrt{\frac{\sum_i (\bar{x}_e - x_i)^2}{N - 1}}. \tag{A.5}$$

Various mathematical models exist which can predict the distribution function that will describe the results of many repetitions of a given radiation measurement[1]. In descending order of complexity these models are:
(a) Binomial Distribution—most general
(b) Poisson Distribution—simplification of binomial model valid if observation time is small compared with half-life of source
(c) Gaussian or Normal Distribution—simplification of Poisson model, valid provided that the total number of measurements (N) is large or $\bar{x}$ is large, widely applicable to many problems in counting statistics.
These models will not be further discussed here, but it should be noted that
 (i) Both the Poisson and the Gaussian/Normal distributions adequately describe the vast majority of situations encountered in the industrial application of radioisotopes.
(ii) Both the Poisson and the Gaussian/Normal models make the fundamentally important prediction that the standard deviation of the distribution is the square root of the mean value which characterizes that same distribution:

$$\sigma = \sqrt{\bar{x}}. \tag{A.6}$$

A.3 Correlation of sets of observations

It may be necessary to assess two or more sets of data to decide whether they come from the same population or whether they should be regarded as significantly different, perhaps corresponding to a malfunction of the counting equipment.

Student's τ test is basically a test for comparing a sample mean with the population mean where the distribution is approximately Gaussian. It can be usefully applied in providing a criterion on which to regard two sets of counting observations as significantly different (or not) as the case may be.

Suppose that the total number of counts in each of two observations are x_1 and x_2 accumulated in times t_1 and t_2 respectively, then the τ of Student's test is given by

$$\tau = \frac{\left| \dfrac{\bar{x}_1}{t_1} - \dfrac{\bar{x}_2}{t_2} \right|}{\sqrt{\dfrac{\bar{x}_1}{t_1^2} + \dfrac{\bar{x}_2}{t_2^2}}} \tag{A.7}$$

Where $\bar{x}_1$ and $\bar{x}_2$ are the population means of the distributions from which x_1 and x_2 are drawn. If x_1 and x_2 are large they can be substituted for $\bar{x}_1$ and $\bar{x}_2$ without serious error. If p is the probability that the observations are in fact from the same population, then p and τ are related as follows:

τ	0.674	1.041	1.155	1.96	2.33	2.58	3.29
p	0.5	0.3	0.1	0.05	0.02	0.01	0.001

The value of p which is taken as indicating a significant difference between the two sets of observations is a matter for judgement in each case, but $\tau \geqslant 2$ corresponding to $p < 0.05$ is usually taken to mean that there is a significant difference.

Similarly, for each set of data we can define the quantity χ^2 given by

$$\chi^2 = \frac{1}{\bar{x}_e} \sum_i (x_i - \bar{x}_e)^2 \tag{A.8}$$

and then, by reference to look-up tables[2,3] we can evaluate how closely each set of data fits Poisson's distribution.

A.4 Precision of a single measurement

If a particular determination requires us to accumulate a set of count readings under identical conditions, then it is usual to report the result of these readings as $\bar{x} \pm \sigma$ where σ is calculated from the readings as in equation (A.5). However, if we have only taken a single count reading the standard deviation cannot be calculated directly but must be estimated by analogy with an appropriate statistical model. Suppose we have a single count reading 'x', and assume that this measurement has been drawn from a population whose theoretical distribution is predicted by either the Poisson or Gaussian models. Then, because we have no other information available, we have no choice other than to assume that the mean of the distribution is equal to our single measurement, i.e. $\bar{x} = x$. From equation (A.6) it follows that

$$\sigma = \sqrt{\bar{x}} = \sqrt{x} \tag{A.9}$$

is our best estimate of the deviation from the true mean which should typify our single measurement x. Consequently, the result of our single count reading is usually expressed as $x \pm \sigma$ or $x \pm \sqrt{x}$. This should be interpreted as implying that the probability that the true mean $\bar{x}$ lies within $x \pm \sqrt{x}$ is 68%. Similarly, the probability that $\bar{x}$ lies between $x \pm 2\sqrt{x}$ is 95%.

Note that we cannot associate the standard deviation σ with the square root of any quantity which is not a directly measured number of counts. For example, the association does not apply to count rates, to sums/differences of counts, or to any derived quantity. In all these cases the quantity is calculated as a function of the number of counts recorded and the error associated with that quantity must be calculated as outlined in section A.6 (error propagation).

A.5 Standard deviation of a ratemeter

In a linear ratemeter, such as is commonly used in nucleonic instruments, each detector pulse adds an element of charge to the integrating circuit and the total charge is indicated on a meter. The contribution of each element of charge to the meter reading depends on its time of arrival. The meter indication then represents a sum of contributions over several time constants. The effective integrating time is twice the time constant[4] and the standard deviation is given by

$$\sigma = \frac{n}{\sqrt{2nT_c}} \tag{A.10}$$

Where n is the countrate and T_c is the time constant ($= RC$).

A.6 Error propagation

If $x, y, z \ldots$ are directly measured counts for which we know the standard deviations $\sigma_x, \sigma_y, \sigma_z \ldots$, then the standard deviation for any quantity u derived from these counts can be calculated from

$$\sigma_u^2 = \left(\frac{\partial u}{\partial x}\right)^2 \sigma_x^2 + \left(\frac{\partial u}{\partial y}\right)^2 \sigma_y^2 + \left(\frac{\partial u}{\partial z}\right)^2 \sigma_z^2 \ldots \tag{A.11}$$

where $u = u(x, y, z..)$ represents the derived quantity.
Simple examples are:
(a) Sums or differences— $u = x + y$ or $u = x - y$

$$\sigma_u = \sqrt{\sigma_x^2 + \sigma_y^2} \tag{A.12}$$

(b) Multiplication or division by a constant— $u = Ax$ or $u = x/B$

$$\sigma_u = A\sigma_x \text{ or } \sigma_u = \frac{\sigma_x}{B} \tag{A.13}$$

(c) Multiplication or division of counts— $u = xy$ or $u = x/y$

$$\left(\frac{\sigma_u}{u}\right)^2 = \left(\frac{\sigma_x}{x}\right)^2 + \left(\frac{\sigma_y}{y}\right)^2 \tag{A.14}$$

(d) Mean value of multiple counts— $u = \sum/N$ where $\sum = x_1 + x_2 \ldots$

$$\sigma_u = \sigma_x = \sqrt{\frac{\bar{x}}{N}} \tag{A.15}$$

Note that the expected standard deviation of any single measurement x_i is $\sigma_{x_i} = \sqrt{x_i}$ and it therefore follows that the mean value based on N independent counts will have an expected error which is smaller by a factor $\sqrt{N}$ than the error associated with any single measurement on which the mean is based.

A.7 Effect of background

The principle of error propagation is often applied in the design of counting experiments. For example, consider the situation where a source is providing a count rate in the presence of a steady-state background.
$S = $ count rate due to source alone (no background).
$B = $ background count rate.
Assume that we count source plus background for a time T_{s+B} and then count background alone for a time T_B. The net rate due to the source alone is then

$$S = \frac{N_1}{T_{s+B}} - \frac{N_2}{T_B} \tag{A.16}$$

where N_1 and N_2 are the total counts in each measurement.
Applying the results disclosed in the previous section it can be shown that

$$\sigma_S = \left[\frac{S+B}{T_{s+B}} + \frac{B}{T_B}\right]^{1/2} \tag{A.17}$$

If a fixed total time $T = T_{s+B} + T_B$ is available to carry out both measurements, the above uncertainty can be minimized by choosing the optimum fraction of T allocated to T_{s+B} (or T_B). Squaring and differentiating equation (A.17) we find that

$$2\sigma_S \, d\sigma_S = \frac{-S+B}{T_{s+B}^2} dT_{s+B} - \frac{B}{T_B^2} dT_B \tag{A.18}$$

Setting $d\sigma_S = 0$, we find that the optimum division of time is obtained by meeting the condition

$$\left.\frac{T_{s+B}}{T_B}\right]_{\text{opt}} = \sqrt{\frac{S+B}{B}}. \tag{A.19}$$

A.8 Statistics of pulse height distributions

When a mono-energetic source of radiation is measured with a proportional or scintillation counter, the observed pulse heights have a Gaussian distribution around the most probable value. The energy resolution of such detectors is expressed in terms of the full width of the pulse height distribution curve measured at half the maximum height and stated as a fraction or percentage of the most probable pulse height H. The full width at half maximum height is given by

$$2.36\,\sigma/H. \tag{A.20}$$

where σ is the standard deviation of the pulse height distribution[5].

In a proportional counter the spread in pulse heights arises from statistical fluctuation in both the initial number of ion pairs formed and in the gas multiplication factor. Thus, the fractional standard deviation of a measured pulse height is the square root of the sum of the squares of the fractional standard deviations of these two quantities. However, the gas multiplication factor is usually so large that its statistical fluctuation is unimportant. Consequently, the fluctuation in the number of ion pairs formed determines the resolution, and the resolution is inversely proportional to the square root of the energy of the ionizing radiation.

In a scintillation counter the statistical fluctuations in pulse height arise from several sources. The conversion of ionizing radiation into light photons in the phosphor, the electron emission at the photocathode and electron multiplication at the dynodes are all subject to statistical error. It can be shown[5] that

$$\sigma = H\sqrt{\delta/E\bar{q}\,\bar{f}\,\bar{p}(\delta - 1)} \tag{A.21}$$

where H is the most probable pulse height for incident energy E keV, $\bar{q}$ is the number of light photons emitted per 1 keV, $\bar{f}$ is the average value of the light collection efficiency at the photocathode, $\bar{p}$ is the mean number of photoelectrons arriving at the first dynode for each photon incident on the photocathode, and δ is the average electron multiplication per dynode.

Substituting typical values into equation (A.21) ($\bar{f} \sim 1$, $\bar{p} \sim 0.1$, $\delta \sim 4$, $\bar{q} \sim 30$ for NaI (Tl)), we can estimate the resolution attainable for the 662 keV Cs[137] γ peak with a sodium iodide scintillation counter. We find that $\sigma/H = 0.026$. The corresponding full width at half maximum is $2.36\,\sigma/H = 0.061$ or 6.1%, which is close to the best resolution obtained experimentally.

A.9 Detector efficiency

The laws of statistics can be used to estimate the efficiency of ion chambers, proportional, Geiger or scintillation counters for ionizing radiation of a given energy by assuming that any incident photon or particle which produces at least one ion pair in the detector also produces a count.

Events occurring within the detector are governed by the Poisson distribution curve. The shape of this curve[6] is described by the formula

$$P(n) = \frac{(n)^n e^{-\bar{n}}}{n!} \qquad (A.22)$$

where $P(n)$ is the probability that n ion pairs will be created and $\bar{n}$ is the average number of ion pairs formed in the same detector for an incident photon of the same energy.

From equation (A.22) it follows that the probability of no ion pairs being formed is given by

$$P(0) = \frac{(n)^0 e^{-\bar{n}}}{0!} = e^{-\bar{n}} \qquad (A.23)$$

Consequently, the probability that at least one ion pair will be created (i.e. a count will be produced) is

$$1 - P(0) = 1 - e^{-\bar{n}} \qquad (A.24)$$

Now $\bar{n}$, the average number of ion pairs created, can be calculated from a knowledge of the incident energy and the detector characteristics. If, for example, $\bar{n} = 7$, then

$1 - e^{-\bar{n}} = 1 - e^{-7} = 99.9\% = $ counter efficiency.

References

1. Knoll, G. F. (1979) *Radiation Detection and Measurement*. Wiley, New York.
2. Fisher, R. A. (1963) *Statistical Methods for Research Workers*. Oliver and Boyd, Edinburgh.
3. Bevington, P. R. (1969) *Data Reduction and Error Analysis for the Physical Sciences*. McGraw-Hill, New York.
4. Price, W. J. (1964) *Nuclear Radiation Detection*. 2nd edn. McGraw-Hill, New York.
5. Kohl, J. *et al.* (1961) *Radioisotope Applications Engineering*. Van Nostrand, Princeton.
6. Feller, W. (1957) *An Introduction to Probability Theory and its Applications*. 2nd edn. Wiley, New York.

Index